Novel Non-Precious Metal Electrocatalysts for Oxygen Electrode Reactions

Novel Non-Precious Metal Electrocatalysts for Oxygen Electrode Reactions

Special Issue Editors

Nicolas Alonso-Vante
Yongjun Feng
Hui Yang

MDPI • Basel • Beijing • Wuhan • Barcelona • Belgrade

MDPI

Special Issue Editors
Nicolas Alonso-Vante
University of Poitiers
France

Yongjun Feng
Beijing University of Chemical Technology
China

Hui Yang
Chinese Academy of Sciences
China

Editorial Office
MDPI
St. Alban-Anlage 66
4052 Basel, Switzerland

This is a reprint of articles from the Special Issue published online in the open access journal *Catalysts* (ISSN 2073-4344) from 2017 to 2019 (available at: https://www.mdpi.com/journal/catalysts/special_issues/metal_electrocatalysts)

For citation purposes, cite each article independently as indicated on the article page online and as indicated below:

LastName, A.A.; LastName, B.B.; LastName, C.C. Article Title. *Journal Name* **Year**, *Article Number*, Page Range.

ISBN 978-3-03921-540-9 (Pbk)
ISBN 978-3-03921-541-6 (PDF)

Contents

About the Special Issue Editors

Nicolas Alonso-Vante has been a Professor at the University of Poitiers since September 1997, teaching graduate courses: Specific Analysis of Solids; Electrocatalysis, Photocatalysis; Conversion and storage of chemical energy. He was a senior scientist at the Hahn-Meitner-Institut-Berlin (now Helmholtz-Zentrum Berlin) from 1985 to 1997. His main research interests at IC2MP UMR CNRS 7285 are (Photo) electrochemistry and (Photo) electrocatalysis of novel materials using various ex-situ and in-situ techniques, fuel generation, interfacial characterization and surface analytical techniques. (http://ic2mp.labo.univ-poitiers.fr/index.php/equipes/samcat/e3-personnels-permanents/alonso-vante-nicolas/). He is the author of over 250 publications, book chapters, editor of a two-volume e-book on electrochemistry in Spanish, author of two books, and six patents. Current h-index 46 with ca. 6780 citations (RG source).

Yongjun Feng (Dr.) has been a full professor and a Ph.D. supervisor at the State Key Laboratory of Chemical Resource Engineering in Beijing University of Chemical Technology (BUCT) since 2014. He got his Ph.D. degree in Chemistry and Materials Science in the Prof. Dr. C. Taviot-Gueho's group at Blaise Pascal University, France, in 2006, and then he spent three and a half years as a postdoctoral fellow with Prof. N. Alonso-Vante at the Laboratory of Catalysis in Organic Chemistry in University of Poitiers, France. Dr. Feng has published more than 80 peer-reviewed papers cited more than 1800 times and two book chapters in English and Chinese, respectively. He has 15 authorized national invention patents and one public U.S. Patent. Dr. Feng is currently working on fundamental, applied fundamental, and engineering research on polymer additives based on layered double hydroxides, non-precious metal electrocatalysts for fuel cells, and porous materials for catalyst support and adsorption. Moreover, he has served as the member of the Editorial Board for Catalysts, Recent Patents on Nanotechnology and Plastics Additives journals.

Hui Yang is currently a professor of Physical Chemistry at Shanghai Advanced Research Institute (SARI), Chinese Academy of Sciences (CAS). He received his Ph.D. (1997) in electrochemistry from Changchun Institute of Applied Chemistry, CAS. After six-years (2000-2005) of postdoctoral research in Japan, France and USA, he joined Shanghai Institute of Microsystem and Information Technology, CAS as a professor of Material Physics and Chemistry and then worked as the director of Center for Energy Storage and Conversion at SARI from 2010. His research interests include the PEMFCs, electrocatalysis, new energy-storage materials and technologies.

catalysts

Editorial

Novel Non-Precious Metal Electrocatalysts for Oxygen Electrode Reactions

Nicolas Alonso-Vante [1,*], Yongjun Feng [2] and Hui Yang [3]

[1] IC2MP-UMR CNRS 7285, University of Poitiers, 86022 Poitiers, France
[2] State Key Laboratory of Chemical Resource Engineering, College of Chemistry, Beijing University of Chemical Technology (BUCT), No. 15, Beisanhuan East Road, Chaoyang District, Beijing 100029, China
[3] Shanghai Advanced Research Institute, Chinese Academy of Sciences, No. 99, Haike Road, Shanghai 201210, China
* Correspondence: nicolas.alonso.vante@univ-poitiers.fr

Received: 16 August 2019; Accepted: 26 August 2019; Published: 29 August 2019

The collection of articles in the Catalyst special issue entitled "Novel Non-Precious Metal Electrocatalysts for Oxygen Electrode Reactions" mirrors the relevance and strengths to address the inevitable increasing demand of energy. This subject matter has stimulated considerable research on alternative energy harvesting technologies, conversion, and storage systems with high efficiency, cost-effective, and environmentally friendly systems, such as fuel cells, rechargeable metal-air batteries, unitized regenerative cells, and water electrolyzers [1–5]. In these devices, the conversion between oxygen and water plays a key step in the development of oxygen electrodes: oxygen reduction reaction (ORR), and oxygen evolution reaction (OER). To date, the state-of-art catalysts for ORR consist of platinum-based materials (Pt), while ruthenium (Ru)- and iridium (Ir)-oxides are the best known OER catalyst materials. The scarcity of the precious metals, their prohibitive cost, and declining activity greatly hamper the practice for large-scale applications. It is thus of paramount practical importance and interest to develop efficient and stable materials for the oxygen electrode based on earth-abundant non-noble metals [6–8]. In this connection, novel non-precious metal electrocatalysts for oxygen electrode reactions have been explored based on the innovative design in chemical compositions, structure, morphology, and supports.

This Special Issue covers recent progress and advances in novel non-precious metal electrocatalysts tailoring with high activity and stability for the catalytic conversion between water and oxygen. Additionally, electrocatalytic activity, selectivity, durability, and the mechanism for single or bifunctional oxygen electrodes, a current key topic in electrocatalysis, is an important subject for this Special Issue.

This special issue comprises a total of 10 scientific articles of which three are review articles and seven are research articles from respected colleagues around the world. Herein, one review paper and five research articles pay special attention to ORR high-performance electrocatalysts. For example, Xiong et al. [9] summed up recent progress on three-dimensional hetero-atom-doped nanocarbon for metal-free ORR electrocatalysis; Schardt et al. [10] carefully investigated the influence of the structure-forming agent on the composition, morphology and ORR performance of Fe-N-C electrocatalysts; Zhu et al. [11] developed a novel metal-organic framework route to embed Co nanoparticles into multi-walled carbon nanotubes for ORR in alkaline media; Liu et al. [12] fabricated N,S co-doped carbon nanofibers derived from bacterial cellulose/poly(methylene blue) hybrid for ORR; Liu et al. [13] prepared porous Fe-N-S/C electrocatalysts for ORR in a Zn-air battery using g-C_3N_4 and 2,4,6-tri(2-pyridyl)-1,3,5-triazine as binary nitrogen precursors; Zeng et al. [14] reported the $Ag_4Bi_2O_5/MnO_2$ corn/cob-like nanomaterial as a superior catalyst for ORR in alkaline media. One review paper and one research article are involved in the OER electrocatalysts. For instance, Li et al. [15] summarized recent advances and perspectives on host-guest engineering of layered double hydroxides (LDH) to manufacture high-performance OER electrocatalysts; Liu et al. [16] engineered mesoporous NiO electrocatalyst with enriched electrophilic Ni^{3+} and O for high-performance OER. Likewise, one review and one research paper are concern bifunctional electrocatalysts. For example, Zhong et al. [17] reviewed recent advance on the design, synthesis and electrocatalytic performance of cobalt-based electrocatalysts for oxygen electrode reactions and hydrogen evolution reaction; and Qiao et al. [18] designed and synthesized cobalt and nitrogen co-doped graphene-carbon nanotube aerogel as an efficient bifunctional electrocatalysts towards ORR and OER.

Summing-up, this special issue covers recent progress on high-performance and non-precious oxygen electrode catalysts providing novel ideas to tailor potential electrocatalytic materials. The Guest Editors really hope that the readers will appreciate the variety of contributions neighboring their own field of research.

Funding: This research received no external funding.

Conflicts of Interest: The authors declare no conflict of interest.

References

1. Dekel, D.R. Review of cell performance in anion exchange membrane fuel cells. *J. Power Sour.* **2018**, *375*, 158–169. [CrossRef]
2. Tahir, M.; Pan, L.; Idrees, F.; Zhang, X.W.; Wang, L.; Zou, J.J.; Wang, Z.L. Electrocatalytic oxygen evolution reaction for energy conversion and storage: A comprehensive review. *Nano Energy* **2017**, *37*, 136–157. [CrossRef]
3. Ghosh, S.; Basu, R.N. Multifunctional nanostructured electrocatalysts for energy conversion and storage: Current status and perspectives. *Nanoscale* **2018**, *10*, 11241–11280. [CrossRef]
4. Omrani, R.; Shabani, B. Review of gas diffusion layer for proton exchange membrane-based technologies with a focus on unitised regenerative fuel cells. *Int. J. Hydrog. Energy* **2019**, *44*, 3834–3860. [CrossRef]
5. Xu, H.M.; Ci, S.Q.; Ding, Y.C.; Wang, G.X.; Wen, Z.H. Recent advances in precious metal-free bifunctional catalysts for electrochemical conversion systems. *J. Mater. Chem. A* **2019**, *7*, 8006–8029. [CrossRef]
6. Liu, D.D.; Tao, L.; Yan, D.F.; Zou, Y.Q.; Wang, S.Y. Recent advances on non-precious metal porous carbon-based electrocatalysts for oxygen reduction reaction. *ChemElectroChem* **2018**, *5*, 1775–1785. [CrossRef]
7. Zhu, Y.P.; Guo, C.X.; Zheng, Y.; Qiao, S.Z. Surface and interface engineering of noble-metal-free electrocatalysts for efficient energy conversion processes. *Acc. Chem. Res.* **2017**, *50*, 915–923. [CrossRef]
8. Wu, G.; More, K.L.; Johnston, C.M.; Zelenay, P. High-performance electrocatalysts for oxygen reduction derived from polyaniline, iron, and cobalt. *Science* **2011**, *332*, 443–447. [CrossRef]
9. Xiong, D.; Li, X.; Fan, L.; Bai, Z. Three-dimensional heteroatom-doped nanocarbon for metal-free oxygen reduction electrocatalysis: A review. *Catalysts* **2018**, *8*, 301. [CrossRef]
10. Schardt, S.; Weidler, N.; Wallace, W.; Martinaiou, I.; Stark, R.; Kramm, U. Influence of the structure-forming agent on the performance of Fe-N-C Catalysts. *Catalysts* **2018**, *8*, 260. [CrossRef]
11. Zhu, H.; Li, K.; Chen, M.; Cao, H.; Wang, F. A novel metal–organic framework route to embed Co nanoparticles into multi-walled carbon nanotubes for effective oxygen reduction in alkaline media. *Catalysts* **2017**, *7*, 364. [CrossRef]
12. Liu, J.; Ji, Y.-G.; Qiao, B.; Zhao, F.; Gao, H.; Chen, P.; An, Z.; Chen, X.; Chen, Y. N,S Co-doped carbon nanofibers derived from bacterial cellulose/poly(methylene blue) hybrids: Efficient electrocatalyst for oxygen reduction reaction. *Catalysts* **2018**, *8*, 269. [CrossRef]
13. Liu, X.; Chen, C.; Cheng, Q.; Zou, L.; Zou, Z.; Yang, H. Binary nitrogen precursor-derived porous Fe-N-S/C catalyst for efficient oxygen reduction reaction in a Zn-air battery. *Catalysts* **2018**, *8*, 158. [CrossRef]
14. Zeng, X.; Pan, J.; Sun, Y. Preparation of $Ag_4Bi_2O_5/MnO_2$ corn/cob like nano material as a superior catalyst for oxygen reduction reaction in alkaline solution. *Catalysts* **2017**, *7*, 379. [CrossRef]
15. Li, J.; Jiang, S.; Shao, M.; Wei, M. Host-guest engineering of layered double hydroxides towards efficient oxygen evolution reaction: Recent advances and perspectives. *Catalysts* **2018**, *8*, 214. [CrossRef]
16. Liu, X.; Zhai, Z.-Y.; Chen, Z.; Zhang, L.-Z.; Zhao, X.-F.; Si, F.-Z.; Li, J.-H. Engineering mesoporous NiO with enriched electrophilic Ni^{3+} and O− toward efficient oxygen evolution. *Catalysts* **2018**, *8*, 310. [CrossRef]

17. Zhong, H.; Campos-Roldán, C.; Zhao, Y.; Zhang, S.; Feng, Y.; Alonso-Vante, N. Recent advances of cobalt-based electrocatalysts for oxygen electrode reactions and hydrogen evolution reaction. *Catalysts* **2018**, *8*, 559. [CrossRef]
18. Qiao, X.; Jin, J.; Fan, H.; Cui, L.; Ji, S.; Li, Y.; Liao, S. Cobalt and nitrogen Co-doped graphene-carbon nanotube aerogel as an efficient bifunctional electrocatalyst for oxygen reduction and evolution reactions. *Catalysts* **2018**, *8*, 275. [CrossRef]

catalysts

MDPI

Review

Host-Guest Engineering of Layered Double Hydroxides towards Efficient Oxygen Evolution Reaction: Recent Advances and Perspectives

Jianming Li [2,†], Shan Jiang [1,†], Mingfei Shao [1,*] and Min Wei [1]

[1] State Key Laboratory of Chemical Resource Engineering, Beijing University of Chemical Technology, Beijing 100029, China; jiangshan3_3@163.com (S.J.); weimin@mail.buct.edu.cn (M.W.)
[2] Petroleum Geology Research and Laboratory Center, Research Institute of Petroleum Exploration & Development (RIPED), PetroChina, Beijing 100083, China; lijm02@petrochina.com.cn
* Correspondence: shaomf@mail.buct.edu.cn
† These authors contribute equally to this work.

Received: 29 March 2018; Accepted: 25 April 2018; Published: 17 May 2018

Abstract: Electrochemical water splitting has great potential in the storage of intermittent energy from the sun, wind, or other renewable sources for sustainable clean energy applications. However, the anodic oxygen evolution reaction (OER) usually determines the efficiency of practical water electrolysis due to its sluggish four-electron process. Layered double hydroxides (LDHs) have attracted increasing attention as one of the ideal and promising electrocatalysts for water oxidation due to their excellent activity, high stability in basic conditions, as well as their earth-abundant compositions. In this review, we discuss the recent progress on LDH-based OER electrocatalysts in terms of active sites, host-guest engineering, and catalytic performances. Moreover, further developments and challenges in developing promising electrocatalysts based on LDHs are discussed from the viewpoint of molecular design and engineering.

Keywords: layered double hydroxide; oxygen evolution reaction; active site; water splitting

1. Introduction

Electrochemical water splitting holds great promise for clean energy resources and has aroused broad study interest in recent years [1–5]. Among all the studies, the development of electrocatalysts for the anode oxygen evolution reaction (OER) is one of the key issues to decrease the overpotential of practical water splitting due to its sluggish four-electron process [6–10]. It is well-known that ruthenium and iridium oxides demonstrate high activity for water oxidation in acid and alkaline electrolytes, respectively [11,12]. However, an efficient alternative is still needed because of the high cost and scarcity of noble metal-based catalysts, which is difficult to meet the large-scale applications. Recently, various transition metal compounds (e.g., oxides, [13–15], hydroxides [16–20], and phosphides [21–24]) have emerged as a new family of OER electrocatalysts. Especially, the homogeneous mixed-transition-metal compounds without phase segregation have been reported with higher OER activity, probably owing to the effectively modulated 3d electronic structures. For instance, NiFe-based electrocatalysts have become a kind of dazzling material attributed to their high OER activity, since first investigated by Corrigan in the 1980s [25], and significantly promoted by Dai in 2013 [26].

Layered double hydroxides (LDHs) are a large class of two-dimensional (2D) intercalated materials which can be described by the general formula $[M^{II}_{1-x}M^{III}_x(OH)_2]^{z+}(A^{n-})_{z/n} \cdot yH_2O$ (M^{II} and M^{III} are divalent and trivalent metals respectively; A^{n-} is the interlayer anion compensating for the positive charge of the brucite-like layers) [27–29]. Recently, LDHs, especially NiFe-LDH, are believed to be one of the ideal and promising electrocatalysts for water splitting due to their excellent OER activity, high stability in basic conditions, and low cost [30–34]. To date, various LDHs, as well as their nanocomposites, have been synthesized for better OER performances. For example, the design and synthesis of LDHs/conducting-material composites can overcome the intrinsically poor conductivity of LDHs and provide a rapid transport of electrons/ions [35–42]. To improve the intrinsic activity, the role of host transition layers and guest interlayer anions in increasing the OER activity have also been considered [43–49]. Boettcher et al. found that the incorporation of Fe shows a more than 30-fold increase in conductivity, as well as a partial-charge transfer activation effect of Fe to Ni sites [50]. Jin et al. detected the presence of Fe^{IV} in NiFe-LDH during steady-state water oxidation by using operando Mössbauer spectroscopy [51], which has important implications for stabilizing the NiOOH lattice. Although, with this progress, how to determine the real active sites, as well as how to rationally design much more efficient electrocatalysts based on LDHs, still remains highly desirable and challenging.

Along with the increasing development of LDH-based OER catalysts, some important reviews on different aspects have already been reported. For example, Dai et al. first gave a mini review about NiFe-based materials (including alloys, oxides, and hydroxides) for OER in 2015 [52], where the related mechanism and applications have been briefly discussed. Our group summarized the development of LDH materials for electrochemical energy storage and conversion [53], in which the OER applications have been discussed from the viewpoint of electrode materials design. Strasser et al. further published a review article focusing on the progresses of NiFe-based (Oxy)hydroxide catalysts [54]. Other excellent reviews also mentioned LDH-based OER electrocatalysts, including the perspective on the OER activity trends and design principles based on transition metal oxides and (oxy)hydroxides by Boettcher et al., [55] and nanocarbon-based electrocatalysts summarized by Zhang et al., [56].

Nevertheless, the discussion on the LDH-based OER catalysts from the viewpoint of their supramolecular intercalated structures is seldom considered. In this review, we will focus on the roles of the host layer and guest interlayer anions in the OER performances of LDHs. The recent advances in the host layer designs will be first discussed in order to provide a systematic digestion of previous achievements in the unveiling of the active nature of metal ions in promoting water oxidation. The interlayer anions also play an indispensable synergistic effect, which will be demonstrated in the next discussion. We also hope to display future efficient OER electrocatalysts based on LDHs from the molecular design and engineering.

2. The Engineering of Host Layer

2.1. The Role of Host Layer Metal Ions

Electrochemical water splitting under both acidic and alkaline conditions has been studied for more than a half century as a means of storing clear energy. It is noted that OER is a four electron-proton coupled reaction, while hydrogen evolution reaction (HER) is only a two electron-transfer reaction, which implies that a higher energy is required for OER process to overcome the reaction barrier. In the past decades, nickel-based electrodes have been widely used as anode catalysts for alkaline water splitting due to the merits of earth-abundance, high activity, and good stability [57–61]. Corrigan found that a low concentration of iron impurity in nickel oxide increased the OER kinetics obviously [25]. Moreover, the oxygen evolution overpotential for the sample with iron impurities of 10–50% in the nickel oxide is substantially lower than that of either nickel oxide or iron oxide. Although without a clear understanding about this interesting phenomenon, a synergetic effect between nickel and iron species for catalyzing the OER process has been rationally assumed. This inspired the intensive investigations to study various NiFe mixed compounds and the catalytic roles for Ni and Fe in order to obtain better OER electrocatalysts. As a family of typical two-dimensional inorganic materials, LDHs consist of brucite-like $[Mg(OH)_2]$ host nanosheets with edge-sharing metal-O_6 octahedra (lateral particle size ranging from nanometer to micrometer-scale). The easy tunability of metal ions without altering the structure, as well as anion exchange properties of LDHs, make them interesting alternatives for applications in electrochemistry. Bell et al. synthesized (Ni,Fe) oxyhydroxides across the entire composition range (Figure 1a,b) [62]. It was found that the addition of 25% Fe to Ni(OH)$_2$ results in up to a 500-fold higher OER current density compared to pure Ni and Fe oxyhydroxide films at the overpotential of 0.3 V. Zhang et al. also investigated the effect of different Ni/Fe molar ratios to their OER performances [63]. It was found that moderate metal substitution into the host hydroxide framework (Fe into Ni or Ni into Fe) substantially enhanced the OER activity with a decrease of both the Tafel slope and overpotential. To further accurately control the Fe content, Boettcher et al. developed a method for purification of KOH electrolyte by using precipitated bulk Ni(OH)$_2$ to absorb Fe impurities [50]. As a result, no significant OER current is observed until >400 mV overpotential for Ni(OH)$_2$ films aged in purified KOH electrolyte. The excellent conductivity of electrocatalysts is important to achieve a fast charge transfer through the catalyst film with a negligible potential drop. To study the conductivity trends of LDHs with different metal ion ratios, various NiFe-LDH have been synthesized by using the purified KOH electrolyte. The conductivity measurement showed that all NiFe-LDH films have low conductivities at low potential, while a sharp increase along with the Ni oxidation (Figure 1c). The NiFe-LDH films with Fe content range from 5% to 25% gives $\sigma \approx 3.5$ to 6.5 mS cm^{-1}, significantly higher than that of Fe-free film ($\sigma \approx 2.5$ mS cm^{-1}), indicating that Fe increases the conductivity of NiOOH. Figure 1d compares the turnover frequency (TOF) as a function of the thickness for Fe-free NiOOH and Ni$_{0.75}$Fe$_{0.25}$OOH deposited on two kinds of substrate (Au and GC). It is concluded that Fe enhances the activity of NiOOH through a Ni−Fe partial-charge-transfer activation process.

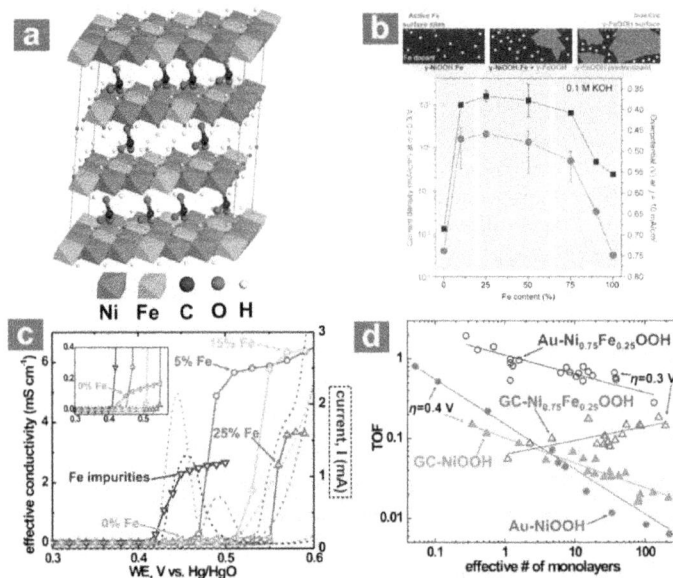

Figure 1. (a) Schematic illustration of the crystal structure of NiFe-LDH. (b) Measured OER activity of mixed Ni−Fe catalysts as a function of Fe content. Reproduced from [62], Copyright American Chemical Society, 2014. (c) Conductivity data for the Ni$_{1-x}$Fe$_x$OOH with various Fe content. (d) TOFs as a function of film thickness for Fe-free NiOOH and Ni$_{0.75}$Fe$_{0.25}$OOH. Reproduced from [50], Copyright American Chemical Society, 2014.

In addition to the enhanced conductivity, the catalytically-active metal redox state of the NiFe-based catalyst has remained under debate. It has long been assumed that Ni is the reactive site for water oxidation in NiFe oxide electrocatalysts on the basis of the high activity of Ni oxide electrocatalysts. For example, Strasser et al. found that the Fe centers consistently remain in the Fe^{III} state regardless of potential and composition [64], which was determined by operando differential electrochemical mass spectrometry (DEMS) and X-ray absorption spectroscopy (XAS) under OER conditions. On the other hand, Ni^{IV} at Fe content below 4% has been detected under catalytic conditions, while Ni atoms stabilized in a low-valent oxidation state when further increasing the content of Fe. This difference in metal valent states mainly depends on the rate of water oxidation—metal-reduction (k_{OER}) and metal oxidation (k_{Mox}) (Figure 2a). The lower k_{Mox}/k_{OER} ratio reflects a dramatically increased rate constant of water oxidation (k_{Mox}), which may exceed the rate of the metal oxidation (k_{OER}). It concluded that high catalytic OER activity of the mixed Ni—Fe catalysts demonstrate a sharply-decreased k_{Mox}/k_{OER} ratio. They further give a catalytic OER cycle (Figure 2b), where the buildup of oxidation equivalents from Ni^{II} to Ni^{IV} sites is followed by the O—O bond formation with the subsequent release of molecular oxygen. However, there is substantial evidence from X-ray absorption and Mössbauer experiments that Ni^{IV} and Fe^{IV} are both found in NiFe-LDH at OER potentials during different studies [51,65,66]. Recently, Stahl et al. detected the Fe^{IV} species (up to 21% of the total Fe) during steady-state water oxidation on NiFe-LDH [51]. The stable presence of Fe^{IV} can be ascribed to the increased electron-donating ability of the π-symmetry lone pairs of the bridging oxygen atoms between Ni and Fe (Figure 2c,d), which makes the NiOOH lattice a more stable environment for high-valent metal ions.

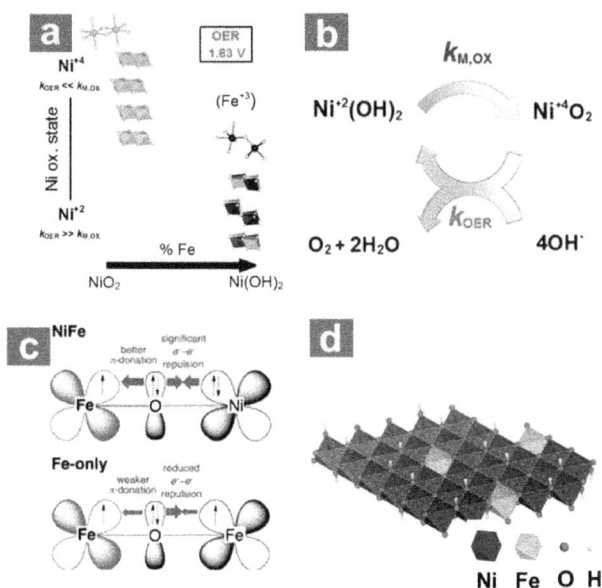

Figure 2. (**a**) XAS-derived structural motifs prevalent during OER catalysis at high and intermediate Ni-content. (**b**) Simplified scheme of the electrochemical water splitting cycle. Reproduced from [64], Copyright American Chemical Society, 2016. (**c**) Electronic effects that could rationalize the observation of Fe^{IV} in NiFe, but not Fe oxide catalysts. (**d**) Schematic representation of a layered NiOOH lattice containing Fe ions in different sites (orange-brown). Reproduced from [51], Copyright American Chemical Society, 2015.

Based on the previous results and the above discussion, the brief conclusions can be obtained: (1) iron impurity in nickel oxide and hydroxide significantly promote the electrocatalytic water oxidation, and suitable Ni/Fe ratio (from 2:1 to 4:1) can further improve OER activity;

(2) the conductivity of $Ni(OH)_2/NiOOH$ increases when combined with iron element at a suitable level (from 5% to 25%) due to the Fe-induced charge transfer; (3) the Ni or Fe as active sites have both been reported, which has been verified by the detected high valence states of Ni^{IV} or Fe^{IV}, while the presence of high valence metal ions probably depends on the metal and water oxidation rate. Therefore, different metal ions introduced into the LDH host layer induces a varied chemical and electronic environment, which thereby varied their OER performances. In addition to NiFe-LDH, other LDHs with various host metal ions (e.g., NiCo, NiMn, ZnCo, and CoAl) have been demonstrated as OER electrocatalysts (Figure 3) [67–91]. However, their activity is still lower than that of NiFe-LDH.

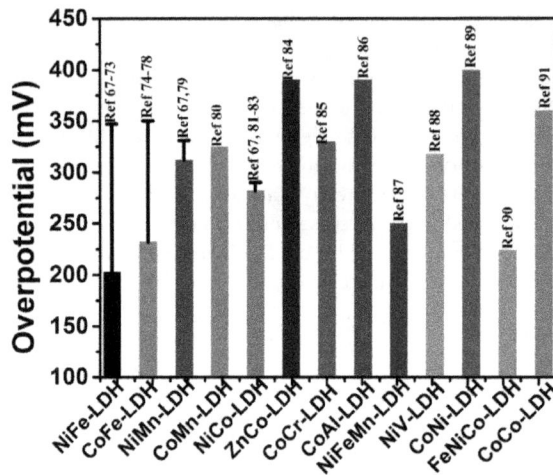

Figure 3. Overpotentials required at $j = 10$ mA cm^{-2} for various LDHs (the error bars indicate a range of overpotentials).

2.2. The Engineering of LDH Host Layers

2.2.1. The Exfoliation of LDHs

The electrochemical properties of an electrocatalyst are affected by its nanostructures [92–95]. For instance, the atomically-thin 2D inorganic materials usually demonstrate unique properties compared with their bulk counterparts [96–100]. LDHs are composed of atomically-thick positive brucite-like host layers and interlayer anions. In practice, LDHs are stacked with several layers, which limits their electrochemical performances due to the inaccessibility to the inner surfaces of the host layers. In the past decades, ultrathin LDH nanosheets with atomic thickness have been synthesized by both "bottom-up" and "top-down" approaches [101–106], which provide opportunities in maximizing the utility of the layers and improving their physicochemical properties (e.g., specific surface area and conductivity). Particularly, the "top-down" delamination method is the most widely developed for producing thin LDH platelets with a thickness of a few atomic layers. Hu et al. first used the exfoliation strategy to promote the OER performances of LDHs [91]. The CoCo, NiCo, and NiFe-LDH with Br⁻ anions are prepared as representative LDH materials, which are exfoliated into single-layer nanosheets in the formamide solution. The OER current densities at an overpotential of 300 mV were enhanced by 2.6-, 3.4-, and 4.5-fold upon exfoliation of CoCo, NiCo, and NiFe-LDH, respectively, compared with their bulk materials. Additionally, the water oxidation activity has the order of NiFe > NiCo > CoCo for both exfoliated nanosheets and bulk LDHs. Following this work, they further synthesized ultrathin CoMn-LDH nanoplatelets (3–5 nm) by a coprecipitation method [80], which gives a current density of 42.5 mA cm^{-2} at η = 350 mV. This value is about 7.6, 22.5, and 2.8 times higher than that of Co(OH)$_2$ + Mn$_2$O$_3$, spinel MnCo$_2$O$_{4+\delta}$, and IrO$_2$, respectively.

Exfoliated ultrathin LDH nanosheets display enhanced active site exposure. However, the LDHs' exfoliation in liquid usually suffers from strong adsorption of solvent molecules, as well as the restacking when removing the surface solvent [107,108]. In addition to liquid exfoliation of LDHs, Wang et al. developed an efficient strategy for the exfoliation of LDHs into stable and clean ultrathin nanosheets by plasma etching [78]. The high-energy plasma destroys the ionic bonds and hydrogen bonds in the interlayers of the bulk LDHs, which interrupted the host-guest charge balance and separated the brucite-like host layers from each other. For instance, the thickness of CoFe-LDH have been successfully decreased from ~20 nm to 0.6 nm by subjecting bulk CoFe-LDH to Ar plasma etching for 60 min. Moreover, the coordination number of the Co-O$_{OH}$ octahedra is lower than that in the bulk CoFe-LDH, suggesting the presence of oxygen vacancies (VO). The as-prepared LDH ultrathin nanosheets demonstrate much-improved OER performance with a low overpotential of 266 mV at 10 mA cm^{-2}. Recently, they further reported the exfoliation of CoFe-LDH by a water plasma-assisted strategy (Figure 4a) [109], which was accompanied with the formation of multi-vacancies, including O, Co, and Fe vacancies. The as-exfoliated ultrathin LDHs nanosheets with multi-vacancies show significantly promoted electrocatalytic activity for water oxidation. As shown in Figure 4b, water-plasma exfoliated CoFe-LDH nanosheets just require a low overpotential of 290 mV to reach 10 mA cm^{-2} while the pristine CoFe-LDH need an overpotential of 332 mV. Therefore, the effective exfoliation, as well as the defect introduction, both promotes the OER activity of LDHs.

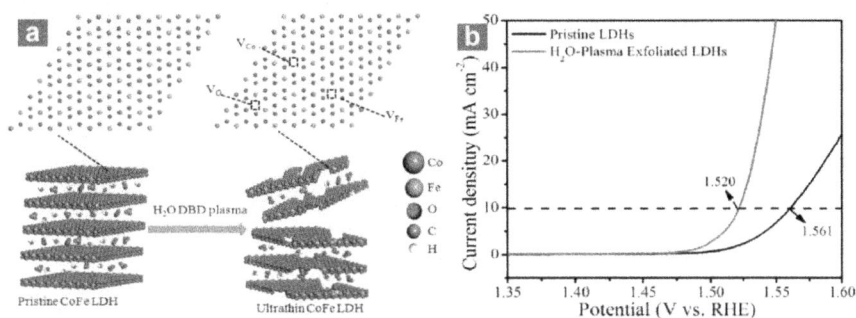

Figure 4. (a) Schematic illustration of the water-plasma-enabled exfoliation of CoFe-LDH nanosheets in a dielectric barrier discharge (DBD) plasma reactor. (b) Linear scan voltammogram (LSV) curves for OER on pristine CoFe-LDH and the water-plasma exfoliated CoFe-LDH nanosheets. Reproduced from [109]. Copyright Wiley, 2017.

2.2.2. Construction of LDH Nanoarrays

LDH nanosheet arrays (NSAs), that have highly-dispersed nanoplatelets, well-uniformed orientation, and improved conductivity compared with LDH powdered samples, have been recently constructed as efficient OER electrocatalysts [110–118]. Various LDH NSAs have been perpendicularly grown on the surface of conducting substrates (metals [115], conducting glasses [116], carbon fibers [117], and papers [118]) by in situ procedures. One of the most effective methods for the fabrication of LDH NSAs is the hydrothermal process [119,120]. To design an highly-active OER electrocatalyst, Huang et al. reported a single-crystalline NiFe-LDH NSA array on a Ni foam with the assistance of a direction agent of NH$_4$F [120]. The top and cross-sectional Scanning Electron Microscope (SEM) images of NiFe-LDH NSAs reveal a highly-oriented flake array nanostructure that is in vertical contact with the substrate (Figure 5a,b), with an edge length of 1–3 μm and a uniform thickness of less than 20 nm. The high-resolution transmission electron microscopy (HRTEM) image and corresponding selected area electron diffraction (SAED) pattern illustrate a single crystal phase of LDH (Figure 5c).

The NiFe-LDH NSAs exhibits superior OER activity compared with the coated NiFe-LDH film, as well as a RuO_2 film electrode, achieving the overpotentials of 210 mV, 240 mV, and 260 mV at the current densities of 10, 50, and 100 mA cm^{-2}, respectively. Moreover, it is found that the single-crystalline NiFe-LDH arrays display smaller overpotentials than that of the reported amorphous NiFe materials and other analogous LDH-based materials. The hexamethylenetetramine ($C_6H_{12}N_4$) also can be used as a direct agent to prepare vertically-aligned LDH NSAs [72]. For example, the NiFe-LDH NSAs grown on the nickel foam were created by an in situ co-precipitation approach using a reaction solution containing $Ni(NO_3)_2$, $Fe(NO_3)_3$, $C_6H_{12}N_4$, and CH_3OH. SEM images of the NiFe-LDH NSAs reveal a three-dimensional (3D) porous architecture with a LDH thickness of about 15 nm. TEM elemental mapping images of as-synthesized NiFe-LDH scratched off nickel foam suggest that Ni, Fe, and O elements are uniformly distributed over the NiFe-LDH (Figure 5f). The OER performances of LDH@nickel foam (NF) NSAs, $Ni(OH)_2$@NF NSAs, and NF were evaluated in a typical three-electrode electrochemical cell in 1.0 M KOH solution at room temperature. Figure 5g displays the OER polarization curves of LDH@NF, $Ni(OH)_2$@NF, and NF. It is clear that NiFe-LDH@NF demonstrates the highest OER activity compared with the contrast samples, with the lowest overpotential of 210 mV at 10 mA cm^{-2}, which are 88, 110, and 161 mV less than those of NiCo-LDH@NF, $Ni(OH)_2$@NF, and NF, respectively. In addition, the well-uniformed NiFe-LDH NSAs also displays promising HER performances in 1.0 mM KOH solution with low overpotential of 133 mV at 10 mA cm^{-2}. The bi-functional electrocatalysts for OER and HER were further used for the overall water splitting in a two-electrode electrolysis cell (Figure 5h), which just needs a cell voltage of 1.59 V to give a water splitting current density of 10 mA cm^{-2} in 1.0 M KOH solution with a scan rate of 2 mV s^{-1} (Figure 5i). It is found that the nanosheet array architecture has increased the electrochemical surface area, which provides more catalytic active sites and favors the efficient adsorption and transfer of reactants. In addition, the well-ordered arrays also benefit the gas evolution reaction and subsequently enhance the electrocatalytic activity [121].

In addition to direct co-precipitation, the template-directed method is another effective strategy for the synthesis of LDH arrays as efficient OER electrocatalysts [122–124]. Sun et al. develop a two-step hydrothermal method to synthesize hierarchical NiCoFe-LDH NSAs [125]. In this process, $Co_2(OH)_2CO_3$ nanowire arrays grown on the Ni foam were first achieved to provide a Co source and to support the growth of NiCoFe-LDH NSAs in the presence of Fe^{III} and urea (Figure 6a). The introduction of Ni in the LDH can be ascribed to the dissolution of the Ni foam substrate at a low pH value (pH = 1.2) due to the hydrolysis of Fe^{III} and precipitation of Ni^{II} in the solution. The density of the LDH NSAs on the nanowires can be tuned by simply changing the molar amount of Fe^{III} during the second hydrothermal step. With a low concentration of Fe^{III} (0.5 mmol), just a few LDH nanoplatelets grew on the nanowires (denoted as H-LDH-0.5, Figure 6b). When the amount of Fe^{III} increased to 1 mmol, the density of LDH NSAs around the nanowires clearly increased and the diameter of the individual nanowire@nanoplatelet expanded to 250 nm (H-LDH-1, Figure 6c). The hierarchical nanoarray architecture benefits the improvement of the electrochemical properties of active materials by exposing more active sites. As a result, a high OER rate of 80 mA cm^{-2} for the H-LDHs could be readily achieved by applying a small overpotential (257 mV for H-LDH-1), much better than that of the $Co_2(OH)_2CO_3$ nanowire arrays (420 mV) and LDH NPs (492 mV). Zhao et al. fabricated CoFe-LDH NSAs by a solution phase cation exchange method at room temperature by dipping the Cu foam loaded with Cu_2O nanoarrays into an aqueous solution of $CoCl_2$ and $FeCl_2$ (Figure 6d) [74]. In this process, OH^- was generated in situ along with the etching of Cu_2O nanowires by $S_2O_3^{2-}$, which then leads to the precipitation of metal ions. The resulting CoFe-LDH NSAs inherited the geometry of the Cu_2O template (Figure 6e,f). The Cu_2O nanoarrays show negligible OER activity, while $Co_{0.70}Fe_{0.30}$-LDH NSAs give an OER onset overpotential as low as 220 mV, a small Tafel slope at 62.4 mV dec^{-1}, as well as excellent long-term durability (>100 h).

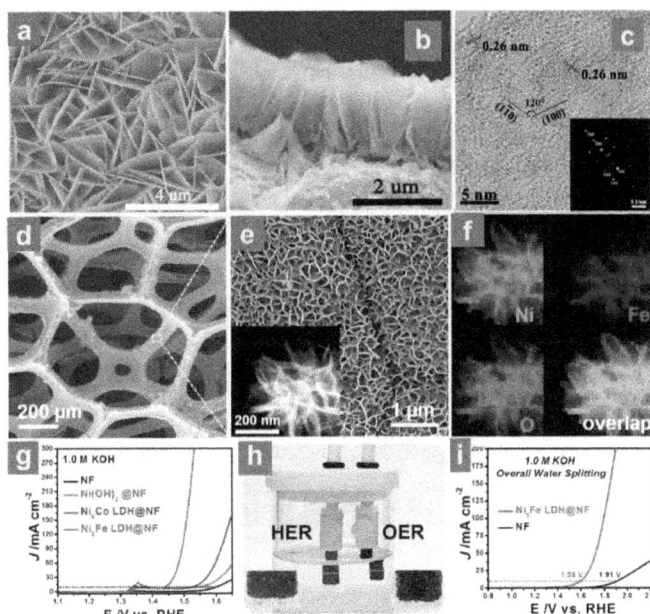

Figure 5. (**a**) SEM image of NiFe-LDH flakes grown on the nickel foam, and (**b**) is the corresponding cross-sectional image. (**c**) HRTEM image and SAED pattern (inset in (**c**)) of a single NiFe-LDH layer. Reproduced from [120], Royal of Society, 2016. (**d,e**) SEM images of Ni_5Fe LDH@NF. Inset in (**e**) is a HAADF-STEM image of the Ni_5Fe-LDH scratched off NF. (**f**) TEM elemental mapping images of Ni_5Fe LDH scratched off NF with blue for Ni, purple for Fe, and red for O. (**g**) OER polarization curves of NF, $Ni(OH)_2$@NF, Ni_5Co-LDH@NF, and Ni_5Fe-LDH@NF. (**h**) Optical images of the Ni_5Fe-LDH@NF electrode and overall water splitting device. (**i**) The steady-state polarization curve for overall water splitting of Ni_5Fe-LDH@NF and NF in a two-electrode configuration. Reproduced from [72], Copyright Wiley, 2017.

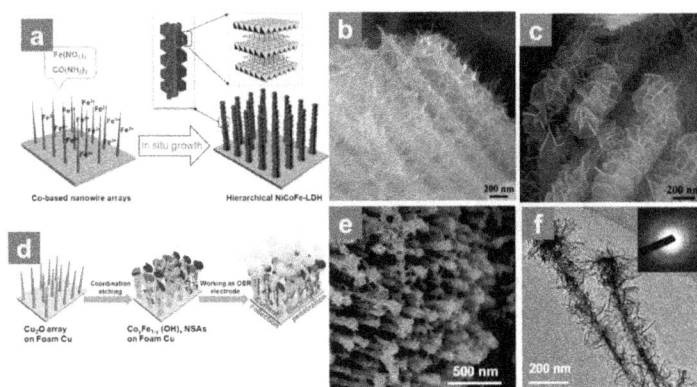

Figure 6. (**a**) Schematic illustration of the hierarchical NiCoFe-LDH nanoarrays via a two-step hydrothermal method. SEM images of the hierarchical NiCoFe-LDH nanoarrays: (**b**) H-LDH-0.5; (**c**) H-LDH-1. Reproduced from [125], Copyright Royal of Society, 2014. (**d**) Schematic illustration of the fabrication process for CoFe-LDH NSAs. (**e**) SEM and (**f**) TEM images of CoFe-LDH NSAs. Reproduced from [74], Copyright Royal of Society, 2017.

A facile method to prepare electrode materials with the merits of fast and one-pot synthesis on the conducting substrates is a critical step when considering the practical operations. The electrosynthesis approach is often used to fabricate electrochemical active films on conducting materials' surfaces [126–128]. Numerous materials have been developed by using the electrosynthesis process for applications in energy storage and conversion, such as transition metal oxides [129] and hydroxides [130], and conducting polymers [131]. The morphology and thickness of as-synthesized films can be manipulated by monitoring electrochemical variables, such as the potential and coulombic charge. Our group developed an electrochemical approach for the fast, precisely-controllable, and economic fabrication of various Fe-containing LDH hierarchical nanoarrays (Figure 7a) [68]. This electrosynthesis process was achieved by the following proposed reduction reaction on the working electrode: $NO_3^- + H_2O + 2e^- \rightarrow NO_2^- + 2OH^-$, in which the resulting OH^- leads to the precipitation of LDHs. The whole electrosynthesis process is finished successively within hundreds of seconds at room temperature and the thickness of the LDH NSAs can be controlled by the deposition time. Figure 7b shows the SEM image of the as-obtained NiFe-LDH NSAs, where ultrathin (8 nm in thickness) and uniform NiFe-LDH platelets were grown perpendicularly to the surface of nickel foam substrate. The TEM image further shows a thin sheet-like nanostructure. In addition, LDHs' NSAs with uniform and homogeneous surface morphology can be synthesized on the foam nickel substrate with different sizes, such as from 2 cm^2 to 100 cm^2 (Figure 7d). This is further adequate for the fabrication of NiFe-LDH NSAs on other conducting substrates, including conducting cloths and glasses (Figure 7e,f). The OER activity of NiFe-LDH NSAs was further studied, which displays the lowest onset potential of OER current and the highest current density at the same overpotential (η) compared with the CoFe- and LiFe-LDH NSAs. Moreover, it shows high energy conversion efficiency from electric energy to chemical energy with a Faradaic efficiency of 99.4% after a testing period of 10 min. The electrosynthesized NiFe-LDH NSAs also gives significantly long-term stability and the current density of OER remains constant at each given potential after 50 h of continued measurement. In addition, the electrosynthesis method can be extended to the various micro-substrates, which gives uniform core-shell nanostructures and improved electrochemical properties [132–136].

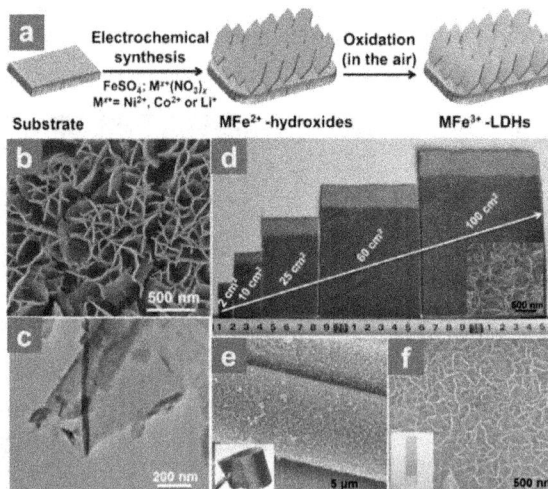

Figure 7. (**a**) Scheme of the synthetic route to MFe-LDH (M=Co, Ni and Li) NSAs. (**b**) SEM and (**c**) TEM image of NiFe-LDH NSAs. (**d**) Photographs of NiFe-LDH NSAs synthesized on the foam nickel substrates at various scales (inset: the SEM image of NiFe-LDH on 100 cm^2 substrate). SEM images for NiFe-LDH nanoplatelet arrays on (**e**) the conducting cloth and (**f**) FTO substrate (inset: their corresponding photographs). Reproduced from [68], Copyright Royal of Society, 2015.

3. The Engineering of Interlayer Guests

Various anions can be intercalated into the interlayer of LDHs, which varies their chemical environment of host layers. Layer-by-layer assembly (LBL) of exfoliated LDH nanosheets with different interlayer anions through electrostatic interaction is a good way to introduce different interlayer anions [137–141]. Our group has reported the assembly of well-ordered CoNi-LDH NS/iron porphyrin (Fe-PP) ultrathin film by means of the LBL strategy, giving rise to an excellent OER performance [142]. Compared with other interlayer anions, like sodium polystyrenesulfonate (PSS) and sodium dodecyl sulfate (SDS), CoNi-LDH NS assembled with Fe-PP shows superior OER activity, which may benefit from the excellent conductivity of Fe-PP. Moreover, LDHs with various host composition has also been combined with Fe-PP, such as CoMn-, CoFe-, and ZnCo-LDH, to form LDH NS/Fe-PP ultrathin films (UTFs). As a result, the OER performance of all these LDH NS/Fe-PP UTFs is significantly improved (Figure 8a), which means the conductive interlayer anions, like Fe-PP can efficiently improve the OER catalytic activities of LDHs. In addition, the increased intersheet spacing would be expected to facilitate OH^-/O_2 transport through the film. Xu et al. synthesized NiFe-LDH by introducing CO_3^{2-} to replace NO_3^-, leading to the reducing of interlayer spacing and poor OER catalytic activity [71]. On the basis of OH^- being the main reactant in alkaline solution, the NO_3^- has a better exchange ability with OH^- in interlamination, which also affects the OER performance of NiFe-LDH. Hunter et al. synthesized NiFe-LDH with 12 different interlayer anions, which shows different activities during OER reactions [143]. As shown in Figure 8b, overpotentials of [NiFe]-LDH materials with different interlayer anions illustrate that the measured overpotentials, which reflects the water oxidation abilities, do not match with the basal spacing. Further study shows that strong correlation can be found between the pKa values of the interlayer anions and overpotentials. As shown in Figure 8c, there is a midpoint of 3.4 ± 0.7. Considering the existence of 1 M OH^- in the electrolyte, it is clear that the di- and trivalent anions outcompeted the hydroxide that presented in the interlayer. This means that di- and trivalent anions can be seen as strongly-bound proton acceptors which can reduce the activation barrier for water oxidation. Interestingly, the DFT calculations suggest that the XPS signal at 405.1 eV correspond to the nitrite bound by its N-atom to edge-site iron, which is a symbol of high water oxidation activity (Figure 8d). In addition, the oxidation of NiFe-LDH in the alkaline media will generate the NiFeOOH along with the detachment of H from the topmost surface of the regular LDH. Zhang et al. found that the presence of interlayer CO_3^{2-} anions stabilized the active sites of LDHs [144]. As discussed above, the distance of interlayer space, as well as the chemical environment induced by the interlayer anions, play an important role in the OER activity of LDHs. Additionally, iron sites at the edges of [NiFe]-LDH nanosheets may be active in water oxidation catalysis through the related studies [145]. We expect that these findings can be potentially used in the engineering of OER catalysts.

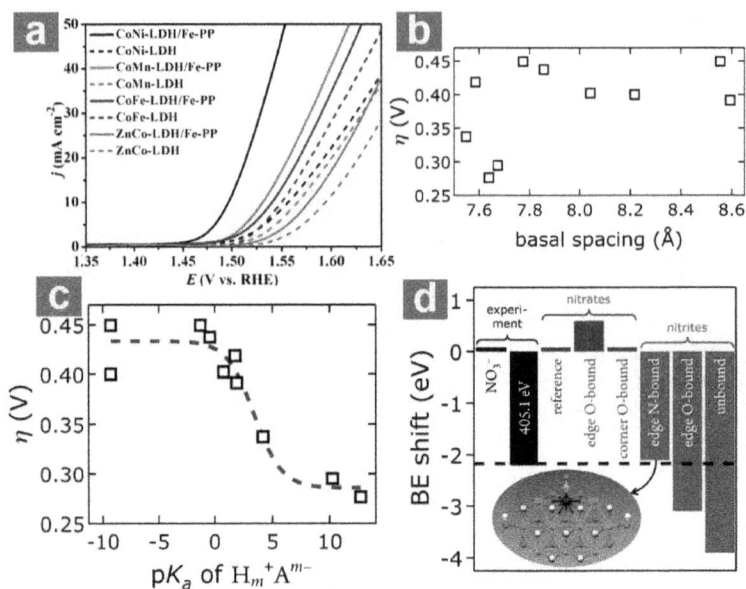

Figure 8. (**a**) LSV curves for various LDHs assembled with Fe-PP. Reproduced from [142], Copyright Royal of Society, 2016. (**b**) Observed overpotentials η as a function of the basal spacing of NiFe-LDH materials with different interlayer anions. (**c**) Overpotentials η of NiFe-LDH materials with different interlayer anions A^{m-}. (**d**) Measured (black) and calculated X-ray photoelectron spectroscopy (XPS) binding energy (BE) core level shifts (is-CLS) for differently-bound nitrates (blue) and nitrites (green) with a calculated structure that is consistent with the feature at 405.1 eV. Atom colors: Ni, green; Fe, maroon; N, blue; O, red; H, white. Reproduced from [143], Copyright Royal of Society, 2016.

4. Conclusions and Perspectives

In summary, we have systematically introduced the recent progress about LDHs, themselves, for electrochemical water oxidation from the view of host and guest engineering. The potential active sites of LDH have been scientifically discussed based on previous experiments and calculation. The metal ion ratio in the LDHs' host layer obviously affects their OER activity. The determination of the valence state of metal ions during water oxidation has been widely used to investigate the active sites of LDH, while this probably depends on the metal and water oxidation rate. As for guest anions, changing the space of the interlayer and the chemical properties (such as acidity and alkalinity) provides a chance to optimize of electrochemical performances for LDHs. Since the nanostructures have a profound impact on electrode materials, it is significant to synthesize uniform and highly-dispersed LDHs with all the active sites exposed to the electrolyte, such as ultrathin nanosheet arrays.

In the future, molecular-level control remains promising to entirely develop the potential of LDH-based OER electrocatalysts. Host-guest interaction of LDHs provides a large space for turning their electrochemical properties for molecule adsorption and catalytic transformation. Additionally, the defect chemistry of LDHs have also been paid increased attention, owing to the significantly-improved surface electronic structures. This also extends the question as to whether the surface or sites show a higher activity for the OER process because of their different defects. Despite the importance of optimizing the material properties, it is still a long way to the synthesis of low-cost and high-quality LDH-based electrocatalysts with precise control over their composition, structure, and morphology. To utilize solar energy in water splitting, how to combine the novel

LDH-based OER catalysts with photocatalysts will be a promising method to achieve an efficient photoelectrochemical process.

Acknowledgments: This work was supported by the National Natural Science Foundation of China, the 973 Program (grant no. 2014CB932102), the Fundamental Research Funds for the Central Universities (buctylkxj01; PYCC1704) and the R and D department of PetroChina.

Conflicts of Interest: The authors declare no conflict of interest.

References

1. Menezes, P.W.; Indra, A.; Das, C.; Walter, C.; Göbel, C.; Vitaly, G.V.; Schmeißer, D.; Driess, M. Uncovering the nature of active species of nickel phosphide catalysts in high–performance electrochemical overall water splitting. *ACS Catal.* **2017**, *7*, 103–109. [CrossRef]
2. Hou, Y.; Qiu, M.; Zhang, T.; Ma, J.; Liu, S.H.; Zhuang, X.D.; Yuan, C.; Feng, X.L. Efficient electrochemical and photoelectrochemical water splitting by a 3D nanostructured carbon supported on flexible exfoliated graphene foil. *Adv. Mater.* **2017**, *29*, 1604480.
3. Ryu, S.J.; Hoffmann, M.R. Mixed-metal semiconductor anodes for electrochemical water splitting and reactive chlorine species generation: Implications for electrochemical wastewater treatment. *Catalysts* **2016**, *6*, 59. [CrossRef]
4. Li, X.M.; Hao, X.G.; Abudula, A.; Guan, G.Q. Nanostructured catalysts for electrochemical water splitting: Current state and prospects. *J. Mater. Chem. A* **2016**, *4*, 11973–12000.
5. Reier, T.; Pawolek, Z.; Cherevko, S.; Bruns, M.; Jones, T.; Teschner, D.; Selve, S.; Bergmann, A.; Nong, H.N.; Schlögl, R.; et al. Molecular insight in structure and activity of highly efficient, low–Ir Ir–Ni oxide catalysts for electrochemical water splitting. *J. Am. Chem. Soc.* **2015**, *137*, 13031–13040. [CrossRef] [PubMed]
6. Liu, S.Y.; Li, L.J.; Ahn, H.S.; Manthiram, A. Delineating the roles of Co_3O_4 and N–doped carbon nanoweb (CNW) in bifunctional Co_3O_4/CNW catalysts for oxygen reduction and oxygen evolution reactions. *J. Mater. Chem. A* **2015**, *3*, 11615–11623. [CrossRef]
7. Li, P.X.; Ma, R.G.; Zhou, Y.; Chen, Y.F.; Zhou, Z.Z.; Liu, G.H.; Liu, Q.; Peng, G.H.; Liang, Z.H.; Wang, J.C. In situ growth of spinel $CoFe_2O_4$ nanoparticles on rod–like ordered mesoporous carbon for bifunctional electrocatalysis of both oxygen reduction and oxygen evolution. *J. Mater. Chem. A* **2015**, *3*, 15598–15606. [CrossRef]
8. Hutchings, G.S.; Zhang, Y.; Li, J.; Yonemoto, B.Y.; Zhou, X.G.; Zhu, K.K.; Jiao, F. In situ formation of cobalt oxide nanocubanes as efficient oxygen evolution catalysts. *J. Am. Chem. Soc.* **2015**, *137*, 4223–4229. [CrossRef] [PubMed]
9. Wu, L.H.; Li, Q.; Wu, C.H.; Zhu, H.Y.; Garcia, A.M.; Shen, B.; Guo, J.H.; Sun, S.H. Stable cobalt nanoparticles and their monolayer array as an efficient electrocatalyst for oxygen evolution reaction. *J. Am. Chem. Soc.* **2015**, *137*, 7071–7074. [CrossRef] [PubMed]
10. He, G.W.; Zhang, W.; Deng, Y.D.; Zhong, C.; Hu, W.B.; Han, X.P. Engineering pyrite-type bimetallic Ni-doped CoS_2 nanoneedle arrays over a wide compositional range for enhanced oxygen and hydrogen electrocatalysis with flexible property. *Catalysts* **2017**, *7*, 366. [CrossRef]
11. Cherevko, S.; Geiger, S.; Kasian, O.; Kulyk, O.; Grote, J.P.; Savan, A.; Shrestha, B.R.; Merzlikin, S.; Breitbach, B.; Ludwig, A.; et al. Oxygen and hydrogen evolution reactions on Ru, RuO_2, Ir, and IrO_2 thin film electrodes in acidic and alkaline electrolytes: A comparative study on activity and stability. *Catalysis Today* **2016**, *262*, 170–180. [CrossRef]
12. Lee, Y.M.; Suntivich, J.; May, K.J.; Perry, E.E.; Yang, S.H. Synthesis and activities of rutile IrO_2 and RuO_2 nanoparticles for oxygen evolution in acid and alkaline solutions. *J. Phys. Chem. Lett.* **2012**, *3*, 399–404. [CrossRef] [PubMed]
13. Bates, M.K.; Jia, Q.Y.; Doan, H.; Liang, W.T.; Mukerjee, S. Charge–transfer effects in Ni–Fe and Ni–Fe–Co mixed–metal oxides for the alkaline oxygen evolution reaction. *ACS Catal.* **2016**, *6*, 155–161. [CrossRef]
14. Weng, B.C.; Xu, F.H.; Wang, C.L.; Meng, W.W.; Grice, C.R.; Yan, Y.F. A layered $Na_{1-x}Ni_yFe_{1-y}O_2$ double oxide oxygen evolution reaction electrocatalyst for highly efficient water–splitting. *Energy Environ. Sci.* **2017**, *10*, 121–128. [CrossRef]

15. Han, X.L.; Yu, Y.F.; Huang, Y.; Liu, D.L.; Zhang, B. Photogenerated carriers boost water splitting activity over transition–metal/semiconducting metal oxide bifunctional electrocatalysts. *ACS Catal.* **2017**, *7*, 6464–6470. [CrossRef]

16. Lu, F.; Zhou, M.; Zhou, Y.X.; Zeng, X.H. First–row transition metal based catalysts for the oxygen evolution reaction under alkaline conditions: Basic principles and recent advances. *Small* **2017**, *13*, 1701931. [CrossRef] [PubMed]

17. Stevens, M.B.; Enman, L.J.; Batchellor, A.S.; Cosby, M.R.; Vise, A.E.; Trang, C.D.M.; Boettcher, S.W. Measurement techniques for the study of thin film heterogeneous water oxidation electrocatalysts. *Chem. Mater.* **2017**, *29*, 120–140. [CrossRef]

18. Han, L.; Dong, S.J.; Wang, E.K. Transition–metal (CO, Ni, and Fe)–based electrocatalysts for the water oxidation reaction. *Adv. Mater.* **2016**, *28*, 9266–9291. [CrossRef] [PubMed]

19. Tan, C.L.; Cao, X.H.; Wu, X.J.; He, Q.Y.; Yang, J.; Zhang, X.; Chen, J.Z.; Zhao, W.; Han, S.K.; Nam, G.H.; et al. Recent advances in ultrathin two-dimensional nanomaterials. *Chem. Rev.* **2017**, *117*, 6225–6331. [CrossRef] [PubMed]

20. Zhu, X.L.; Tang, C.; Wang, H.F.; Zhang, Q.; Yang, C.H.; Wei, F. Dual-sized NiFe layered double hydroxides in situ grown on oxygen-decorated self-dispersal nanocarbon as enhanced water oxidation catalysts. *J. Mater. Chem. A* **2015**, *3*, 24540–24546. [CrossRef]

21. Zhou, L.; Shao, M.F.; Li, J.B.; Jiang, S.; Wei, M.; Duan, X. Two–dimensional ultrathin arrays of CoP: Electronic modulation toward high performance overall water splitting. *Nano Energy* **2017**, *41*, 583–590. [CrossRef]

22. Zhang, G.; Wang, G.C.; Liu, Y.; Liu, H.J.; Qu, J.H.; Li, J.H. Highly active and stable catalysts of phytic acid–derivative transition metal phosphides for full water splitting. *J. Am. Chem. Soc.* **2016**, *138*, 14686–14693. [CrossRef] [PubMed]

23. Tan, Y.W.; Wang, H.; Liu, P.; Shen, Y.H.; Cheng, C.; Hirata, A.; Fujita, T.; Tang, Z.; Chen, M.W. Versatile nanoporous bimetallic phosphides towards electrochemical water splitting. *Energy Environ. Sci.* **2016**, *9*, 2257–2261. [CrossRef]

24. Li, W.; Zhang, S.L.; Fan, Q.N.; Zhang, F.Z.; Xu, S.L. Hierarchically scaffolded CoP/CoP$_2$ nanoparticles: Controllable synthesis and their application as a well-matched bifunctional electrocatalyst for overall water splitting. *Nanoscale* **2017**, *9*, 5677–5685. [CrossRef] [PubMed]

25. Corrigan, D.A. The catalysis of the oxygen evolution reaction by iron impurities in thin film nickel oxide electrodes. *J. Electrochem. Soc.* **1987**, *134*, 377–384. [CrossRef]

26. Gong, M.; Li, Y.G.; Wang, H.L.; Liang, Y.Y.; Wu, J.Z.; Zhou, J.G.; Wang, J.; Regier, T.; Wei, F.; Dai, H.J. An advanced Ni–Fe layered double hydroxide electrocatalyst for water oxidation. *J. Am. Chem. Soc.* **2013**, *135*, 8452–8455. [CrossRef] [PubMed]

27. Shao, M.F.; Ning, F.Y.; Zhao, J.W.; Wei, M.; Evans, D.G.; Duan, X. Preparation of Fe$_3$O$_4$@SiO$_2$@layered double hydroxide core–shell microspheres for magnetic separation of proteins. *J. Am. Chem. Soc.* **2012**, *134*, 1071–1077. [CrossRef] [PubMed]

28. Evans, D.G.; Duan, X. Preparation of layered double hydroxides and their applications as additives in polymers, as precursors to magnetic materials and in biology and medicine. *Chem. Commun.* **2006**, *0*, 485–496. [CrossRef] [PubMed]

29. Wang, Q.; O'Hare, D. Recent advances in the synthesis and application of layered double hydroxide (LDH) nanosheets. *Chem. Rev.* **2012**, *7*, 4124–4155. [CrossRef] [PubMed]

30. Ning, F.Y.; Shao, M.F.; Xu, S.M.; Fu, Y.; Zhang, R.K.; Wei, M.; Evans, D.G.; Duan, X. TiO$_2$/Graphene/ NiFe-layered double hydroxide nanorod array photoanodes for efficient photoelectrochemical water splitting. *Energy Environ. Sci.* **2016**, *9*, 2633–2643. [CrossRef]

31. Zhang, H.J.; Li, X.P.; Hahnel, A.; Naumann, V.; Lin, C.; Azimi, S.; Schweizer, S.L.; Maijenburg, A.W.; Wehrspohn, R.B. Bifunctional heterostructure assembly of NiFe LDH nanosheets on NiCoP nanowires for highly efficient and stable overall water splitting. *Adv. Funct. Mater.* **2018**, *28*, 1706847. [CrossRef]

32. Lu, Z.Y.; Xu, W.W.; Zhu, W.; Yang, Q.; Lei, X.D.; Liu, J.F.; Li, Y.P.; Sun, X.M.; Duan, X. Three-dimensional NiFe layered double hydroxide film for high- efficiency oxygen evolution reaction. *Chem. Commun.* **2014**, *50*, 6479–6482. [CrossRef] [PubMed]

33. Hou, Y.; Lohe, M.R.; Zhang, J.; Liu, S.H.; Zhuang, X.D.; Feng, X.L. Vertically oriented cobalt selenide/NiFe layered-double-hydroxide nanosheets supported on exfoliated graphene foil: An efficient 3D electrode for overall water splitting. *Energy Environ. Sci.* **2016**, *9*, 478–483. [CrossRef]

34. Wang, Q.; Shang, L.; Shi, R.; Zhang, X.; Zhao, Y.F.; Waterhouse, G.I.N.; Wu, L.Z.; Tung, C.H.; Zhang, T.Z. NiFe layered double hydroxide nanoparticles on Co., N-codoped carbon nanoframes as efficient bifunctional catalysts for rechargeable zinc–air batteries. *Adv. Energy Mater.* **2017**, *7*, 1700467. [CrossRef]

35. Li, Y.G.; Gong, M.; Liang, Y.Y.; Feng, J.; Kim, J.E.; Wang, H.L.; Hong, G.S.; Zhang, B.; Dai, H.J. Advanced zinc-air batteries based on high-performance hybrid electrocatalysts. *Nat. Commun.* **2013**, *4*, 1805. [CrossRef] [PubMed]

36. Tang, D.; Liu, J.; Wu, X.Y.; Liu, R.H.; Han, X.; Han, Y.Z.; Huang, H.; Liu, Y.; Kang, Z.H. Carbon quantum dot/NiFe layered double-hydroxide composite as a highly efficient electrocatalyst for water oxidation. *ACS Appl. Mater. Interfaces* **2014**, *6*, 7918–7925. [CrossRef] [PubMed]

37. Ma, W.; Ma, R.Z.; Wang, C.X.; Liang, J.B.; Liu, X.H.; Zhou, K.C.; Sasaki, T. A superlattice of alternately stacked Ni–Fe hydroxide nanosheets and graphene for efficient splitting of water. *ACS Nano* **2015**, *9*, 1977–1984. [CrossRef] [PubMed]

38. Tang, C.; Wang, H.S.; Wang, H.F.; Zhang, Q.; Tian, G.L.; Nie, J.Q.; Wei, F. Spatially confined hybridization of nanometer-sized NiFe hydroxides into nitrogen-doped graphene frameworks leading to superior oxygen evolution reactivity. *Adv. Mater.* **2015**, *27*, 4516–4522. [CrossRef] [PubMed]

39. Tang, D.; Han, Y.Z.; Ji, W.B.; Qiao, S.; Zhou, X.; Liu, R.H.; Han, X.; Huang, H.; Liu, Y.; Kang, Z.H. A high-performance reduced grapheme oxide/ZnCo layered double hydroxide electrocatalyst for efficient water oxidation. *Dalton Trans.* **2014**, *43*, 15119–15125. [CrossRef] [PubMed]

40. Long, X.; Li, J.K.; Xiao, S.; Yan, K.Y.; Wang, Z.L.; Chen, H.N.; Yang, S.H. A strongly coupled graphene and FeNi double hydroxide hybrid as an excellent electrocatalyst for the oxygen evolution reaction. *Angew. Chem. Int. Ed.* **2014**, *53*, 7584–7588. [CrossRef] [PubMed]

41. Ma, W.; Ma, R.Z.; Wu, J.H.; Sun, P.Z.; Liu, X.H.; Zhou, K.C.; Sasaki, T. Development of efficient electrocatalysts via molecular hybridization of NiMn layered double hydroxide nanosheets and grapheme. *Nanoscale* **2016**, *8*, 10425–10432. [CrossRef] [PubMed]

42. Vargas, G.; Vazquez, S.J.; Oliver, T.M.A.; Ramos, S.L.; Flores, M.G.; Reguera, E. Influence on the electrocatalytic water oxidation of M^{2+}/M^{3+} cation arrangement in NiFe LDH: Experimental and theoretical DFT evidences. *Electrocatalysis* **2017**, *8*, 383–391. [CrossRef]

43. Han, N.; Zhao, F.P.; Li, Y.G. Ultrathin nickel–iron layered double hydroxide nanosheets intercalated with molybdate anions for electrocatalytic water oxidation. *J. Mater. Chem. A* **2015**, *3*, 16348–16353. [CrossRef]

44. Gao, R.; Yan, D.P. Fast formation of single-unit-cell-thick and defect-rich layered double hydroxide nanosheets with highly enhanced oxygen evolution reaction for water splitting. *Nano Res.* **2018**, *11*, 1883–1894. [CrossRef]

45. Luo, M.; Cai, Z.; Wang, C.; Bi, Y.M.; Qian, L.; Hao, Y.C.; Li, L.; Kuang, Y.; Li, Y.P.; Lei, X.D.; et al. Phosphorus oxoanion-intercalated layered double hydroxides for high-performance oxygen evolution. *Nano Res.* **2017**, *10*, 1732–1739. [CrossRef]

46. Ge, X.; Gu, C.D.; Wang, X.L.; Tu, J.P. Ionothermal synthesis of cobalt iron layered double hydroxides (LDHs) with expanded interlayer spacing as advanced electrochemical materials. *J. Mater. Chem. A* **2014**, *2*, 17066–17076. [CrossRef]

47. Jin, H.Y.; Mao, S.J.; Zhan, G.P.; Xu, F.; Bao, X.B.; Wang, Y. Fe incorporated α-Co(OH)$_2$ nanosheet with remarkably improved activity towards oxygen evolution reaction. *J. Mater. Chem. A* **2017**, *5*, 1078–1084. [CrossRef]

48. Wang, D.Y.; Costa, F.R.; Vyalikh, A.; Leuteritz, A.; Scheler, U.; Jehnichen, D.; Wagenknecht, U.; Häussler, L.; Heinrich, G. One-step synthesis of organic LDH and its comparison with regeneration and anion exchange method. *Chem. Mater.* **2009**, *21*, 4490–4497. [CrossRef]

49. Xu, H.J.; Wang, B.K.; Shan, C.F.; Xi, P.X.; Liu, W.S.; Tang, Y. Ce-doped NiFe-layered double hydroxide ultrathin nanosheets/nanocarbon hierarchical nanocomposite as an efficient oxygen evolution catalyst. *ACS Appl. Mater. Interfaces* **2018**, *10*, 6336–6345. [CrossRef] [PubMed]

50. Trotochaud, L.; Young, S.L.; Ranney, J.K.; Boettcher, S.W. Nickel−iron oxyhydroxide oxygen-evolution electrocatalysts: The role of intentional and incidental iron incorporation. *J. Am. Chem. Soc.* **2014**, *136*, 6744–6753. [CrossRef] [PubMed]

51. Chen, J.Y.C.; Dang, L.; Liang, H.F.; Bi, W.L.; Gerken, J.B.; Jin, S.; Alp, E.E.; Stahl, S.S. Operando analysis of NiFe and Fe oxyhydroxide electrocatalysts for water oxidation: Detection of Fe^{4+} by Mössbauer spectroscopy. *J. Am. Chem. Soc.* **2015**, *137*, 15090–15093. [CrossRef] [PubMed]

52. Gong, M.; Dai, H.J. A mini review of NiFe-based materials as highly active oxygen evolution reaction electrocatalysts. *Nano Res.* **2015**, *8*, 23–39. [CrossRef]
53. Shao, M.F.; Zhang, R.K.; Li, Z.H.; Wei, M.; Evans, D.G.; Duan, X. Layered double hydroxides toward electrochemical energy storage and conversion: Design, synthesis and applications. *Chem. Commun.* **2015**, *51*, 15880–15893. [CrossRef] [PubMed]
54. Dionigi, F.; Strasser, P. NiFe-based (oxy)hydroxide catalysts for oxygen evolution reaction in non-acidic electrolytes. *Adv. Energy Mater.* **2016**, *6*, 1600621. [CrossRef]
55. Burke, M.S.; Enman, L.J.; Batchellor, A.S.; Zou, S.H.; Boettcher, S.W. Oxygen evolution reaction electrocatalysis on transition metal oxides and (oxy)hydroxides: Activity trends and design principles. *Chem. Mater.* **2015**, *27*, 7549–7558. [CrossRef]
56. Tang, C.; Titirici, M.M.; Zhang, Q. A review of nanocarbons in energy electrocatalysis: Multifunctional substrates and highly active sites. *J. Energy Chem.* **2017**, *26*, 1077–1093. [CrossRef]
57. Suen, N.T.; Hung, S.F.; Quan, Q.; Zhang, N.; Xu, Y.J.; Chen, H.M. Electrocatalysis for the oxygen evolution reaction: recent development and future perspectives. *Chem. Soc. Rev.* **2017**, *46*, 337–365. [CrossRef] [PubMed]
58. Li, Y.B.; Zhao, C. Iron-doped nickel phosphate as synergistic electrocatalyst for water oxidation. *Chem. Mater.* **2016**, *28*, 5659–5666. [CrossRef]
59. Xi, W.; Ren, Z.Y.; Kong, L.J.; Wu, J.; Du, S.C.; Zhu, J.Q.; Xue, Y.Z.; Meng, H.Y.; Fu, H.G. Dual-valence nickel nanosheets covered with thin carbon as bifunctional electrocatalysts for full water splitting. *J. Mater. Chem. A* **2016**, *4*, 7297–7304. [CrossRef]
60. Pfrommer, J.; Azarpira, A.; Steigert, A.; Olech, K.; Menezes, P.W.; Duarte, R.F.; Liao, X.X.; Wilks, R.G.; Bär, M.; Thomas, S.N.; et al. Active and stable nickel-based electrocatalysts based on the ZnO:Ni system for water oxidation in alkaline media. *ChemCatChem* **2017**, *9*, 672–676. [CrossRef]
61. Luo, P.; Zhang, H.J.; Liu, L.; Zhang, Y.; Deng, J.; Xu, C.H.; Hu, N.; Wang, Y. Targeted synthesis of unique nickel sulfide (NiS, NiS$_2$) microarchitectures and the applications for the enhanced water splitting system. *ACS Appl. Mater. Interfaces* **2017**, *9*, 2500–2508. [CrossRef] [PubMed]
62. Friebel, D.; Louie, M.W.; Bajdich, M.; Sanwald, K.E.; Cai, Y.; Wise, A.M.; Cheng, M.J.; Sokaras, D.; Weng, T.C.; Mori, R.A.; et al. Identification of highly active Fe sites in (Ni,Fe)OOH for electrocatalytic water splitting. *J. Am. Chem. Soc.* **2015**, *137*, 1305–1313. [CrossRef] [PubMed]
63. Tang, C.; Wang, H.F.; Wang, H.S.; Wei, F.; Zhang, Q. Guest–host modulation of multi-metallic (oxy)hydroxides for superb water oxidation. *J. Mater. Chem. A* **2016**, *4*, 3210–3216. [CrossRef]
64. Gorlin, M.; Chernev, P.; Araujo, J.F.; Reier, T.; Dresp, S.; Paul, B.; Krahnert, R.; Dau, H.; Strasser, P. Oxygen evolution reaction dynamics, faradaic charge efficiency, and the active metal redox states of Ni–Fe oxide water splitting electrocatalysts. *J. Am. Chem. Soc.* **2016**, *138*, 5603–5614. [CrossRef] [PubMed]
65. Liu, H.J.; Zhou, J.; Wu, C.Q.; Wang, C.D.; Zhang, Y.K.; Liu, D.B.; Lin, Y.X.; Jiang, H.L.; Song, L. Integrated flexible electrode for oxygen evolution reaction: layered double hydroxide coupled with single-walled carbon nanotubes film. *ACS Sustainable Chem. Eng.* **2018**, *6*, 2911–2915. [CrossRef]
66. Yang, Y.; Dang, L.; Shearer, M.J.; Sheng, H.Y.; Li, W.J.; Chen, J.; Xiao, P.; Zhang, Y.H.; Hamers, R.J.; Jin, S. Highly active trimetallic NiFeCr layered double hydroxide electrocatalysts for oxygen evolution reaction. *Adv. Energy Mater.* **2018**, 1703189. [CrossRef]
67. Sun, X.H.; Shao, Q.; Pi, Y.C.; Guo, J.; Huang, X.Q. A general approach to synthesise ultrathin NiM(M=Fe, Co, Mn) hydroxide nanosheets as high performance low-cost electrocatalysts for overall water splitting. *J. Mater. Chem. A* **2017**, *5*, 7769–7775. [CrossRef]
68. Li, Z.H.; Shao, M.F.; An, H.L.; Wang, Z.X.; Xu, S.M.; Wei, M.; Evans, D.G.; Duan, X. Fast electrosynthesis of Fe-containing layered double hydroxide arrays toward highly efficient electrocatalytic oxidation reactions. *Chem. Sci.* **2015**, *6*, 6624–6631. [CrossRef] [PubMed]
69. Li, X.M.; Hao, X.G.; Wang, Z.D.; Abudula, A.; Guan, G.Q. In-situ intercalation of NiFe LDH materials: An efficient approach to improve electrocatalytic activity and stability for water splitting. *J. Power Sources* **2017**, *347*, 193–200. [CrossRef]
70. Yang, H.D.; Luo, S.; Bao, Y.; Luo, Y.T.; Jin, J.; Ma, J.T. In situ growth of ultrathin Ni–Fe LDH nanosheets for high performance oxygen evolution reaction. *Inorg. Chem. Front.* **2017**, *4*, 1173–1181. [CrossRef]

71. Xu, Y.Q.; Hao, Y.C.; Zhang, G.X.; Lu, Z.Y.; Han, S.; Li, Y.P.; Sun, X.M. Room-temperature synthetic NiFe layered double hydroxide with different anions intercalation as an excellent oxygen evolution catalyst. *RSC Adv.* **2015**, *5*, 55131–55135. [CrossRef]

72. Zhang, Y.; Shao, Q.; Pi, Y.C.; Guo, J.; Huang, X.Q. A cost-effiient bifunctional ultrathin nanosheets array for electrochemical overall water splitting. *Small* **2017**, *13*, 1700355. [CrossRef] [PubMed]

73. Tian, X.Q.; Liu, Y.H.; Xiao, D.; Sun, J. Ultrafast and large scale preparation of superior catalyst for oxygen evolution reaction. *J. Power Sources* **2017**, *365*, 320–326. [CrossRef]

74. Zhou, T.T.; Cao, Z.; Wang, H.; Gao, Z.; Lia, L.; Mab, H.Y.; Zhao, Y.F. Ultrathin Co–Fe hydroxide nanosheet arrays for improved oxygen evolution during water splitting. *RSC Adv.* **2017**, *7*, 22818–22824. [CrossRef]

75. Feng, L.X.; Li, A.R.; Li, Y.X.; Liu, J.; Wang, L.D.Y.; Huang, L.Y.; Wang, Y.; Ge, X.B. A highly active CoFe layered double hydroxide for water splitting. *ChemPlusChem* **2017**, *82*, 483–488. [CrossRef]

76. Zhou, P.; Wang, Y.Y.; Xie, C.; Chen, C.; Liu, H.W.; Chen, R.; Huo, J.; Wang, S.Y. Acid-etched layered double hydroxides with rich defects for enhancing the oxygen evolution reaction. *Chem. Commun.* **2017**, *53*, 11778–11781. [CrossRef] [PubMed]

77. Liu, P.F.; Yang, S.; Zhang, B.; Yang, H.G. Defect-rich ultrathin cobalt–iron layered double hydroxide for electrochemical overall water splitting. *ACS Appl. Mater. Interfaces* **2016**, *8*, 34474–34481. [CrossRef] [PubMed]

78. Wang, Y.Y.; Zhang, Y.Q.; Liu, Z.J.; Xie, C.; Feng, S.; Liu, D.D.; Shao, M.F.; Wang, S.Y. Layered double hydroxide nanosheets with multiple vacancies obtained by dry exfoliation as highly efficient oxygen evolution electrocatalysts. *Angew. Chem. Int. Ed.* **2017**, *56*, 5867–5871. [CrossRef] [PubMed]

79. Sumboja, A.; Chen, J.W.; Zong, Y.; Lee, P.S.; Liu, Z.L. NiMn layered double hydroxide as efficient electrocatalyst for oxygen evolution reaction and its application in rechargeable Zn air batteries. *Nanoscale* **2017**, *9*, 774–780. [CrossRef] [PubMed]

80. Song, F.; Hu, X.L. Ultrathin cobalt–manganese layered double hydroxide is an efficient oxygen evolution catalyst. *J. Am. Chem. Soc.* **2014**, *136*, 16481–16484. [CrossRef] [PubMed]

81. Jiang, J.; Zhang, A.L.; Li, L.L.; Ai, L.H. Nickel–cobalt layered double hydroxide nanosheets as high performance electrocatalyst for oxygen evolution reaction. *J. Power Sources* **2015**, *278*, 445–451. [CrossRef]

82. Liu, W.J.; Bao, J.; Guan, M.L.; Zhao, Y.; Lian, J.B.; Qiu, J.X.; Xu, L.; Huang, Y.P.; Qian, J.; Li, H.M. Nickel–cobalt–layered double hydroxide nanosheet arrays on Ni foam as a bifunctional electrocatalyst for overall water splitting. *Dalton Trans.* **2017**, *46*, 8372–8376. [CrossRef] [PubMed]

83. Liang, H.F.; Meng, F.; Acevedo, M.C.; Li, L.C.; Forticaux, A.; Xiu, L.C.; Wang, Z.C.; Ji, S. Hydrothermal continuous flow synthesis and exfoliation of NiCo layered double hydroxide nanosheets for enhanced oxygen evolution catalysis. *Nano Lett.* **2015**, *15*, 1421–1427. [CrossRef] [PubMed]

84. Qiao, C.; Zhang, Y.; Zhu, Y.Q.; Cao, C.B.; Bao, X.H.; Xu, J.Q. One–step synthesis of zinc–cobalt layered double hydroxide (Zn–Co–LDH) nanosheets for high efficiency oxygen evolution reaction. *J. Mater. Chem. A* **2015**, *3*, 6878–6883. [CrossRef]

85. Dong, C.L.; Yuan, X.T.; Wang, X.; Liu, X.Y.; Dong, W.J.; Wang, R.Q.; Duan, Y.H.; Huang, F.Q. Rational design of cobalt–chromium layered double hydroxide as a highly efficient electrocatalyst for water oxidation. *J. Mater. Chem. A* **2016**, *4*, 11292–11298. [CrossRef]

86. Yang, F.K.; Sliozberg, K.R.; Sinev, I.; Antoni, H.; Bähr, A.; Ollegott, K.; Xia, W.; Mas, J.; Grünert, W.; Cuenya, B.R.; et al. Synergistic effect of cobalt and iron in layered double hydroxide catalysts for the oxygen evolution reaction. *ChemSusChem* **2017**, *10*, 156–165. [CrossRef] [PubMed]

87. Lu, Z.Y.; Qian, L.; Tian, Y.; Li, Y.P.; Sun, X.M.; Duan, X. Ternary NiFeMn layered double hydroxides as highly-efficient oxygen evolution catalysts. *Chem. Commun.* **2016**, *52*, 90–911. [CrossRef] [PubMed]

88. Fan, K.; Chen, H.; Ji, Y.F.; Huang, H.; Claesson, P.M.; Daniel, Q.; Bertrand Philippe, B.; Rensmo, H.; Li, F.S.; Luo, Y.; et al. Nickel–vanadium monolayer double hydroxide for efficient electrochemical water oxidation. *Nat. Commun.* **2016**, *7*, 11981. [CrossRef] [PubMed]

89. Yoon, S.H.; Yun, J.Y.; Lim, J.H.; Yoo, B.Y. Enhanced electrocatalytic properties of electrodeposited amorphous cobalt-nickel hydroxide nanosheets on nickel foam by the formation of nickel nanocones for the oxygen evolution reaction. *J. Alloys Compd.* **2017**, *693*, 964–969. [CrossRef]

90. Long, X.; Xiao, S.; Wang, Z.L.; Zheng, X.L.; Yang, S.H. Co intake mediated formation of ultrathin nanosheets of transition metal LDH—an advanced electrocatalyst for oxygen evolution reaction. *Chem. Commun.* **2015**, *51*, 1120–1123. [CrossRef] [PubMed]

91. Song, F.; Hu, X.L. Exfoliation of layered double hydroxides for enhanced oxygen evolution catalysis. *Nat. Commun.* **2014**, *5*, 4477. [CrossRef] [PubMed]

92. Yan, D.F.; Li, Y.X.; Huo, J.; Chen, R.; Dai, L.M.; Wang, S.Y. Defect chemistry of nonprecious-metal electrocatalysts for oxygen reactions. *Adv. Mater.* **2017**, *29*, 1606459. [CrossRef] [PubMed]

93. Dogan, F.; Long, B.R.; Croy, J.R.; Gallagher, K.G.; Iddir, H.; Russell, J.T.; Balasubramanian, M.; Key, B. Re-entrant lithium local environments and defect driven electrochemistry of Li- and Mn-rich Li-ion battery cathodes. *J. Am. Chem. Soc.* **2015**, *6*, 2328–2335. [CrossRef] [PubMed]

94. Gu, Y.Q.; Xu, K.; Wu, C.Z.; Zhao, J.Y.; Xie, Y. Surface chemical-modification for engineering the intrinsic physical properties of inorganic two-dimensional nanomaterials. *Chem. Soc. Rev.* **2015**, *44*, 637–646. [CrossRef] [PubMed]

95. Hartmann, P.; Brezesinski, T.; Sann, J.; Lotnyk, A.; Eufinger, J.P.; Kienle, L.; Janek, J. Defect chemistry of oxide nanomaterials with high surface area: Ordered mesoporous thin films of the oxygen storage catalyst CeO_2–ZrO_2. *ACS Nano* **2013**, *7*, 2999–3013. [CrossRef] [PubMed]

96. Dou, L.; Wong, A.B.; Yu, Y.; Lai, M.L.; Kornienko, N.; Eaton, S.W.; Fu, A.; Bischak, C.G.; Ma, J.; Ding, T.; et al. Atomically thin two-dimensional organic-inorganic hybrid perovskites. *Science* **2015**, *349*, 1518–1521. [CrossRef] [PubMed]

97. Smith, R.J.; King, P.J.; Lotya, M.; Wirtz, C.; Khan, U.; De, S.; O'Neill, A.; Duesberg, G.C.; Grunlan, J.C.; Moriarty, G.; et al. Large–scale exfoliation of inorganic layered compounds in aqueous surfactant solutions. *Adv. Mater.* **2011**, *23*, 3944–3948. [CrossRef] [PubMed]

98. Zhou, L.; Shao, M.F.; Zhang, C.; Zhao, J.W.; He, S.; Rao, D.M.; Wei, M.; Evans, D.G.; Duan, X. Hierarchical CoNi-sulfie nanosheet arrays derived from layered double hydroxides toward effiient hydrazine electrooxidation. *Adv. Mater.* **2017**, *29*, 1604080. [CrossRef] [PubMed]

99. Subbaiah, Y.P.V.; Saji, K.J.; Tiwari, A. Atomically thin MoS_2: A versatile nongraphene 2D material. *Adv. Funct. Mater.* **2016**, *26*, 2046–2069. [CrossRef]

100. Zhang, R.K.; Shao, M.F.; Li, Z.H.; Ning, F.Y.; Wei, M.; Evans, D.G.; Duan, X. Photo electrochemical catalysis toward selective anaerobic oxidation of alcohols. *Chem. Eur. J.* **2017**, *23*, 8142–8147. [CrossRef] [PubMed]

101. Zhao, Y.F.; Zhang, X.; Jia, X.D.; Waterhouse, G.I.N.; Shi, R.; Zhang, X.R.; Zhan, F.; Tao, Y.; Wu, L.Z.; Tung, C.H.; et al. Sub-3 nm Ultrafine monolayer layered double hydroxide nanosheets for electrochemical water oxidation. *Adv. Energy Mater.* **2018**. [CrossRef]

102. Ma, W.L.; Wang, L.; Xue, J.Y.; Cui, H.T. Ultra-large scale synthesis of Co–Ni layered double hydroxides monolayer nanosheets by a solvent-free bottom-up strategy. *J. Alloys Compd.* **2016**, *662*, 315–319. [CrossRef]

103. Hu, G.; Wang, N.; O'Hare, D.; Davis, J. One-step synthesis and AFM imaging of hydrophobic LDH monolayers. *Chem. Commun.* **2006**, *0*, 287–289. [CrossRef] [PubMed]

104. Zhao, Y.F.; Li, B.; Wang, Q.; Gao, W.; Wang, C.J.; Wei, M.; Evans, D.G.; Duan, X.; O'Hare, D. NiTi-Layered double hydroxides nanosheets as efficient photocatalysts for oxygen evolution from water using visible light. *Chem. Sci.* **2014**, *5*, 951–958. [CrossRef]

105. Ma, R.Z.; Liu, R.P.; Li, L.; Iyia, N.; Sasak, T. Exfoliating layered double hydroxides in formamide: A method to obtain positively charged nanosheets. *J. Mater. Chem.* **2006**, *16*, 3809–3813. [CrossRef]

106. Mao, N.; Zhou, C.H.; Tong, D.S.; Yu, W.H.; Lind, C.X.C. Exfoliation of layered double hydroxide solids into functional nanosheets. *Appl. Clay Sci.* **2017**, *144*, 60–78. [CrossRef]

107. Yu, J.; Wang, Q.; O'Hare, D.; Sun, L.Y. Preparation of two dimensional layered double hydroxide nanosheets and their applications. *Chem. Soc. Rev.* **2017**, *46*, 5950–5974. [CrossRef] [PubMed]

108. Huang, S.; Cen, X.; Peng, H.D.; Guo, S.Z.; Wang, W.Z.; Liu, T.X. Heterogeneous ultrathin films of poly (vinyl alcohol)/layered double hydroxide and montmorillonite nanosheets via layer-by-layer assembly. *J. Phys. Chem. B* **2009**, *46*, 15225–15230. [CrossRef] [PubMed]

109. Liu, R.; Wang, Y.Y.; Liu, D.D.; Zou, Y.Q.; Wang, S.Y. Water-plasma-enabled exfoliation of ultrathin layered double hydroxide nanosheets with multivacancies for water oxidation. *Adv. Mater.* **2017**, *29*, 1701546. [CrossRef] [PubMed]

110. Zhang, R.K.; Shao, M.F.; Xu, S.M.; Ning, F.Y.; Zhou, L.; Wei, M. Photo-assisted synthesis of zinc-iron layered double hydroxides/TiO_2 nanoarrays toward highly-efficient photoelectrochemical water splitting. *Nano Energy* **2017**, *33*, 21–28. [CrossRef]

111. Zhou, L.; Jiang, S.; Liu, Y.K.; Shao, M.F.; Wei, M.; Duan, X. Ultrathin CoNiP@layered double hydroxides core–shell nanosheets arrays for largely enhanced overall water splitting. *ACS Appl. Energy Mater.* **2018**, *1*, 623–631. [CrossRef]

112. Lin, J.H.; Jia, H.N.; Liang, H.Y.; Cheng, S.L.; Cai, Y.F.; Qi, J.L.; Qu, C.Q.; Cao, J.; Fei, W.D.; Feng, J.C. Hierarchical CuCo$_2$S$_4$@NiMn-layered double hydroxide core-shell hybrid arrays as electrodes for supercapacitors. *Chem. Eng. J.* **2018**, *336*, 562–569. [CrossRef]

113. Wang, J.H.; Cui, W.; Liu, Q.; Xing, Z.C.; Asiri, A.M.; Sun, X.P. Recent progress in cobalt–based heterogeneous catalysts for electrochemical water splitting. *Adv. Mater.* **2016**, *28*, 215–230. [CrossRef] [PubMed]

114. Chi, H.M.; Qin, B.; Fu, L.; Jia, J.; Yi, B.L.; Shao, Z.G. Vertically aligned FeOOH/NiFe layered double hydroxides electrode. *ACS Appl. Mater. Interfaces* **2017**, *9*, 464–471. [CrossRef] [PubMed]

115. Cho, S.; Jang, J.W.; Park, Y.B.; Kim, J.Y.; Magesh, G.; Kim, J.H.; Seol, M.; Yong, K.J.; Lee, K.H.; Lee, J.S. An exceptionally facile method to produce layered double hydroxides on a conducting substrate and their application for solar water splitting without an external bias. *Energy Environ. Sci.* **2014**, *7*, 2301–2307. [CrossRef]

116. Yu, C.; Liu, Z.B.; Han, X.T.; Huang, H.W.; Zhao, C.T.; Yang, J.; Qiu, J. NiCo-layered double hydroxides vertically assembled on carbon fiber papers as binder-free high-active electrocatalysts for water oxidation. *Carbon* **2016**, *110*, 1–7. [CrossRef]

117. Zhao, M.Q.; Zhang, Q.; Huang, J.Q.; Wei, F. Hierarchical nanocomposites derived from nanocarbons and layered double hydroxides–properties, synthesis, and applications. *Adv. Funct. Mater.* **2012**, *22*, 675–694. [CrossRef]

118. Ma, Y.; Wang, Y.C.; Xie, D.H.; Gu, Y.; Zhang, H.M.; Wang, G.Z.; Zhang, Y.X.; Zhao, H.J.; Wong, P.K. NiFe-layered double hydroxide nanosheet arrays supported on carbon cloth for highly sensitive detection of nitrite. *ACS Appl. Mater. Interfaces* **2018**, *10*, 6541–6551. [CrossRef] [PubMed]

119. Ning, F.Y.; Shao, M.F.; Zhang, C.L.; Xu, S.M.; Wei, M.; Duan, X. Co$_3$O$_4$@ layered double hydroxide core/shell hierarchical nanowire arrays for enhanced supercapacitance performance. *Nano Energy* **2014**, *7*, 134–142. [CrossRef]

120. Li, X.Y.; Wang, X.; Yuan, X.T.; Dong, W.J.; Huang, F.Q. Rational composition and structural design of in situ grown nickel-based electrocatalysts for efficient water electrolysis. *J. Mater. Chem. A* **2016**, *4*, 167–172. [CrossRef]

121. Lu, Z.Y.; Sun, M.; Xu, T.H.; Li, Y.J.; Xu, W.W.; Chang, Z.; Ding, Y.; Sun, X.M.; Jiang, L. Superaerophobic electrodes for direct hydrazine fuel cells. *Adv. Mater.* **2015**, *27*, 2361–2366. [CrossRef] [PubMed]

122. Zhang, C.; Shao, M.F.; Zhou, L.; Li, Z.H.; Xiao, K.M.; Wei, M. Hierarchical NiFe layered double hydroxide hollow microspheres with highly-efficient behavior toward oxygen evolution reaction. *ACS Appl. Mater. Interfaces* **2016**, *8*, 33697–33703. [CrossRef] [PubMed]

123. Liu, J.P.; Li, Y.Y.; Fan, H.J.; Zhu, Z.H.; Jiang, J.; Ding, R.M.; Hu, Y.Y.; Huang, X.T. Iron oxide-based nanotube arrays derived from sacrificial template-accelerated hydrolysis: Large-area design and reversible lithium storage. *Chem. Mater.* **2010**, *22*, 212–217. [CrossRef]

124. Tang, Y.Q.; Fang, X.Y.; Zhang, X.; Fernandes, G.; Yan, Y.; Yan, D.P.; Xiang, X.; He, J. Space–Confined earth–abundant bifunctional electrocatalyst for high-efficiency water splitting. *ACS Appl. Mater. Interfaces* **2017**, *9*, 36762–36771. [CrossRef] [PubMed]

125. Yang, Q.; Li, T.; Lu, Z.Y.; Sun, X.M.; Liu, J.F. Hierarchical construction of an ultrathin layered double hydroxide nanoarray for highly-efficient oxygen evolution reaction. *Nanoscale* **2014**, *6*, 11789–11794. [CrossRef] [PubMed]

126. Blanchard, P.; Huchet, L.; Lillain, E.; Roncali, J. ation template assisted electrosynthesis of a highly π-conjugated polythiophene containing oligooxyethylene segments. *Electrochem. Commun.* **2000**, *2*, 1–5. [CrossRef]

127. Surendranath, Y.; Dincă, M.; Nocera, D.G. Electrolyte-dependent electrosynthesis and activity of cobalt-based water oxidation catalysts. *J. Am. Chem. Soc.* **2009**, *131*, 2615–2620. [CrossRef] [PubMed]

128. Francke, R.; Little, R.D. Three-dimensional hierarchical metal oxide–carbon electrode materials for highly efficient microbial electrosynthesis. *Chem. Soc. Rev.* **2014**, *43*, 2492–2521. [CrossRef] [PubMed]

129. Hamid, H.; Shiria, M.; Ehsani, S. Electrosynthesis of neodymium oxide nanorods and its nanocomposite with conjugated conductive polymer as a hybrid electrode material for highly capacitive pseudocapacitors. *J. Colloid Interface Sci.* **2017**, *495*, 102–110.

130. Qiao, X.Y.; Wei, M.C.; Tian, D.; Xia, F.Q.; Chen, P.P.; Zhou, C.L. One-step electrosynthesis of cadmium/aluminum layered double hydroxides composite as electrochemical probe for voltammetric detection of anthracene. *J. Electroanal. Chem.* **2018**, *808*, 35–40. [CrossRef]

131. Yamabe, K.; Goto, H. Electrosynthesis of conducting polymers in lecith in liquid crystal reaction field. *Fibers Polym.* **2018**, *19*, 248–253. [CrossRef]

132. Shao, M.F.; Li, Z.H.; Zhang, R.K.; Ning, F.Y.; Wei, M.; Evans, D.G.; Duan, X. Hierarchical conducting polymer@clay core–shell arrays for flexible all-solid-state supercapacitor devices. *Small* **2015**, *11*, 3530–3538. [CrossRef] [PubMed]

133. Wu, S.X.; Hui, K.S.; Hui, K.N. One-dimensional core–shell architecture composed of silver nanowire@hierarchical nickel–aluminum layered double hydroxide nanosheet as advanced electrode materials for pseudocapacitor. *J. Phys. Chem. C* **2015**, *119*, 23358–23365. [CrossRef]

134. Serrà, A.; Gómez, E.; Vallés, E. Novel electrodeposition media to synthesize CoNi-Pt core@shell stable mesoporous nanorods with very high active surface for methanol electro-oxidation. *Electrochim. Acta* **2015**, *174*, 630–639. [CrossRef]

135. Yanilkina, V.V.; Nastapovaa, N.V.; Nasretdinovaa, G.R.; Fazleevaa, R.R.; Toropchinaa, A.V.; Osin, Y.N. Methylviologen mediated electrochemical reduction of AgCl—A new route to produce a silica core/Ag shell nanocomposite material in solution. *Electrochem. Commun.* **2015**, *59*, 60–63. [CrossRef]

136. Li, Z.H.; Shao, M.F.; Zhou, L.; Zhang, R.K.; Zhang, C.; Wei, M.; Evans, D.G.; Duan, X. A flexible all-solid-state micro-supercapacitor based on hierarchical CuO@layered double hydroxide core–shell nanoarrays. *Nano Energy* **2016**, *20*, 294–304. [CrossRef]

137. Dong, X.Y.; Wang, L.; Wang, D.; Li, C.; Jin, J. Layer-by-layer engineered Co–Al hydroxide nanosheets/graphene multilayer films as flexible electrode for supercapacitor. *Langmuir* **2012**, *28*, 293–298. [CrossRef] [PubMed]

138. Han, J.B.; Lu, J.; Wei, M.; Wang, Z.L.; Duan, X. Heterogeneous ultrathin films fabricated by alternate assembly of exfoliated layered double hydroxides and polyanion. *Chem. Commun.* **2008**, 5188–5190. [CrossRef] [PubMed]

139. Yan, D.P.; Lu, J.; Wei, M.; Han, J.B.; Li, F.; Evans, D.G.; Duan, X. Ordered poly(p-phenylene)/layered double hydroxide ultrathin films with blue luminescence by layer-by-layer assembly. *Angew. Chem.* **2009**, *121*, 3119–3122. [CrossRef]

140. Han, J.B.; Xu, X.Y.; Rao, X.Y.; Wei, M.; Evans, D.G.; Duan, X. Layer-by-layer assembly of layered double hydroxide/cobalt phthalocyanine ultrathin film and its application for sensors. *J. Mater. Chem.* **2011**, *21*, 2126–2130. [CrossRef]

141. Liang, R.Z.; Tian, R.; Shi, W.Y.; Liu, Z.H.; Yan, D.P.; Wei, M.; Evans, D.G.; Duan, X. Temperature sensor based on CdTe quantum dots–layered double hydroxide ultrathin films via layer-by-layer assembly. *Chem. Commun.* **2013**, *49*, 969–971. [CrossRef] [PubMed]

142. Zhang, C.; Zhao, J.W.; Zhou, L.; Li, Z.H.; Shao, M.F.; Wei, M. Layer-by-layer assembly of exfoliated layered double hydroxide nanosheets for enhanced electrochemical oxidation of water. *J. Mater. Chem. A* **2016**, *4*, 11516–11523. [CrossRef]

143. Hunter, B.M.; Hieringer, W.; Winkler, J.R.; Gray, H.B.; Muller, A.M. Effect of interlayer anions on [NiFe]-LDH nanosheet water oxidation activity. *Energy Environ. Sci.* **2016**, *9*, 1734–1743. [CrossRef]

144. Zhang, J.F.; Liu, J.Y.; Xi, L.F.; Yu, Y.F.; Chen, N.; Sun, S.H.; Wang, W.C.; Lange, K.M.; Zhang, B. Single-atom Au/NiFe layered double hydroxide electrocatalyst: Probing the origin of activity for oxygen evolution reaction. *J. Am. Chem. Soc.* **2018**, *140*, 3876–3879. [CrossRef] [PubMed]

145. Ahn, H.S.; Bard, A.J. Surface interrogation scanning electrochemical microscopy of $Ni_{1-x}Fe_xOOH$ ($0 < x < 0.27$) oxygen evolving catalyst: Kinetics of the "fast" iron sites. *J. Am. Chem. Soc.* **2016**, *138*, 313–318. [PubMed]

catalysts

MDPI

Article

N,S Co-Doped Carbon Nanofibers Derived from Bacterial Cellulose/Poly(Methylene Blue) Hybrids: Efficient Electrocatalyst for Oxygen Reduction Reaction

Jing Liu [1], Yi-Gang Ji [2], Bin Qiao [1], Fengqi Zhao [3], Hongxu Gao [3], Pei Chen [1,*], Zhongwei An [1], Xinbing Chen [1] and Yu Chen [1]

[1] Key Laboratory of Applied Surface and Colloid Chemistry (MOE); Shaanxi Key Laboratory for Advanced Energy Devices; Shaanxi Engineering Lab for Advanced Energy Technology, School of Materials Science and Engineering, Shaanxi Normal University, Xi'an 710119, China; liujing551177@gmail.com (J.L.); qiaobin0229@gmail.com (B.Q.); gmecazw@163.com (Z.A.); chenxinbing@snnu.edu.cn (X.C.); ndchenyu@gmail.com (Y.C.)

[2] Jiangsu Key Laboratory of Biofuction Molecule, Department of Life Sciences and Chemistry, Jiangsu Second Normal University, Nanjing 210013, China; ygji@jssnu.edu.cn

[3] National Key Laboratory of Science and Technology on Combustion and Explosion, Xi'an Modern Chemistry Research Institute, 168 East Zhangba Road, Xi'an 710065, China; zhaofqi@163.com (F.Z.); gordon888@163.com (H.G.)

* Correspondence: chenpei@snnu.edu.cn; Tel.: +86-029-8153-0719

Received: 26 May 2018; Accepted: 27 June 2018; Published: 30 June 2018

Abstract: Exploring inexpensive and highly efficient electrocatalyst to decrease the overpotential of oxygen reduction reaction (ORR) is one of the key issues for the commercialization of energy conversion and storage devices. Heteroatom-doped carbon materials have attracted increasing attention as promising electrocatalysts. Herein, we prepared a highly active electrocatalyst, nitrogen, sulfur co-doped carbon nanofibers (N/S-CNF), via in situ chemical oxidative polymerization of methylene blue on the bacterial cellulose nanofibers, followed by carbonization process. It was found that the type of nitrogen/sulfur source, methylene blue and poly(methylene blue), has significantly influence on the catalytic activity of the resultant carbon nanofibers. Benefiting from the porous structure and high surface area (729 m^2/g) which favors mass transfer and exposing of active N and S atoms, the N/S-CNF displays high catalytic activity for the ORR in alkaline media with a half-wave potential of about 0.80 V, and better stability and stronger methanol tolerance than that of 20 wt % Pt/C, indicating great potential application in the field of alkaline fuel cell.

Keywords: nitrogen sulfur co-doped carbon nanofibers; bacterial cellulose/poly(methylene blue) hybrids; oxygen reduction reaction; electrocatalyst

1. Introduction

The development of low-cost and efficient energy conversion and storage technologies is of vital importance in alleviating the energy crisis and environmental protection. In recent years, some novel fuel cells and metal-air batteries, a class of devices that convert the chemical energy directly into electricity by electrochemical reactions, have attracted increasing attention [1–4]. In these devices, the oxygen reduction reaction (ORR) on the cathode is very slow kinetically, and thus requires platinum (Pt) as electrocatalyst. As the high price and unsatisfactory methanol tolerance of Pt have become a bottleneck of its extensive application in the fuel cells and metal-air batteries, the development of cheap and steady non-platinum catalysts is a practical and urgent issue [5–8]. In such conditions, nitrogen doped carbon nanomaterials (N-Cs) has recently been expanding rapidly because of excellent catalytic activity and high stability for the ORR in alkali media, and many techniques have been developed to prepare high active N-Cs as electrocatalyst for the ORR [9–14].

It is very well-known that the catalytic activity of N-Cs is closely related to their structure, meanwhile, the structural characteristics, such as the type and number of active centers, the porous structures [15] and the graphitizing extent [16,17], are controlled by the preparation method. Generally, the most used method to prepare the N-Cs is carbonizing various nitrogen-containing precursors, including (i) the heat-treatment of the existing carbon materials (such as graphene) using N-containing compounds (ammonia etc.); (ii) the pyrolysis of the N,C-containing precursor. In these processes, nitrogen source has great influence on the structure characteristics of the resultant N-Cs, and thus has a positive or negative impact on the catalytic activity for the ORR [18,19]. Understanding the impact of nitrogen source in molecular scale will provide fundamental knowledge for the rational design of N-Cs with high catalytic activity for the ORR.

In addition to the nitrogen source, the carbon source should be paid much attention to. Most of the reported N-Cs are prepared through chemical reagents or pre-synthesized precursor [20,21]. Considering the mass production in practical application, the cheaper raw and more convenient procedures are desired. Biomass is an attractive raw material due to its low cost, abundance and environmental friendly. Recently, various biomass, such as soybean shells [22], poplar catkins [23], biomass lysine [24] and soybean [25], are used to prepare the N-Cs as electrocatalyst for the ORR. Compare to these materials, cellulose, as the most abundant polymer on earth, is an excellent precursor for producing various carbon-based catalyst [23,26]. Especially, bacterial cellulose (BC), a biomass material produced by microbial industrial fermentation process at a very low price, possess a interconnected three dimension porous network structure consisting of cellulose nanofibers, and thus, is an ideal material to prepare of three dimension carbon-based functional nanomaterials [27,28].

Methylene blue (MB), as a cationic phenothiazines dye, contains not only N but also S element. Both can incorporate into the carbon matrix through facile carbonization, and the synergistic effect of N and S further enhances the catalysis performance for the ORR [29]. In addition, MB has good adsorption on the phenolic group of BC through various mechanisms such as electrostatic attractions [30], and it is easy to obtain the hybrid of BC/MB. Though carbonizing it, the nitrogen, sulfur-co-doped carbon nanofibers (N/S-CNF) has been facilely achieved [31]. However, compared with the comical Pt/C, the catalytic activity of N/S-CNF for the ORR is unsatisfied, and there is still no rational explanation for this result, due to the complicated carbonation process [32].

Compared with MB which is a small molecule compound, poly(methylene blue) (PMB) can gradually reduce and release the N/S-containing gas products during the carbonization process. These gas products readily react with the carbonization product and incorporate into it with higher N/S doping amount. Hence, we speculate that poly(methylene blue) (PMB) may be more appropriate as nitrogen source than MB. Luckily, the chemical oxidative polymerization of MB can occur at room temperature using the common oxidant, such as Au^{3+} and ammonium persulfate ($(NH_4)_2S_2O_8$) [33,34].

Therefore, in this work, using BC as carbon source, MB and PMB as nitrogen source respectively, the N/S-CNF was prepared. After characterizing the microstructure and evaluating the catalytic activity of the N/S-CNF derived from the hybrids of BC/MB and BC/PMB respectively, it is found that the activity for the ORR can be tuned by varying the type of nitrogen precursor. The N/S-CNF, prepared via in situ chemical oxidative polymerization of MB on the BC followed by carbonization process, displays high catalytic activity for the ORR in alkaline media with a half-wave potential of about 0.80 V, and better stability and stronger methanol tolerance than that of 20 wt % Pt/C.

2. Results

2.1. Characterization of the BC/MB and BC/PMB

As presented in Figure 1, the N/S-CNF were prepared by three steps. First, the MB was absorbed on the surface of BC at 100 °C, driven by the electrostatic interaction or hydrogen bond. Secondly, the chemical oxidative polymerization of MB was initiated by $(NH_4)_2S_2O_8$ and the formed PMB enwrapped evenly the nanofibers of BC [34]. Finally, the obtained BC/PMB hybrid was carbonized to form the N/S-CNF.

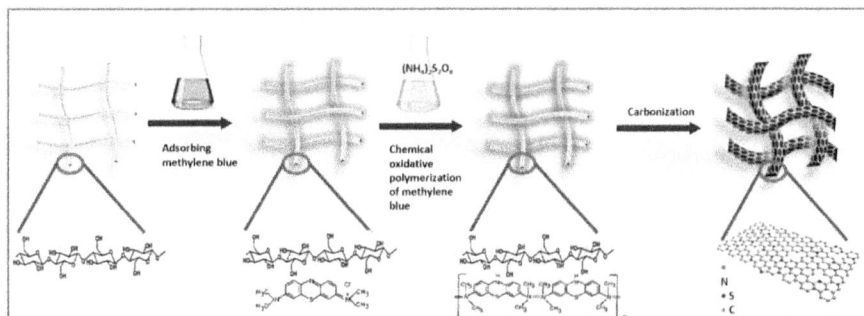

Figure 1. Synthetic procedure of the N/S-CNF.

To identify the formation of PMB on the surface of BC, the samples of BC, BC/MB and BC/PMB were characterized. Scan electron microscopy (SEM) image (Figure 2a) shows that the BC consists of the intertwining nanofibers with a dimension of about 100 nm. After adsorbing MB, some MB particle aggregations are deposited on the surface of BC (Figure 2b). However, after polymerization, these aggregations disappear and some smooth joints gumming the nanofiber together are observed from the BC/PMB (Figure 2c), suggesting the dissolution/reprecipitation process happened during the chemical oxidation polymerization of MB. The EDS results show that N, S and Cl are detected in the BC/MB and BC/PMB, and the contents of N and S in these two samples are similar (Figure 2b,c and Figure S1). However, the content of Cl in the BC/PMB is much lower than that of the BC/MB. This result indicates that, the MB cations are adsorbed on the surface of BC by static electric attractive, after in situ oxidation polymerization, Cl^- dissolved into solution and the electroneutral PMB was formed. From the FTIR spectra of BC, PMB, and BC/PMB in Figure 2d, the typical peaks belonging to the PMB and BC, which are in agreement with the reported [35–38], are observed. The peak at 1600 cm^{-1} assigned to the stretching vibration of the –C=N group of the PMB is detected from the BC/PMB, demonstrating that the PMB has successfully loaded on the BC. In addition, the survey spectra of X-ray photoelectron spectroscopy (XPS) (Figure 2e) further prove the presence of S and N in both the BC/MB and BC/PMB. Similar to the results of EDS, the peak (198.9 eV) ascribed to Cl^- is detected from the BC/MB while not the BC/PMB. The XPS fine spectra of N1s in Figure 2f further demonstrates the formation of PMB in the BC/PMB due to the appearance of PMB characteristic peak at 400.1 eV [39]. Furthermore, the peaks of pyridinic N (399.7 eV) and protonated amine N (401.6 eV) of the PMB shift to low energy direction, indicating the PMB tightly enwrap the nanofibers of BC, which results in the electron of the skeleton carbon atoms in the BC shifting toward the N atom of PMB due to the difference in electronegativity between them. All these results reveal that the hybrid of BC/PMB has been successfully prepared. Further study (Figure S2) reveals that, without the BC, the prepared PMB are blocks with irregular morphology, testifying the important role of BC in inhibiting the agglomeration of PMB.

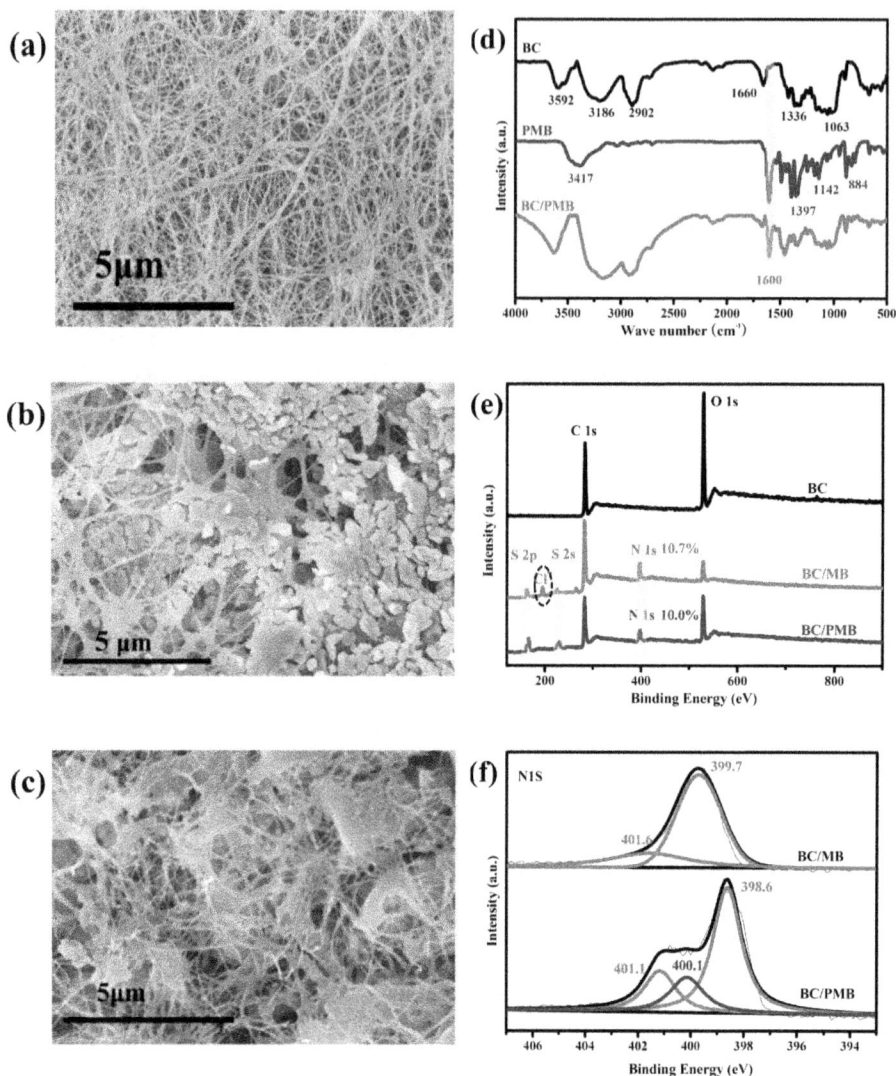

Figure 2. SEM images of (**a**) BC; (**b**) BC/MB and (**c**) BC/PMB; (**d**) FTIR spectra of BC, PMB and BC/PMB; (**e**) XPS survey spectra of BC, BC/MB and BC/PMB; and (**f**) fine XPS spectra of N1S for the BC/MB and BC/PMB.

Thermogravimetric analysis (TGA) was further carried out to verify the thermal decomposition behavior of the BC, MB, PMB, BC/MB and BC/PMB, and the corresponding TG curves are shown in Figure 3. The MB exhibits the first mass loss step below 250 °C with a mass loss of about 13%, and the second one centers at 250–400 °C with a mass loss of about 26%. With further increasing the temperature, the mass loss increases slowly, and the carbon product at 800 °C is about 52%. Compared with the MB, PMB shows much better thermal stability, and the onset decomposition temperature is up to ~250 °C, but the carbon yield has no change. The BC has a sharp mass loss in the range of 250–375 °C, and a low carbon yield of 11%. For the BC/MB, the thermal decomposition behavior in the low temperature (<250 °C) is similar to that of MB, and the carbon yield is ~40%. No surprise, the BC/PMB displays higher thermal stability below 250 °C, and the carbon yield is close to that of BC/MB and of 37%. Based on these results, it can be suspected that the high decomposition temperature of PMB should be helpful for its decomposition products to participate in the carbonization reaction, and thus promote the formation of N/S co-doped carbon materials.

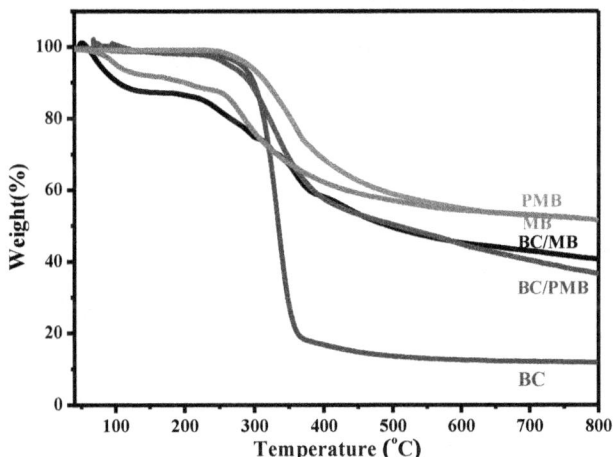

Figure 3. TG curves of the PMB, MB, BC, BC/MB and BC/PMB.

2.2. Characterization of the Carbonization Products of BC, BC/MB and BC/PMB

After carbonizing the BC, BC/MB and BC/PMB, the resultant products are named by C-BC, C-BC/MB and N/S-CNF, and their morphologies were investigated by SEM and TFM. Compared the images of BC (Figure 2a) and C-BC (Figure 4a,c), no obvious change is observed, except that the nanofibers are fluffier and thinner in the case of C-BC. However, compared with the BC/MB (Figure 2b), the C-BC/MB (Figure 4b) exhibits quite different morphology, which is similar to that of C-BC, seeming that MB particles have been completely decomposed. Nevertheless, the corresponding transmission electron microscopy (TEM) image (Figure 4e) shows that, except for the nanofibers derived from the BC, there are some isolated nanoparticles with a dimension of about 30 nm. Based on the different Z-contrast between S and C elements, these nanoparticles should be the sulfur-rich materials, which are proved by the element mapping results of energy dispersive spectrometer (EDS) (Figure S3). Obviously, these nanoparticles are the carbonized products coming from the incomplete decomposition of MB, due to the absence of S in the BC. Considering the similar morphology of the carbon nanofibers in cases of C-BC and C-BC/MB, it is deduced that the adsorbed MB has little influence on the decomposition of BC, probably due to the weak interface interaction between them, just as shown the TG curve in Figure 3. It is interesting to note that, in contrast to the C-BC and C-BC/MB, the N/S-CNF derived from the BC/PMB, consists of some short and wide nanobelts (Figure 4c), and their diameter and length (Figure S4) are about 70 and 400 nm, respectively. This result demonstrates the great influence of PMB on the carbonization of BC. TEM image in Figure 4e shows that the contrast of the nanobelts is uniform, indicating the absence of sulfur-rich domains in the N/S-CNF. In addition, elemental mappings from energy dispersive spectroscopy (EDS) (Figure 4h,i) confirm the nitrogen and sulfur are successfully incorporated and uniformly distribute in the carbon matrix. Moreover, EDS results (insets in Figure 4b,c) show that the contents of N and S are slightly higher in the case of N/S-CNF than that of C-BC/MB, further proving that PMB as N/S source is helpful for the incorporation of N/S into the carbon framework. No Cl was found by EDS, probably because of the formation of Cl-containing gas.

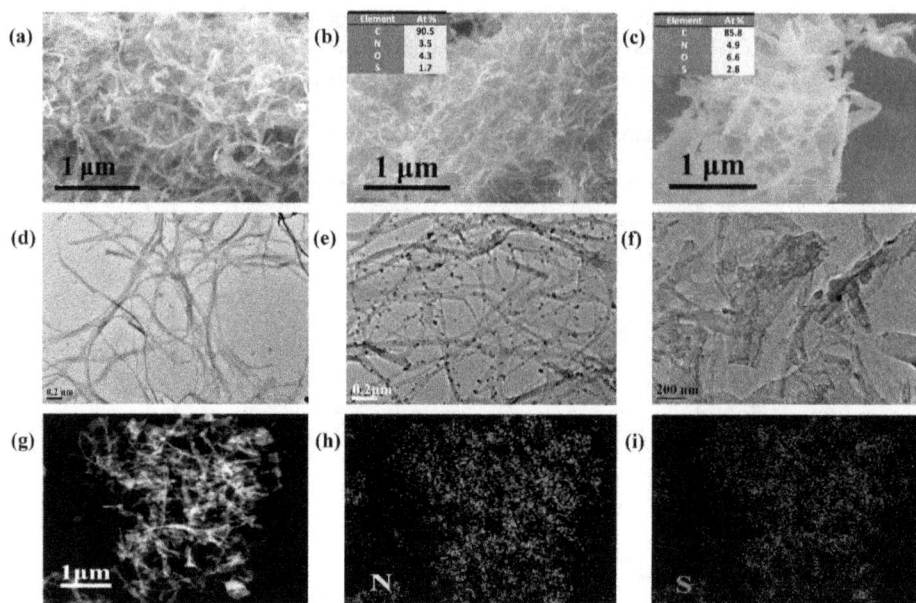

Figure 4. SEM and TEM images of the (**a,d**) C-BC; (**b,e**) C-BC/MB and (**c,f**) N/S-CNF; and the element maps (**g–i**) of N,S-CNF. The insets in (**b,c**) are EDS data.

To gain more insight, the porous structure of N/S-CNF was characterized. As shown in Figure 5a, with respect to the C-BC and C-BC/MB, the N/S-CNF shows a type I isotherm due to the accomplishment of the predominant adsorption of N_2 below the relative pressure (P/P$_0$) of 0.02, implying the presence of micro-pores. In addition, a hysteresis loop at P/P$_0$ from 0.40 to 1.0 is also observed, which is a characteristic of mesoporous materials. The pore size distribution (PSD) and surface area were calculated with the slit/cylinder model of quenched solid density functional theory using the adsorption branch. The PSD curve further confirms that both the micro-pore centering at 1.0 nm and the meso-pores with various sizes in 2–35 nm coexist in the N/S-CNF. The calculated surface area is 729 g·cm^{-2}. The C-BC has similar PSD but smaller surface area, while the C-BC/MB belongs to the mesoporous material and has the smallest surface area. These results suggest that the MB or its decomposition product destroyed the micro-pores of C-BC, but the PMB promoted to form more micro-pores in the C-BC, probably by chemical etching the carbon matrix. The hierarchical porous structure and high surface area of N/S-CNF are beneficial to the exposure of more active sites and the diffusion of reactants.

To elucidate the crystallinity of N/S-CNF, X-ray powder diffraction (XRD) and Raman spectroscopic investigation were conducted. The XRD patterns of the N/S-CNF and the control samples show a broad peak at approximately $2\theta = 24°$ and a very weak peak at $2\theta = 42°$ (Figure 5b), which are the characteristics of graphitic carbon materials with low graphitization degree [40]. Raman spectra (Figure 5c) further reveal that both amorphous and crystalline carbon coexist in these samples. The intensity ratios of the D to G bands (I_D/I_G) are nearly same for these samples, reflecting their similar graphitization degree [41].

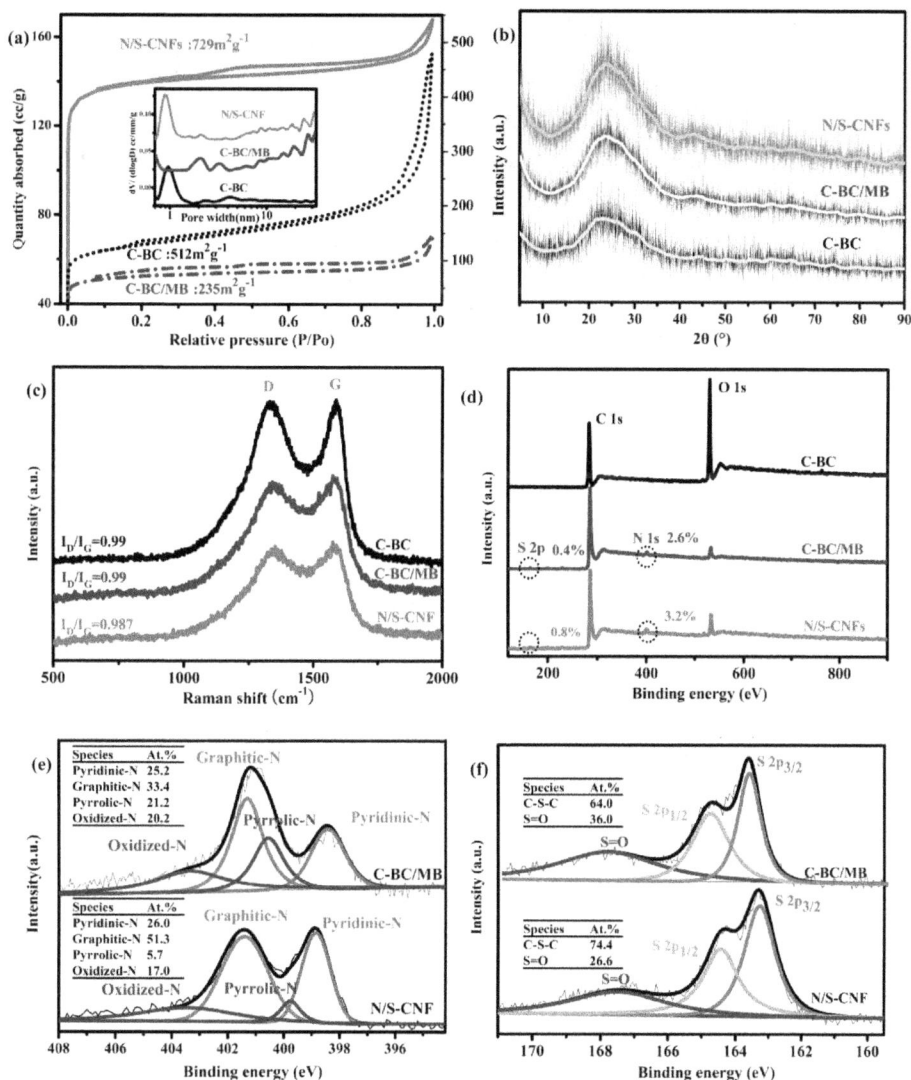

Figure 5. (a) N_2 adsorption-desorption isotherms and the inset is the pore size distribution curve; (b) XRD patterns; (c) Raman spectra and (d) XPS full spectra of C-BC, C-BC/MB and N/S-CNF; (e) N1s and (f) S2p fine XPS spectra of C-BC/MB and N/S-CNF.

To further confirm the chemical state of N and S, X-ray photo electron spectroscopy (XPS) measurement was carried out. As shown the XPS survey spectra in Figure 5d, N and S are detected from the N/S-CNF and C-BC/MB, confirming that the N and S atoms have been successfully introduced into the carbon matrix, which is in agreement with the EDS results. However, the contents of N and S in the N/S-CNF are 3.2% and 0.8%, respectively, which are higher than that (2.6% and 0.4%) of the C-BC/MB. Generally, It is believed that the N and S content play a key role for the improved ORR catalytic activity [13,42], and thus the N/S-CNF should have better catalysis performance than that of C-BC/MB. In addition, it is reported that the N and S species proportion also is crucial for the

ORR catalytic activity of electrocatalyst [43]. Form the high-resolution N1s spectra (Figure 5e) of the N/S-CNF and C-BC/MB, four peaks can be deconvoluted into, which are assigned to the pyridinic N (398.7 eV), pyrrolic N (399.8 eV), graphitic N (401.2 eV) and (403.2eV), respectively. Usually, it is accepted that the pyridinic N and graphitic N are active species for the ORR [43]. Compared with the C-BC/MP, the N/S-CNF has a higher amount (77.3 at. %) of the two species and thus should display better catalysis performance for the ORR. In addition, the high-resolution S2p spectra (Figure 5f) show the peaks of $P_{1/2}$ and $P_{3/2}$ at binding energy of 163.5 and 164.3 eV which attribute to C–S–C bonds, and the ones at 166.0–170 eV are associated with the oxidized-S species that are chemically inactive for the ORR [30]. Similar to the case of N, the N/S-CNF also possesses higher amount (74.4%) of the active S species than that (64%) of C-BC/MP. Therefore, combining all the above results of SEM, PSD, XRD, XPS, and so on, a conclusion can be drawn that the PMB is more suitable as N and S source to prepare efficient elecrocatalyst for the ORR than the MB.

2.3. Electrocatalytic Activity of the N/S-CNF for the ORR

The electrocatalytic activity of the N/S-CNF for the ORR was evaluated by cyclic voltammetry (CV) measurements in 0.1 M KOH electrolyte. Being compared with the smooth CV curve obtained from the N_2-saturated electrolyte (Figure 6a), the CV curve in O_2-saturated electrolyte shows a cathodic peak at 0.78 V, implying the electrocatalytic activity of N/S-CNF for the ORR in alkaline media. The ORR performance of the N/S-CNF was further measured with the rotating disk electrode (RDE) using linear sweep voltammetry (LSV) technique. As control subjects, the ORR activities of the C-BC, C-BC/MB and Pt/C electrocatalyst were also tested under the same experimental conditions (Figure 6b). Unsurprisingly, the C-BC shows the worst catalytic activity due to lacking active centres derived from N/S doping. With the benefit of N/S doping, the C-BC/MB exhibits a significant performance boost over the C-BC. Especially, with the aid of the large surface area and high amount of active N and S species, the N/S-CNF shows the best catalytic performance for the ORR, featuring with a comparable $E_{1/2}$ value (0.80 V) to the commercial 20 wt % Pt/C (0.83 V). In addition, the $E_{1/2}$ value (0.80 V) is also more positive than some of the reported N,S-co-doped carbon-based electrocatalysts (Table 1). Especially, the $E_{1/2}$ (0.80 V) in this work is 170 mV higher than that reported N-S-CNF-800 (MB) [31], which was prepared through BC physically absorbing MB and followed the carbonization process. We think the novel preparation method should be responsible for the improved catalytic activity. Firstly, based on the adsorption kinetic of MB on the cellulose [44] a harsh adsorption condition, heating the saturated solution of MB containing the dried BC at 100 °C for 4.5 h with autoclave, was employed to increase the adsorption amount of MB. Secondly, the N/S-CNF was obtained by carbonizing the BC/PMB hybrid that derived from the in situ polymerization of MB on the BC surface, but not by directly carbonizing the BC/MB hybrid as mentioned in the literature [31]. Obviously, compared with the small molecule compound, N,S-containing polymer is much more suitable as N/S source for the synthesis of carbon-based electrocatalyst with high ORR catalytic activity. To clarify the influence of synthesis parameters on the catalytic activity of N/S-CNF, a series of samples were prepared. The corresponding LSV curves for the ORR (Figure S5) reveal that the catalytic activity of N/S-CNF is sensitive to the synthesis condition, and the optimized experimental parameters are critical for achieving the N/S-CNF with high catalytic activity.

The catalysis kinetic of the N/S-CNF for the ORR was further investigated. It is normal that the limiting diffusion current increases with the rotation speed due to the thinned diffusion layer (Figure 6c). The transferred electron number (n) per oxygen molecule involved in the ORR process was calculated with Koutecky–Levich equation, which was to be ca. 3.89 in the potential range of 0.4 to 0.6 V, demonstrating an approximate four-electron pathway. To elucidate the electron transfer mechanism, the hydrogen peroxide yields were measured with rotating ring-disk electrode (RRDE). As shown in Figure 6f, the ring current originating from the oxidation of hydrogen peroxide ions (HO_2^-) is low. The calculated percentage of HO_2^- is below 17% over the potential range from 0.2 to 0.8 V, which corresponds to a transfer number of ~3.88. This is agreement with the results obtained from the Koutecky–Levich plots, again illuminating a nearly $4e^-$ pathway for the ORR catalyzed by the N/S-CNF.

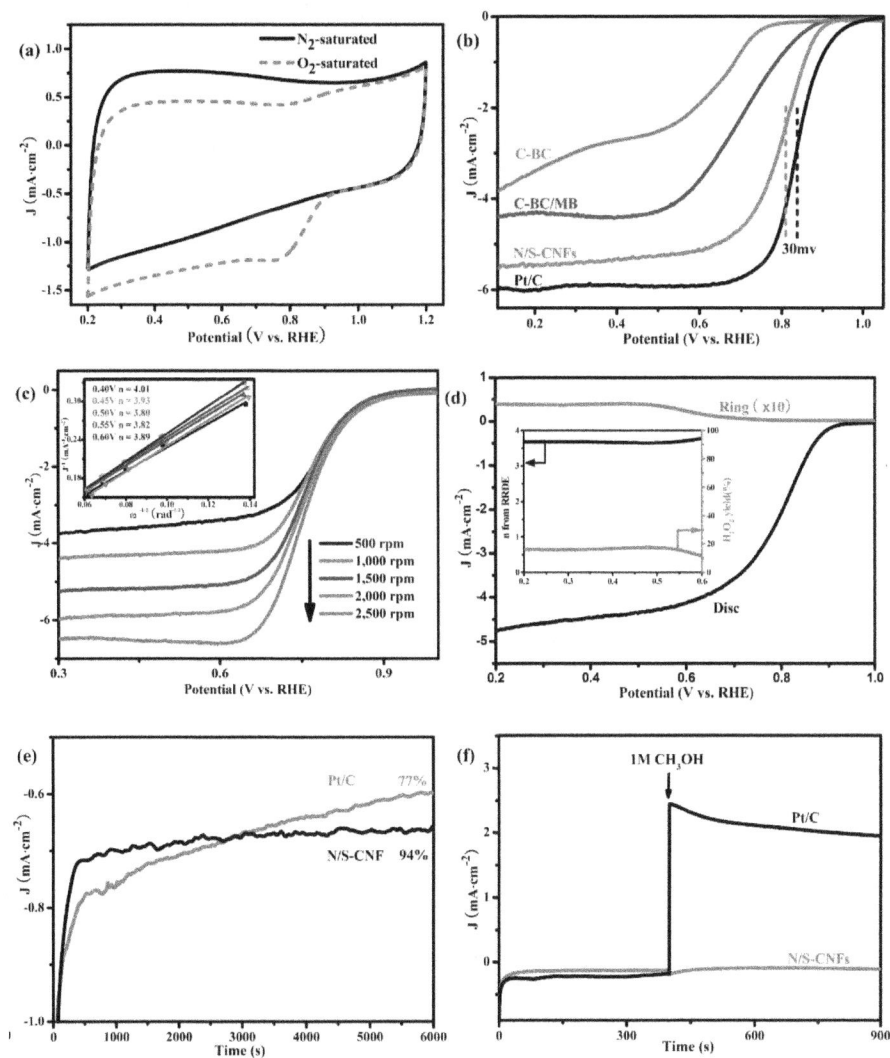

Figure 6. (**a**) CV curves of the N/S-CNF in N_2/O_2-saturated 0.1 M KOH solution at scan rate of 50 mV·s^{-1}; (**b**) ORR polarization curves of the C-BC, C-BC/MB, N/S-CNF and Pt/C electrocatalysts at scan rate of 10 mV·s^{-1} and a rotation rate of 1600 rpm; (**c**) LSV curves of N/S-CNF at different rotation rates, and the inset is the corresponding Koutecky–Levich plots; (**d**) RRDE measurements for N/S-CNF electrode; chronoamperometry curves of the N/S-CNF and Pt/C obtained at 0.75 V for (**e**) the stability test; and (**f**) methanol tolerance test before and after adding 1 M methanol. (**b–f**) were obtained in O_2-saturated 1 M KOH solution.

Table 1. The difference of half wave potential between the N/S-co-doped carbon-based elecrocatalyst and 20 wt % Pt/C for the ORR in 0.1 M KOH solution.

Sample	$\Delta E_{\frac{1}{2}}^{1}$ (mV)	Ref.
N/S-CNFs	30	This work
N-S-CNF-800 (MB)	200	[31]
NS-3DrGO-950	64	[43]
N,S-PGN-800	30	[45]
NSC800	35.4	[46]
NSC-A2	80	[47]
PAC-5S	48	[48]
S1N5-OMC	65	[49]
N/S-2DPC-60	27	[50]
N,S-hcs-900 °C	40	[51]
NS-G	72	[52]
N,S-CN	50	[53]
CNx/CSx-GNRs 1000° for 2.5 h	50	[54]

Subsequently, the chronoamperometric responses is used to evaluate the electrocatalytic activity and stability of the N/S-CNF. As shown in Figure 6e, after a brief transient period, the oxygen reduction current at the N/S-CNF electrode remains stable for the long time (6000 s) of polarization, while the current at the Pt/C electrode reduced to about 77% during the same test period, implying the excellent durability of the N/S-CNF.

Methanol-tolerance is an important benchmark for the electrocatalyst used in fuel cells, the catalysis performance of the N/S-CNF in KOH electrolyte containing methanol was investigated by chronoamperometry. As displayed in Figure 6f, after adjusting the concentration of methanol to 1 M in the O_2-saturated 0.1 M KOH electrolyte, the ORR current for Pt/C electrocatalyst shows a drastic surge and cannot be recovered to the initial level. Conversely, the current level of the N/S-CNF remains virtually unchanged, indicating its excellent methanol-tolerance.

3. Materials and Methods

3.1. Materials

BC membranes were supplied by Hainan Yida Food Co., Ltd., (Hainan, China), stored in acetic acid solution. Before use, BC membranes were immersed into a 0.1 M sodium hydroxide solution at 100 °C for 60 min to remove the residual cells, and then thoroughly washed with deionized water until pH reached neutral. To facilitate experimental operation, the BC membranes were cut to quadrate pieces with size of 3 × 3 cm, and followed by freeze drying at −50 °C. All other chemicals were purchased from Aladdin Reagent Co. Ltd., (Shanghai, China) and used as received.

3.2. Preparation of BC/MB and BC/PMB

6 mL of the saturated solution of MB was poured into the autoclave with volume of 15 mL, and the dried BC membranes of 15 mg were soaked into it. To accelate adsorption of MB on BC membrane, the autoclave was sealed and heated at 100 °C for 4.5 h. The blued BC membranes were washed by water to remove the physically adsorbed day molecules, and then immsered into 10 mL of ammonium peroxydisulphate solution (0.1 M) for 1.5 h to make the methylene blue polymerize. The purple BC membranes were freeze-dried, and transferred into a tubular furnace for carbonization under a flowing N_2 atmosphere at 800 °C for 2 h with a heating rate of 2 °C·min^{-1}. After cooled to room temperature, the black carbon powder coded as N/S-CNF was achieved. To compare, the controlled samples were also prepared with the BC membranes adsorbing and unadsorbing MB by the similar process.

3.3. Electrochemical Measurements

Electrochemical experiments were carried out by computer-controlled CHI 760E (Chenhua Instrument, Shanghai, China) electrochemical workstation equipped with RDE apparatus at room temperature, using a standard three-electrode system. The electrocatalyst-modified working electrode was prepared with the elelctrocatalyst ink, which was obtained by mixing 10 mg of electrocatalyst and 5 mL of water/Nafion solution, and sonicating the mixture for 60 min. After the working electrode loaded 10 μL of electrocatalyst ink was dried at room temperature, the CV and LSV tests were measured in N_2- or O_2-saturated 0.1 M KOH solution, by using a saturated calomel and carbon rod as reference electrode and counter electrode, respectively. All potentials in this work were reported with respect to reversible hydrogen electrode (RHE). The electron transfer number (n) was determined by the Koutecky–Levich equation:

$$1/j = 1/j_k + 1/B\omega^{0.5} \tag{1}$$

Here, j_k is the kinetic current density, and B is expressed by the following expression:

$$B = 0.2 \times n \times F \times (D_{O2})^{2/3} \times v^{-1/6} \times C_{O2} \tag{2}$$

where n represents the number of electrons transferred per oxygen molecule; F is the Faraday constant ($F = 96,485\ C·mol^{-1}$); D_{O2} is the diffusion coefficient of O_2 in 0.1 M KOH ($1.9 \times 10^{-5}\ cm^2·s^{-1}$); v is the kinematic viscosity of the electrolyte solution ($0.01\ cm^2·s^{-1}$); C_{O2} is the concentration of dissolved O_2 ($1.2 \times 10^{-6}\ mol·cm^{-3}$). The constant 0.2 is adopted when the rotation speed is expressed in rpm.

The n and H_2O_2 yield (%) were measured with RRDE, and calculated with the followed equations:

$$n = 4I_d/(I_d + I_r/N) \tag{3}$$

$$H_2O_2\ (\%) = 100 \times (4 - n)/2 \tag{4}$$

where I_d is the disk current density, I_r is the ring current density, and N = 0.37 is the current collection efficiency.

3.4. Characterization

The crystalline structure, morphology and surface composition of sample were physically characterized by XRD (D/max-rC) (Rigalcu, Tokyo, Japan), TEM (JEM-2100F) (JEOL, Tokyo, Japan) equipped with EDS, SEM (SU-8020) (Hitachi High-Technologies Corp., Japan), Raman Spectrometer (In Via Reflex) (Renishaw PLC, Wotton-under-Edge, UK) and XPS (AXIS ULTRA) (Kratos Analytical Ltd., Hadano, Japan). The surface area and pore volume of the samples were measured on a physical adsorption instrument (ASAP 2400) (Norcross, GA, USA), the functional groups were analyzed by using Fourier transform infrared spectroscopy (FTIR, Nicolet Is10) (Thermo Fisher Scientific, Waltham, MA, USA), and the thermal decomposition behavior was measured by TGA (Q600) (TA Instruments, New Castle, DE, USA).

4. Conclusions

We developed a facile method to fabricate superior ORR catalysts. By simply in situ polymerization of MB on the surface of BC, followed by carbonization, the N/S-CNF nanomaterials was obtained. It was found that, compared with the small molecular substance, the polymer as nitrogen/sulfur source can significantly enhance the catalytic activity of the resultant N,S-co-doped carbon nanofibers. Benefited from the synergistic effect of the multiple active sites, as well as the enriched porosity and high surface area, the N/S-CNF revealed excellent ORR activity in alkaline media with a half-wave potential of about 0.80 V. Moreover, the N/S-CNF showed better long-term stability and methanol tolerance than that commercial 20 wt % Pt/C, demonstrating its potential application in fuel cells and metal-air batteries as alternative Pt electrocatalysts.

Supplementary Materials: The following are available online at http://www.mdpi.com/2073-4344/8/7/269/s1, Figure S1: EDS spectra of (**a**) the BC/MB and (**b**) BC/PMB, Figure S2: SEM image of the PMB, Figure S3: The dark field TEM image of the C-BC/MB and the corresponding element mapping images of S, N, O and C, Figure S4: The diameter (**b**) and length (**c**) distribution of the nanofibers obtained from TEM image (**a**) of the sample N/S-CNF, Figure S5: Linear sweep voltammetry curves of the samples of N/S-CNF prepared at different condition. Fixed the other experiment parameters, (**a**) the concentration of MB in the adsorption process of MB on BC was varied from 0.01 to 0.1 mol/L; (**b**) the heat treatment temperature in the adsorption process of MB on BC was varied from 60 to 150 °C; (**c**) the polymerization time of MB was varied from 0.5 to 6 h. All these linear sweep voltammetry curves were obtained from O_2-saturated 0.1 M KOH solution at scan rate of 10 mv/s and a rotation rate of 1600 rpm.

Author Contributions: J.L. analyzed and interpreted the experiment data; Y.-G.J., B.Q., F.Z. and H.G. performed the experiments and collected data; P.C. designed the experiments and drafted the article; Y.C., and Z.A. critically revised the article; X.C. given final approval of the version to be published.

Acknowledgments: This research was supported by National Science Foundation Committee of China (51773112), Program for Science & Technology Innovation Team of Shaanxi Province (2018TD-030), the Fundamental Research Funds for the Central Universities (GK201801001, GK201703030).

Conflicts of Interest: The authors declare no conflict of interest.

References

1. Fu, J.; Cano, Z.P.; Park, M.G.; Yu, A.; Fowler, M.; Chen, Z. Electrically rechargeable zinc-air batteries: Progress, challenges, and perspectives. *Adv. Mater.* **2017**, *29*, 1604685. [CrossRef] [PubMed]
2. Tang, C.; Zhang, Q. Nanocarbon for oxygen reduction electrocatalysis: Dopants, edges, and defects. *Adv. Mater.* **2017**, *29*, 1604103. [CrossRef] [PubMed]
3. Shao, M.; Chang, Q.; Dodelet, J.-P.; Chenitz, R. Recent advances in electrocatalysts for oxygen reduction reaction. *Chem. Rev.* **2016**, *116*, 3594–3657. [CrossRef] [PubMed]
4. Pan, J.; Xu, Y.Y.; Yang, H.; Dong, Z.; Liu, H.; Xia, B.Y. Advanced architectures and relatives of air electrodes in Zn-air batteries. *Adv. Sci.* **2018**, *5*, 1700691:1–1700691:30. [CrossRef] [PubMed]
5. Xu, H.; Yan, B.; Li, S.; Wang, J.; Wang, C.; Guo, J.; Du, Y. Facile construction of N-doped graphene supported hollow ptag nanodendrites as highly efficient electrocatalysts toward formic acid oxidation reaction. *ACS Sustain. Chem. Eng.* **2018**, *6*, 609–617. [CrossRef]
6. Cheng, H.; Li, M.L.; Su, C.Y.; Li, N.; Liu, Z.Q. Cu-Co bimetallic oxide quantum dot decorated nitrogen-doped carbon nanotubes: A high-efficiency bifunctional oxygen electrode for Zn-air batteries. *Adv. Funct. Mater.* **2017**, *27*, 1701833. [CrossRef]
7. Liu, X.; Amiinu Ibrahim, S.; Liu, S.; Pu, Z.; Li, W.; Ye, B.; Tan, D.; Mu, S. H_2O_2-assisted synthesis of porous N-doped graphene/molybdenum nitride composites with boosted oxygen reduction reaction. *Adv. Mater. Interfaces* **2017**, *4*, 1601227. [CrossRef]
8. Su, C.Y.; Cheng, H.; Li, W.; Liu, Z.Q.; Li, N.; Hou, Z.; Bai, F.Q.; Zhang, H.X.; Ma, T.Y. Atomic modulation of FeCo-nitrogen-carbon bifunctional oxygen electrodes for rechargeable and flexible all-solid-state zinc-air battery. *Adv. Energy Mater.* **2017**, *7*, 1602420. [CrossRef]
9. Cui, H.; Zhou, Z.; Jia, D. Heteroatom-doped graphene as electrocatalysts for air cathodes. *Mater. Horiz.* **2017**, *4*, 7–19. [CrossRef]
10. Zhu, C.; Li, H.; Fu, S.; Du, D.; Lin, Y. Highly efficient nonprecious metal catalysts towards oxygen reduction reaction based on three-dimensional porous carbon nanostructures. *Chem. Soc. Rev.* **2016**, *45*, 517–531. [CrossRef] [PubMed]
11. Lee Won, J.; Lim, J.; Kim Sang, O. Nitrogen dopants in carbon nanomaterials: Defects or a new opportunity? *Small Methods* **2016**, *1*, 1600014.
12. Xiang, Z.; Cao, D.; Huang, L.; Shui, J.; Wang, M.; Dai, L. Nitrogen-doped holey graphitic carbon from 2D covalent organic polymers for oxygen reduction. *Adv. Mater.* **2014**, *26*, 3315–3320. [CrossRef] [PubMed]
13. Yu, H.; Shang, L.; Bian, T.; Shi, R.; Waterhouse Geoffrey, I.N.; Zhao, Y.; Zhou, C.; Wu, L.Z.; Tung, C.H.; Zhang, T. Nitrogen-doped porous carbon nanosheets templated from g-C_3N_4 as metal-free electrocatalysts for efficient oxygen reduction reaction. *Adv. Mater.* **2016**, *28*, 5080–5086. [CrossRef] [PubMed]
14. Chatterjee, K.; Ashokkumar, M.; Gullapalli, H.; Gong, Y.; Vajtai, R.; Thanikaivelan, P.; Ajayan, P.M. Nitrogen-rich carbon nano-onions for oxygen reduction reaction. *Carbon* **2018**, *130*, 645–651. [CrossRef]

15. Geng, D.; Chen, Y.; Chen, Y.; Li, Y.; Li, R.; Sun, X.; Ye, S.; Knights, S. High oxygen-reduction activity and durability of nitrogen-doped graphene. *Energy Environ. Sci.* **2011**, *4*, 760–764. [CrossRef]

16. Li, L.; Dai, P.; Gu, X.; Wang, Y.; Yan, L.; Zhao, X. High oxygen reduction activity on a metal-organic framework derived carbon combined with high degree of graphitization and pyridinic-N dopants. *J. Mater. Chem. A* **2017**, *5*, 789–795. [CrossRef]

17. Zhao, J.; Liu, Y.; Quan, X.; Chen, S.; Yu, H.; Zhao, H. Nitrogen-doped carbon with a high degree of graphitization derived from biomass as high-performance electrocatalyst for oxygen reduction reaction. *Appl. Surf. Sci.* **2017**, *396*, 986–993. [CrossRef]

18. Zhan, Y.; Yu, X.; Cao, L.; Zhang, B.; Wu, X.; Xie, F.; Zhang, W.; Chen, J.; Xie, W.; Mai, W.; et al. The influence of nitrogen source and doping sequence on the electrocatalytic activity for oxygen reduction reaction of nitrogen doped carbon materials. *Int. J. Hydrogen Energy* **2016**, *41*, 13493–13503. [CrossRef]

19. Higgins, D.; Chen, Z.; Chen, Z. Nitrogen doped carbon nanotubes synthesized from aliphatic diamines for oxygen reduction reaction. *Electrochim. Acta* **2011**, *56*, 1570–1575. [CrossRef]

20. Lin, C.Y.; Zhang, D.; Zhao, Z.; Xia, Z. Covalent organic framework electrocatalysts for clean energy conversion. *Adv. Mater.* **2017**, *30*, 1703646. [CrossRef] [PubMed]

21. Xu, Q.; Tang, Y.; Zhang, X.; Oshima, Y.; Chen, Q.; Jiang, D. Template conversion of covalent organic frameworks into 2d conducting nanocarbons for catalyzing oxygen reduction reaction. *Adv. Mater.* **2018**, *30*, 1706330. [CrossRef] [PubMed]

22. Zhou, H.; Zhang, J.; Amiinu, I.S.; Zhang, C.; Liu, X.; Tu, W.; Pan, M.; Mu, S. Transforming waste biomass with an intrinsically porous network structure into porous nitrogen-doped graphene for highly efficient oxygen reduction. *Phys. Chem. Chem. Phys.* **2016**, *18*, 10392–10399. [CrossRef] [PubMed]

23. Gao, S.; Li, X.; Li, L.; Wei, X. A versatile biomass derived carbon material for oxygen reduction reaction, supercapacitors and oil/water separation. *Nano Energy* **2017**, *33*, 334–342. [CrossRef]

24. Zheng, X.; Cao, X.; Li, X.; Tian, J.; Jin, C.; Yang, R. Biomass lysine-derived nitrogen-doped carbon hollow cubes via a nacl crystal template: An efficient bifunctional electrocatalyst for oxygen reduction and evolution reactions. *Nanoscale* **2017**, *9*, 1059–1067. [CrossRef] [PubMed]

25. Lin, G.; Ma, R.; Zhou, Y.; Liu, Q.; Dong, X.; Wang, J. Koh activation of biomass-derived nitrogen-doped carbons for supercapacitor and electrocatalytic oxygen reduction. *Electrochim. Acta* **2018**, *261*, 49–57. [CrossRef]

26. Chen, W.; Yu, H.; Lee, S.-Y.; Wei, T.; Li, J.; Fan, Z. Nanocellulose: A promising nanomaterial for advanced electrochemical energy storage. *Chem. Soc. Rev.* **2018**, *47*, 2837–2872. [CrossRef] [PubMed]

27. Wu, Z.-Y.; Liang, H.-W.; Chen, L.-F.; Hu, B.-C.; Yu, S.-H. Bacterial cellulose: A robust platform for design of three dimensional carbon-based functional nanomaterials. *Acc. Chem. Res.* **2016**, *49*, 96–105. [CrossRef] [PubMed]

28. Liang, H.-W.; Wu, Z.-Y.; Chen, L.-F.; Li, C.; Yu, S.-H. Bacterial cellulose derived nitrogen-doped carbon nanofiber aerogel: An efficient metal-free oxygen reduction electrocatalyst for zinc-air battery. *Nano Energy* **2015**, *11*, 366–376. [CrossRef]

29. Pei, Z.; Li, H.; Huang, Y.; Xue, Q.; Huang, Y.; Zhu, M.; Wang, Z.; Zhi, C. Texturing in situ: N,S-enriched hierarchically porous carbon as a highly active reversible oxygen electrocatalyst. *Energy Environ. Sci.* **2017**, *10*, 742–749. [CrossRef]

30. Rafatullah, M.; Sulaiman, O.; Hashim, R.; Ahmad, A. Adsorption of methylene blue on low-cost adsorbents: A review. *J. Hazard. Mater.* **2010**, *177*, 70–80. [CrossRef] [PubMed]

31. Wu, Z.-Y.; Liang, H.-W.; Li, C.; Hu, B.-C.; Xu, X.-X.; Wang, Q.; Chen, J.-F.; Yu, S.-H. Dyeing bacterial cellulose pellicles for energetic heteroatom doped carbon nanofiber aerogels. *Nano Res.* **2014**, *7*, 1861–1872. [CrossRef]

32. Guo, D.; Shibuya, R.; Akiba, C.; Saji, S.; Kondo, T.; Nakamura, J. Active sites of nitrogen-doped carbon materials for oxygen reduction reaction clarified using model catalysts. *Science* **2016**, *351*, 361–365. [CrossRef] [PubMed]

33. Tang, Z.; Wang, L.; Ma, Z. Triple sensitivity amplification for ultrasensitive electrochemical detection of prostate specific antigen. *Biosens. Bioelectron.* **2017**, *92*, 577–582. [CrossRef] [PubMed]

34. Lv, X.; Zhang, P.; Li, M.; Guo, Z.; Zheng, X. Synthesis of polymethylene blue nanoparticles and their application to label-free DNA detection. *Anal. Lett.* **2016**, *49*, 2728–2740. [CrossRef]

35. Li, X.; Zhong, M.; Sun, C.; Luo, Y. A novel bilayer film material composed of polyaniline and poly(methylene blue). *Mater. Lett.* **2005**, *59*, 3913–3916. [CrossRef]

36. Shan, J.; Wang, L.; Ma, Z. Novel metal-organic nanocomposites: Poly(methylene blue)-Au and its application for an ultrasensitive electrochemical immunosensing platform. *Sensor. Actuators B Chem.* **2016**, *237*, 666–671. [CrossRef]

37. Xiao, X.; Zhou, B.; Tan, L.; Tang, H.; Zhang, Y.; Xie, Q.; Yao, S. Poly(methylene blue) doped silica nanocomposites with crosslinked cage structure: Electropolymerization, characterization and catalytic activity for reduction of dissolved oxygen. *Electrochim. Acta* **2011**, *56*, 10055–10063. [CrossRef]

38. Yang, S.; Li, G.; Zhao, J.; Zhu, H.; Qu, L. Electrochemical preparation of ag nanoparticles/poly(methylene blue) functionalized graphene nanocomposite film modified electrode for sensitive determination of rutin. *J. Electroanal. Chem.* **2014**, *717–718*, 225–230. [CrossRef]

39. Rincón Rosalba, A.; Artyushkova, K.; Mojica, M.; Germain Marguerite, N.; Minteer Shelley, D.; Atanassov, P. Structure and electrochemical properties of electrocatalysts for nadh oxidation. *Electroanalysis* **2010**, *22*, 799–806. [CrossRef]

40. Zheng, F.; Yang, Y.; Chen, Q. High lithium anodic performance of highly nitrogen-doped porous carbon prepared from a metal-organic framework. *Nat. Commun.* **2014**, *5*, 5261. [CrossRef] [PubMed]

41. Chen, L.-F.; Lu, Y.; Yu, L.; Lou, X.W. Designed formation of hollow particle-based nitrogen-doped carbon nanofibers for high-performance supercapacitors. *Energy Environ. Sci.* **2017**, *10*, 1777–1783. [CrossRef]

42. Klingele, M.; Pham, C.; Vuyyuru, K.R.; Britton, B.; Holdcroft, S.; Fischer, A.; Thiele, S. Sulfur doped reduced graphene oxide as metal-free catalyst for the oxygen reduction reaction in anion and proton exchange fuel cells. *Electrochem. Commun.* **2017**, *77*, 71–75. [CrossRef]

43. Li, Y.; Yang, J.; Huang, J.; Zhou, Y.; Xu, K.; Zhao, N.; Cheng, X. Soft template-assisted method for synthesis of nitrogen and sulfur co-doped three-dimensional reduced graphene oxide as an efficient metal free catalyst for oxygen reduction reaction. *Carbon* **2017**, *122*, 237–246. [CrossRef]

44. Chan, C.H.; Chia, C.H.; Zakaria, S.; Sajab, M.S.; Chin, S.X. Cellulose nanofibrils: a rapid adsorbent for the removal of methylene blue. *RSC Adv.* **2015**, *5*, 18204–18212. [CrossRef]

45. Wu, M.; Liu, Y.; Zhu, Y.; Lin, J.; Liu, J.; Hu, H.; Wang, Y.; Zhao, Q.; Lv, R.; Qiu, J. Supramolecular polymerization-assisted synthesis of nitrogen and sulfur dual-doped porous graphene networks from petroleum coke as efficient metal-free electrocatalysts for the oxygen reduction reaction. *J. Mater. Chem. A* **2017**, *5*, 11331–11339. [CrossRef]

46. Xu, L.; Fan, H.; Huang, L.; Xia, J.; Li, S.; Li, M.; Ding, H.; Huang, K. Chrysanthemum-derived n and s co-doped porous carbon for efficient oxygen reduction reaction and aluminum-air battery. *Electrochim. Acta* **2017**, *239*, 1–9. [CrossRef]

47. Yang, S.; Mao, X.; Cao, Z.; Yin, Y.; Wang, Z.; Shi, M.; Dong, H. Onion-derived n, s self-doped carbon materials as highly efficient metal-free electrocatalysts for the oxygen reduction reaction. *Appl. Surf. Sci.* **2018**, *427*, 626–634. [CrossRef]

48. You, C.; Jiang, X.; Han, L.; Wang, X.; Lin, Q.; Hua, Y.; Wang, C.; Liu, X.; Liao, S. Uniform nitrogen and sulphur co-doped hollow carbon nanospheres as efficient metal-free electrocatalysts for oxygen reduction. *J. Mater. Chem. A* **2017**, *5*, 1742–1748. [CrossRef]

49. Jiang, T.; Wang, Y.; Wang, K.; Liang, Y.; Wu, D.; Tsiakaras, P.; Song, S. A novel sulfur-nitrogen dual doped ordered mesoporous carbon electrocatalyst for efficient oxygen reduction reaction. *Appl. Catal. B Environ.* **2016**, *189*, 1–11. [CrossRef]

50. Su, Y.; Yao, Z.; Zhang, F.; Wang, H.; Mics, Z.; Cánovas, E.; Bonn, M.; Zhuang, X.; Feng, X. Sulfur-enriched conjugated polymer nanosheet derived sulfur and nitrogen co-doped porous carbon nanosheets as electrocatalysts for oxygen reduction reaction and zinc-air battery. *Adv. Funct. Mater.* **2016**, *26*, 5893–5902. [CrossRef]

51. Wu, Z.; Liu, R.; Wang, J.; Zhu, J.; Xiao, W.; Xuan, C.; Lei, W.; Wang, D. Nitrogen and sulfur co-doping of 3D hollow-structured carbon spheres as an efficient and stable metal free catalyst for the oxygen reduction reaction. *Nanoscale* **2016**, *8*, 19086–19092. [CrossRef] [PubMed]

52. Pan, F.; Duan, Y.; Zhang, X.; Zhang, J. A facile synthesis of nitrogen/sulfur co-doped graphene for the oxygen reduction reaction. *ChemCatChem* **2015**, *8*, 163–170. [CrossRef]

53. Qu, K.; Zheng, Y.; Dai, S.; Qiao, S.Z. Graphene oxide-polydopamine derived n, s-codoped carbon nanosheets as superior bifunctional electrocatalysts for oxygen reduction and evolution. *Nano Energy* **2016**, *19*, 373–381. [CrossRef]
54. Zehtab Yazdi, A.; Roberts, E.P.L.; Sundararaj, U. Nitrogen/sulfur co-doped helical graphene nanoribbons for efficient oxygen reduction in alkaline and acidic electrolytes. *Carbon* **2016**, *100*, 99–108. [CrossRef]

catalysts

MDPI

Article

Binary Nitrogen Precursor-Derived Porous Fe-N-S/C Catalyst for Efficient Oxygen Reduction Reaction in a Zn-Air Battery

Xiao Liu [1,2], Chi Chen [1], Qingqing Cheng [1,2], Liangliang Zou [1], Zhiqing Zou [1,*] and Hui Yang [1,*]

[1]   Shanghai Advanced Research Institute, Chinese Academy of Sciences, Shanghai 201210, China;
      liuxiao@sari.ac.cn (X.L.); chenchi@sari.ac.cn (C.C.); chengqinghc@163.com (Q.C.); zoull@sari.ac.cn (L.Z.)
[2]   University of the Chinese Academy of Sciences, Beijing 100039, China
*   Correspondence: zouzq@sari.ac.cn (Z.Z.); yangh@sari.ac.cn (H.Y.); Tel.: +86-21-2032-4112 (H.Y.)

Received: 2 March 2018; Accepted: 19 March 2018; Published: 13 April 2018

Abstract: It is still a challenge to synthesize non-precious-metal catalysts with high activity and stability for the oxygen reduction reaction (ORR) to replace the state-of-the art Pt/C catalyst. Herein, a Fe, N, S co-doped porous carbon (Fe-NS/PC) is developed by using g-C_3N_4 and 2,4,6-tri(2-pyridyl)-1,3,5-triazine (TPTZ) as binary nitrogen precursors. The interaction of binary nitrogen precursors not only leads to the formation of more micropores, but also increases the doping amount of both iron and nitrogen dispersed in the carbon matrix. After a second heat-treatment, the best Fe/NS/C-g-C_3N_4/TPTZ-1000 catalyst exhibits excellent ORR performance with an onset potential of 1.0 V vs. reversible hydrogen electrode (RHE) and a half-wave potential of 0.868 V (RHE) in alkaline medium. The long-term durability is even superior to the commercial Pt/C catalyst. In the meantime, an assembled Zn-air battery with Fe/NS/C-g-C_3N_4/TPTZ-1000 as the cathode shows a maximal power density of 225 mW·cm^{-2} and excellent durability, demonstrating the great potential of practical applications in energy conversion devices.

Keywords: non-precious metal catalyst; oxygen reduction reaction; binary nitrogen precursors; g-C_3N_4; 2,4,6-tri(2-pyridyl)-1,3,5-triazine

1. Introduction

The oxygen reduction reaction (ORR) plays an important role in the energy efficiency of polymer electrolyte membrane fuel cells (PEMFCs) and metal-air batteries (MABs). So far, platinum (Pt)-based materials are still the most effective catalysts for the ORR due to its sluggish kinetics. However, the high price and scarcity of Pt severely hinder the large-scale applications of PEMFCs and MABs. Therefore, extensive efforts have been devoted to develop low-cost and earth-abundant non-precious metal catalysts with efficient ORR performance. The transition metal (M=Fe, Co, etc.) and nitrogen co-doped carbon materials (M-N-C), such as graphene [1,2], nanotube [3,4], and porous carbon [5,6], have shown great progress in ORR electrocatalysis, especially for Fe-N-C materials, which have been considered as the most promising catalysts for substituting the expensive Pt catalysts [7–9].

The excellent ORR activity usually depends on two main factors, namely, the high intrinsic activity of single sites and high density of active sites. Therefore, the M-N-C electrocatalysts for efficient ORR require high heteroatom doping contents, high surface area, porous structure, and good conductivity [10–12]. Heteroatom doping is an effective method to tailor the electronic structure of electroneutral carbon matrix, which would facilitate the adsorption of O_2. High surface area and porous structure are beneficial to increase the number of accessible active sites and facilitate the mass transport of the ORR relevant species approaching the internal active sites of catalysts [13,14]. Heat-treatment at high temperature is a vital process to form the ORR active centers, therefore, the precursors should be chosen carefully. Recently, the use of binary nitrogen precursors has been developed as an effective synthetic strategy to improve the porosity and heteroatoms doping contents, hence to improve the ORR activity. Wu et al. [15] synthesized a Fe-N-C catalyst derived from polyaniline (PANI) and dicyandiamide (DCDA) as binary nitrogen precursors, which possessed higher ORR activity than the individual PANI or DCDA-derived ones. The superior ORR activity can be ascribed to the increased content of pyridinic nitrogen doped into the carbon matrix. The combination of PANI and DCDA could enhance the porosity and increase the surface area, thanks to the different decomposition temperatures. Chen et al. [16] synthesized a class of Fe-N-C catalysts with 3D nanoporous structure using phenanthroline (Phen) and PANI as dual nitrogen sources. The Phen played the role of pore-creating agent due to its lower thermostability. A similar strategy was also reported by Zelenay et al. [17]. Moreover, it has been proved that the additional doping of S atoms in the Fe-N-C catalyst will remarkably enhance the ORR activity [18]. However, although the development of non-precious metal catalysts has achieved great progress, the application of these materials in practical devices, such as PEMFCs and MABs, is still far from satisfactory, especially for the long-term stability [19].

In this work, we have developed a facile method to synthesize Fe, N, S co-doped porous carbon materials (Fe-NS/PC) as efficient ORR catalysts with g-C$_3$N$_4$ and 2,4,6-tri(2-pyridyl)-1,3,5-triazine (TPTZ) as binary nitrogen precursors. The TPTZ is able to coordinate with Fe^{3+} [20], which could contribute to the uniformly-dispersed metal-containing species located at the N-doped carbon skeleton [21,22]. The addition of g-C$_3$N$_4$ sheets could inhibit the sintering of TPTZ during carbonization. The interaction of g-C$_3$N$_4$ and TPTZ is beneficial to increase the doping amount of iron and nitrogen, and to facilitate the mass transfer of ORR relevant species. As a result, the binary nitrogen precursor-derived Fe-NS/PC catalyst exhibited better ORR performance than the single nitrogen precursor-derived ones. After second heat treatment, the Fe/NS/C-g-C$_3$N$_4$/TPTZ catalyst shows enhanced activity and long-term durability. The best ORR activity of Fe/NS/C-g-C$_3$N$_4$/TPTZ-1000 is even superior to that of the state-of-the art Pt/C catalyst. A Zn-air battery with the Fe/NS/C-g-C$_3$N$_4$/TPTZ-1000 cathode exhibits a maximal power density of 225 mW·cm^{-2} at room temperature and superior stability with only 4.03% loss of output voltage at a current density of 20 mA·cm^{-2} after 20,000 s.

2. Results and Discussion

Figure 1a–c show the scanning electron microscope (SEM) images of the as-synthesized Fe-NS/PC catalysts using g-C$_3$N$_4$ and TPTZ separated as single nitrogen precursor and together as binary nitrogen precursors, denoted as Fe/NS/C-g-C$_3$N$_4$, Fe/NS/C-TPTZ, and Fe/NS/C-g-C$_3$N$_4$/TPTZ, respectively. The Fe/NS/C-g-C$_3$N$_4$ has a fluffy morphology with high porosity (Figure 1a). By contrast, Fe/NS/C-TPTZ shows a denser morphology (Figure 1b), probably because of the collapse of the carbon skeleton during heat-treatment. Through the combination of both g-C$_3$N$_4$ and TPTZ with various thermal stability (Figure S1), the Fe/NS/C-g-C$_3$N$_4$/TPTZ preserves the fluffy morphology (Figure 1c), probably due to the fact that the mixed g-C$_3$N$_4$ sheets prevent the sintering of TPTZ.

The effects of N precursors on BET surface areas and pore structures were studied by N$_2$ adsorption-desorption isotherms. As can be seen in Figure 1d,f, the Fe/NS/C-g-C$_3$N$_4$ exhibits a highest BET surface area of 928 m^2/g, the most of which are external surface area. The micropore area is only 82 m^2/g. The Fe/NS/C-TPTZ shows a slightly lower BET surface area of 849 m^2/g, but a much larger micropore area of 317 m^2/g. However, the amounts of mesopores and macropores are relatively few (Figure 1e), which are consistent with the dense structure displayed by SEM image (Figure 1b). After the combination of g-C$_3$N$_4$ and TPTZ, although the BET surface area slightly decreases (759 m^2/g), the Fe/NS/C-g-C$_3$N$_4$/TPTZ catalyst integrates the micropores of Fe/NS/C-TPTZ and the fluffy structure of Fe/NS/C-g-C$_3$N$_4$ (Figure 1f), which might facilitate the mass transfer and the ORR catalytic activity [23,24].

X-ray diffraction (XRD) was carried out to characterize the crystal structure of the Fe-NS/PC catalysts. According to the XRD patterns in Figure 1g, the Fe/NS/C-TPTZ exhibits two main diffraction peaks at around 25.5° and 43°, associated to the (002) and (100) planes of graphitic carbon, respectively. The (002) diffraction peak of the Fe/NS/C-g-C_3N_4 shifts to a higher angle of 29.8°, which can be associated with the carbon nitride (PDF-#78-1747). Noteworthy, the binary nitrogen precursor-derived Fe/NS/C-g-C_3N_4/TPTZ displays a broad peak at around 25.8° corresponding to the (002) diffraction of graphitic carbon, and a weak swell at 29.8° corresponding to the carbon nitride, which clearly indicates the interaction between two nitrogen precursors during pyrolysis. No other diffraction peaks can be observed, demonstrating the absence of any other Fe-containing crystalline phases.

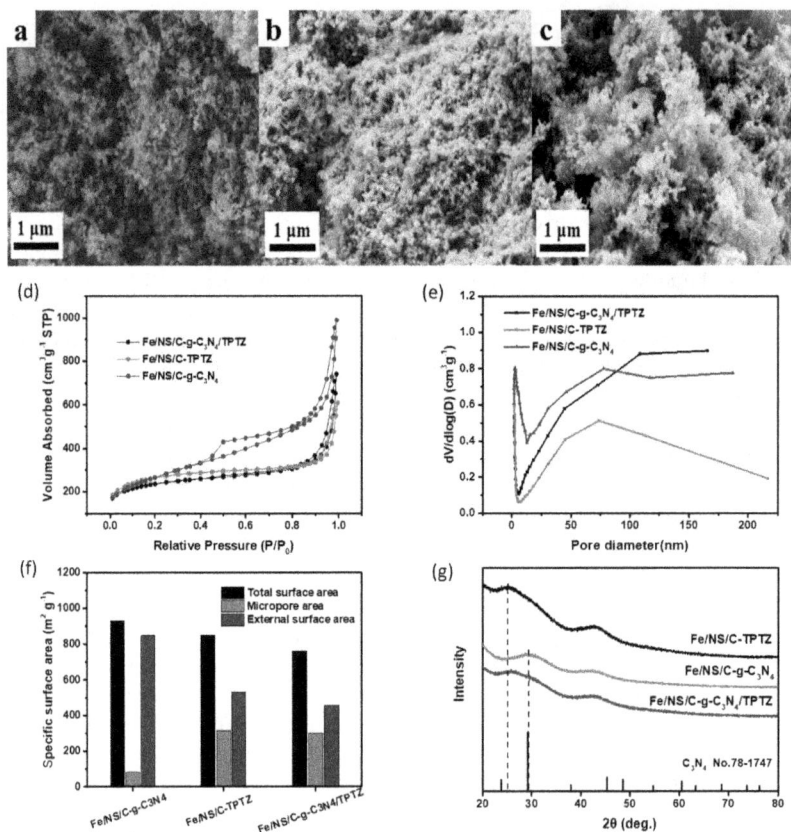

Figure 1. Scanning electron microscope (SEM) images of (**a**) Fe/NS/C-g-C_3N_4, (**b**) Fe/NS/C-TPTZ (TPTZ: 2,4,6-tri(2-pyridyl)-1,3,5-triazine), and (**c**) Fe/NS/C-g-C_3N_4/TPTZ; (**d**) N_2 adsorption-desorption isotherms; (**e**) corresponding pore size distributions; (**f**) comparison of BET surface areas; and (**g**) X-ray diffraction (XRD) patterns of Fe/NS/C-g-C_3N_4, Fe/NS/C-TPTZ, and Fe/NS/C-g-C_3N_4/TPTZ.

X-ray photoelectron spectroscopy (XPS) was implemented to investigate the states of each component within the Fe-NS/PC catalysts. The survey XPS spectra (Figure 2a) reveal that the main elements of Fe-NS/PC catalysts consist of Fe, N, C, O, and S. The elemental compositions are summarized in Table S1. Figure 2b–d display the high resolution N 1s spectra of the three Fe-NS/PC catalysts. All the spectra can be deconvoluted into four peaks corresponding to pyridinic N (N1,

398.1–398.7 eV), pyrrolic N (N2, 399.78–400.7 eV), graphitic N (N3, 400.99–401.3 eV), and oxidized N (N4, 402–404.27 eV) [25,26], respectively. Previous reports have demonstrated that both pyridinic N and graphitic N may participate the oxygen reduction reaction [27–29]. These two N species (N1 + N3) account for 4.03 at% of all the elements in Fe/NS/C-g-C$_3$N$_4$/TPTZ, remarkably higher than that of Fe/NS/C-g-C$_3$N$_4$ (1.88%) and Fe/NS/C-TPTZ (3.46%), as shown in Table S1. Notably, the contents of Fe and N in Fe/NS/C-g-C$_3$N$_4$/TPTZ are 0.29 at% and 6.67 at%, respectively, both of which are the highest among the three catalysts. In addition, the co-doping of S element in Fe-N-C catalysts would further improve the ORR activity, probably due to the structural defects and electron distribution induced by S atoms [30,31]. Based on the consideration of the high heteroatoms doping contents combined with the porous structure, the high ORR activity could be expected for the Fe/NS/C-g-C$_3$N$_4$/TPTZ catalyst.

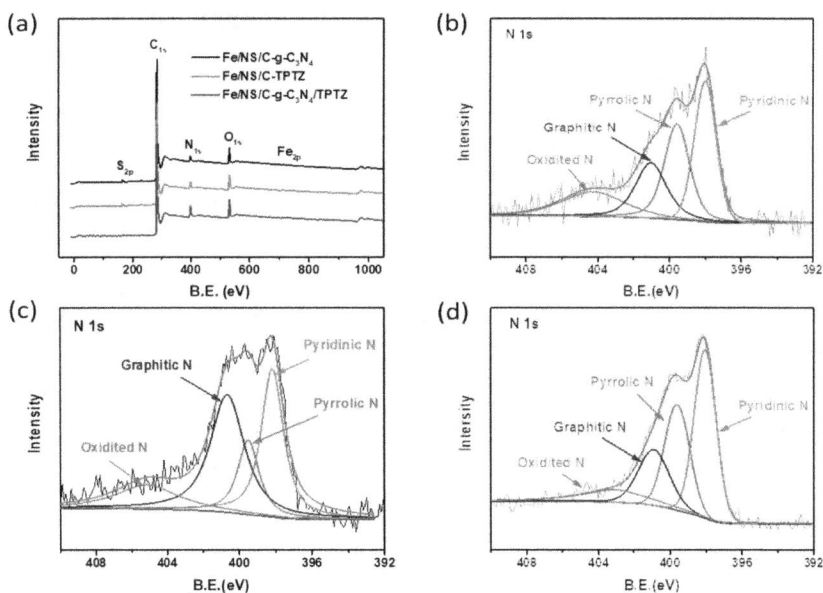

Figure 2. (a) X-ray photoelectron spectroscopy (XPS) survey spectra of Fe/NS/C-g-C$_3$N$_4$, Fe/NS/C-TPTZ, and Fe/NS/C-g-C$_3$N$_4$/TPTZ catalysts; high-resolution N 1s XPS spectra of (b) Fe/NS/C-g-C$_3$N$_4$, (c) Fe/NS/C-TPTZ, and (d) Fe/NS/C-g-C$_3$N$_4$/TPTZ.

The ORR activity of the Fe-NS/PC catalysts were evaluated by rotating disk electrode (RDE) test in O$_2$-saturated 0.1 M KOH solution. As displayed in Figure 3, the Fe/NS/C-g-C$_3$N$_4$/TPTZ catalyst exhibits the best ORR activity with an onset (E_0) and half-wave ($E_{1/2}$) potential of 0.95 V and 0.853 V (RHE), respectively, higher than that of Fe/NS/C-g-C$_3$N$_4$ catalyst (E_0 = 0.946 V and $E_{1/2}$ = 0.843 V). By sharp contrast, the Fe/NS/C-TPTZ shows inferior ORR activity with E_0 = 0.917 V (RHE) and much smaller diffusion-limited current, probably due to the agglomerations of sintered carbon that are difficult to disperse obstruct the transfer of ORR-related species, in spite of the high BET surface area. The presence of g-C$_3$N$_4$ in dual nitrogen precursors might avoid the sintering of TPTZ, meanwhile maintain the fluffy structure and increase the Fe, N doping content, thus resulting in the full exposure of active sites, which could be responsible for the high ORR activity of the Fe/NS/C-g-C$_3$N$_4$/TPTZ catalyst.

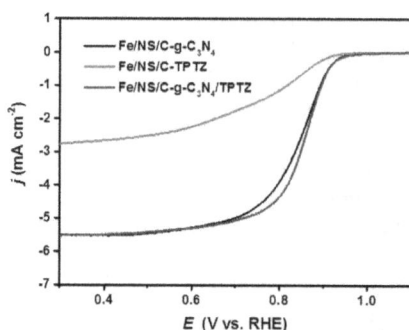

Figure 3. Oxygen reduction reaction (ORR) polarization curves of Fe/NS/C-g-C$_3$N$_4$, Fe/NS/C-TPTZ, and Fe/NS/C-g-C$_3$N$_4$/TPTZ catalysts in O$_2$-saturated 0.1 M KOH with a rotating speed of 1600 rpm and at a scan rate of 10 mV s^{-1}. RHE: reversible hydrogen electrode.

To further improve the ORR performance, a secondary heat treatment was conducted to the best Fe/NS/C-g-C$_3$N$_4$/TPTZ catalyst at the range of 800–1000 °C, denoted as Fe/NS/C-g-C$_3$N$_4$/TPTZ-T (T = 800, 900, 1000). Figure 4a–h present the transmission electron microscope (TEM) images of Fe/NS/C-g-C$_3$N$_4$/TPTZ, and Fe/NS/C-g-C$_3$N$_4$/TPTZ-T (T = 800, 900, 1000) at diversed magnifications. All these samples show the morphological characteristics of agglomerations of amorphous carbon nanoparticles with the diameter of 20–50 nm. No crystalline iron-containing nanoparticles can be observed for all three catalysts, indicating no agglomerations of iron formed during the second heat treatment. The energy dispersive X-ray spectroscopy (EDX) mapping analysis of Fe/NS/C-g-C$_3$N$_4$/TPTZ-1000 was also carried out to observe the elemental distributions. As can be seen in Figure 4i, the TEM image and the corresponding elemental mapping reveal that all doping heteroatoms, which are regarded as the components of the ORR active sites, are uniformly distributed throughout the carbon matrix, leading to the full exposure of the active sites to the ORR related species.

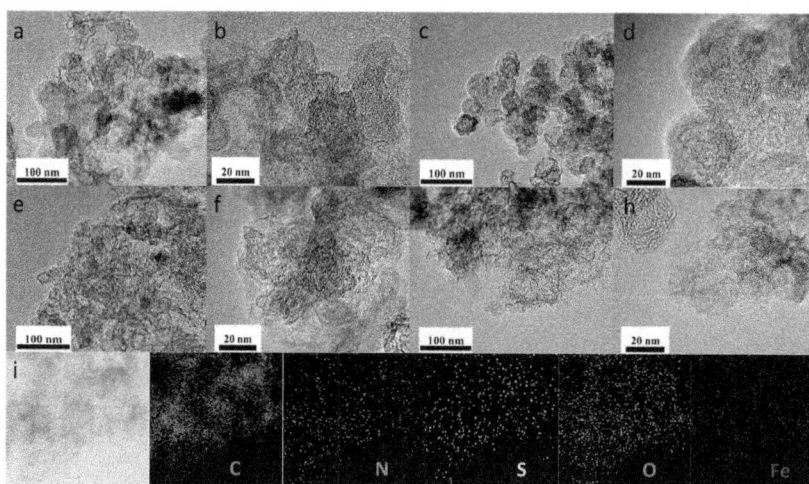

Figure 4. Transmission electron microscope (TEM) images of (**a,b**) Fe/NS/C-g-C$_3$N$_4$/TPTZ; (**c,d**) Fe/NS/C-g-C$_3$N$_4$/TPTZ-800; (**e,f**) Fe/NS/C-g-C$_3$N$_4$/TPTZ-900; (**g,h**) Fe/NS/C-g-C$_3$N$_4$/TPTZ-1000; and (**i**) TEM-EDX (EDX: energy dispersive X-ray spectroscopy) mapping analysis of C, N, S, O, and Fe of Fe/NS/C-g-C$_3$N$_4$/TPTZ-1000.

After a second heat treatment at different temperatures, all three Fe/NS/C-g-C$_3$N$_4$/TPTZ-T (T = 800, 900, 1000) catalysts show two main diffraction peaks at around 29.8° and 43° (Figure 5a), similar to Fe/NS/C-g-C$_3$N$_4$. There are no other diffraction peaks appear, further demonstrating the absence of crystalline iron-containing phases, which is consistent with the TEM results (Figure 4). Figure 5b displays the N$_2$ adsorption-desorption isotherms of the Fe/NS/C-g-C$_3$N$_4$/TPTZ-T (T = 800, 900, 1000) catalysts. The BET suface areas of Fe/NS/C-g-C$_3$N$_4$/TPTZ-800, Fe/NS/C-g-C$_3$N$_4$/TPTZ-900, and Fe/NS/C-g-C$_3$N$_4$/TPTZ-1000 are 876.8 m^2 g^{-1}, 1026.4 m^2 g^{-1}, 1138.9 m^2 g^{-1}, respectively. The pore size distributions of the Fe/NS/C-g-C$_3$N$_4$/TPTZ-T (T = 800, 900, 1000) indicate that the three catalysts all possess a good porous structure (Figure 5c).

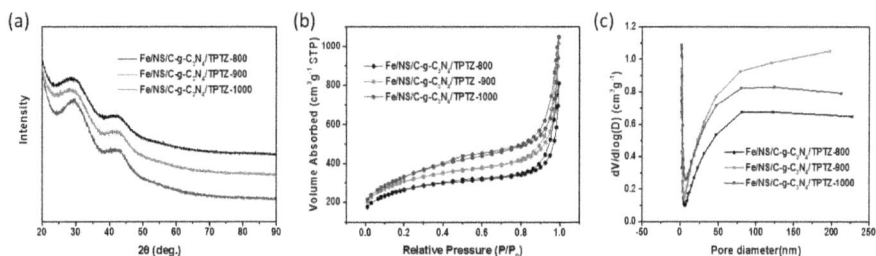

Figure 5. (**a**) XRD patterns; (**b**) N$_2$ adsorption-desorption isotherms; and (**c**) corresponding pore size distributions of Fe/NS/C-g-C$_3$N$_4$/TPTZ-800, Fe/NS/C-g-C$_3$N$_4$/TPTZ-900, and Fe/NS/C-g-C$_3$N$_4$/TPTZ-1000.

The XPS survey spectra of Fe/NS/C-g-C$_3$N$_4$/TPTZ and Fe/NS/C-g-C$_3$N$_4$/TPTZ-T (T = 800, 900, 1000) were collected to measure the elemental compositions, as shown in Figure 6a and Table S2. The contents of doped Fe and N elements reduce along with the secondary heat treatment temperature elevates. The N doping contents of Fe/NS/C-g-C$_3$N$_4$/TPTZ-T (T = 800, 900, 1000) are 4.48%, 2.57%, and 1.40%, respectively, all lower than that of Fe/NS/C-g-C$_3$N$_4$/TPTZ, due to the formation of gaseous N-containing phases. The deconvoluted high-resolution N 1s spectra reveal that the pyridinic N and graphitic N predominate in all three catalysts, as shown in Figure 6b–d and summarized in Table S2. The high-resolution S 2p spectrum of Fe/NS/C-g-C$_3$N$_4$/TPTZ-1000 can be deconvoluted into three peaks, as shown in Figure 6e. The two peaks at 164.0 and 165.2 eV can be described to S 2p$_{3/2}$ and S 2p$_{1/2}$ of thiophene-like C-S-C structure, respectively [32,33], while the third peak at 167.3 eV corresponds to sulfate species. The synergetic effects of N, S co-doping would significantly improve the ORR activity by reducing the electron localization around the Fe centers, and improve the interaction with oxygen, facilitating the four-electron pathway [34]. The high-resolution Fe 2p spectrum of Fe/NS/C-g-C$_3$N$_4$/TPTZ-1000 presents two major peaks at around 711 and 724 eV, corresponding to Fe 2p$_{1/2}$ and Fe 2p$_{3/2}$, respectively (Figure 6f) [15,35]. The dominant peak at 711 eV can be assigned to Fe^{3+} or Fe^{2+} coordinated with N, which are suggested to be the ORR active centers [15,36]. The peak at 718.5 eV is a satellite peak indicating the co-existence of Fe^{3+} and Fe^{2+} in the Fe/NS/C-g-C$_3$N$_4$/TPTZ-1000 [35].

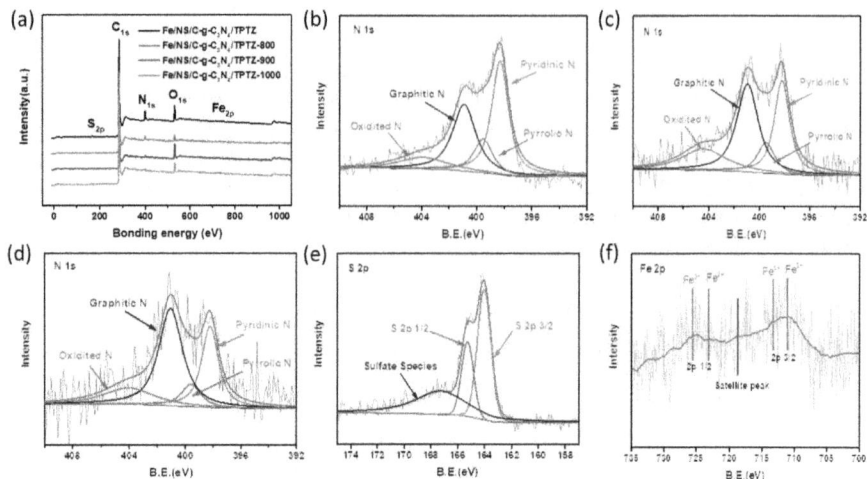

Figure 6. (**a**) XPS survey spectra of Fe/NS/C-g-C$_3$N$_4$/TPTZ and Fe/NS/C-g-C$_3$N$_4$/TPTZ-T (T = 800, 900, 1000); high-resolution N 1s spectra of (**b**) Fe/NS/C-g-C$_3$N$_4$/TPTZ-800, (**c**) Fe/NS/C-g-C$_3$N$_4$/TPTZ-900 and (**d**) Fe/NS/C-g-C$_3$N$_4$/TPTZ-1000; (**e**) high-resolution S 2p spectrum and (**f**) Fe 2p spectrum of Fe/NS/C-g-C$_3$N$_4$/TPTZ-1000.

The ORR activities of Fe/NS/C-g-C$_3$N$_4$/TPTZ-T (T = 800, 900, 1000) catalysts were measured in O$_2$-saturated 0.1 M KOH solution. For a comparison, commercial Pt/C (20 wt %) catalyst was also evaluated. As displayed in Figure 7a, after secondary heat treatment, the ORR activity of the Fe/NS/C-g-C$_3$N$_4$/TPTZ is further improved. Fe/NS/C-g-C$_3$N$_4$/TPTZ-800 and Fe/NS/C-g-C$_3$N$_4$/TPTZ-900 exhibit similar activity with the half-wave potential ($E_{1/2}$) of 0.863 V and 0.864 V (RHE), respectively. The best ORR activity is achieved at Fe/NS/C-g-C$_3$N$_4$/TPTZ-1000 with the onset and half-wave potential of 1.0 V and 0.868 V (RHE), respectively, higher than that of Pt/C catalyst (E_0 = 0.97 V, $E_{1/2}$ = 0.841 V). To evaluate the intrinsic catalytic activity, the mass activity (j_m) was calculated based on the Koutecky-Levich equation (Table S3). The Fe/NS/C-g-C$_3$N$_4$/TPTZ-1000 catalyst shows the best mass activity of 5.73 A g^{-1} at 0.9 V, reaching up to 48.8% of that of Pt/C (11.73 A g^{-1}).

To explore the origin of the high ORR activity, we have probed the possible role of Fe and S within the Fe-NS/PC catalysts. As shown in Figure S2, Fe/N/C-g-C$_3$N$_4$/TPTZ-1000 without sulfur doping exhibits lower onset and half-wave potentials for the ORR than Fe/NS/C-g-C$_3$N$_4$/TPTZ-1000, indicating the promoting effects of sulfur co-doping. By sharp comparison, the NS/C-g-C$_3$N$_4$/TPTZ-1000 without Fe doping presents inferior activity, with the half-wave potential 64 mV lower than that of Fe/NS/C-g-C$_3$N$_4$/TPTZ-1000. These results definitely reflect the indispensable roles of Fe and S doping in Fe-NS/PC for high ORR activity. It is well known that SCN$^-$ can strongly coordinate with Fe atoms. As shown in Figure S3, the Fe/NS/C-g-C$_3$N$_4$/TPTZ-1000 catalyst shows obvious ORR activity degradation after injecting 5 mM SCN$^-$, the remaining ORR activity probably results from the N, S co-doped carbon materials. Interestingly, the ORR activity of the Fe/NS/C-g-C$_3$N$_4$/TPTZ-1000 can almost recover after rinsing and replacing with fresh electrolyte, indicating that Fe atoms are at least part of the ORR active sites.

Figure 7b shows the hydrogen peroxide yield (H$_2$O$_2$ %) and the electron transfer number (n) of Fe/NS/C-g-C$_3$N$_4$/TPTZ-1000 calculated from disk current (I_d) and ring current (I_r) obtained by rotating ring-disk electrode (RRDE) test. The H$_2$O$_2$ yield is lower than 1.0% over the whole potential range. The corresponding electron transfer number is calculated to be larger than 3.975. Notably, the H$_2$O$_2$ starts to generate at the potential lower than 0.8 V, where the ORR polarization curve has reached the diffusion-limited current. These results indicate a direct four-electron pathway of ORR on the Fe/NS/C-g-C_3N_4/TPTZ-1000. The Tafel plots of Fe/NS/C-g-C_3N_4/TPTZ-1000 depicted in Figure 7c show the Tafel slope of 68 mV dec^{-1}, closed to that of Pt/C catalyst (73 mV dec^{-1}).

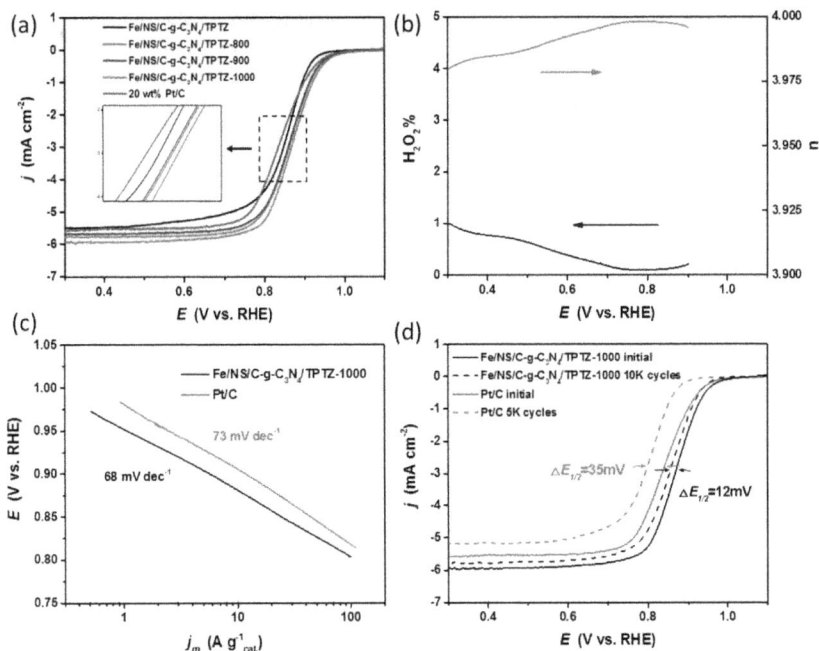

Figure 7. (**a**) ORR polarization curves of Fe/NS/C-g-C_3N_4/TPTZ, Fe/NS/C-g-C_3N_4/TPTZ-T (T = 800, 900, 1000) and Pt/C catalyst in O_2-saturated 0.1 M KOH solution with a rotational speed of 1600 rpm and at a scan rate of 10 mV/s; (**b**) hydrogen peroxide yield and electron transfer number of Fe/NS/C-g-C_3N_4/TPTZ-1000 catalyst, (**c**) Tafel plots of Fe/NS/C-g-C_3N_4/TPTZ-1000 and Pt/C; and (**d**) ORR polarization curves of Fe/NS/C-g-C_3N_4/TPTZ-1000 (before and after 10,000 potential cycles) and Pt/C (before and after 5000 potential cycles).

An accelerated durability test (ADT) was carried out to assess the durability of Fe/NS/C-g-C_3N_4/TPTZ-1000 catalyst by potential-cycling between 0.6 and 1.1 V at a scan rate of 50 mV s^{-1} in O_2-saturated 0.1 M KOH. As shown in Figure 7d, after 10,000 cycles, the $E_{1/2}$ of Fe/NS/C-g-C_3N_4/TPTZ-1000 slightly decreases by 12 mV, demonstrating the superior durability of the Fe/NS/C-g-C_3N_4/TPTZ-1000 catalyst, due to the robust structure of Fe and N anchored in carbon matrix and the low yield of corrosive H_2O_2. In sharp contrast, the $E_{1/2}$ of Pt/C remarkably decreases by 35 mV after only 5000 cycles. The insufficient durability of the Pt/C might suffer from the aggregation of Pt nanoparticles.

To further evaluate the potential for practical application, the Fe/NS/C-g-C_3N_4/TPTZ-1000 was assembled into a homemade primary Zn-air battery as cathode catalyst. Figure 8a presents the polarization curves of Zn-air batteries with Fe/NS/C-g-C_3N_4/TPTZ-1000 and Pt/C as cathodes, respectively. The Zn-air battery with Fe/NS/C-g-C_3N_4/TPTZ-1000 catalyst exhibits an open-circuit voltage (OCV) of 1.385 V and a maximal power density of 225 mW cm^{-2} at a temperature of ca. 25 °C, which is quite comparable to that of Pt/C with the OCV of 1.411 V and the maximal power density of 246 mW cm^{-2}. This performance outperforms the most of Zn-Air batteries utilizing analogous Fe-N-C cathode reported so far [37–40]. In the meantime, the long-term stability of a Zn-air battery with Fe/NS/C-g-C$_3$N$_4$/TPTZ-1000 was also tested by recording the galvanostatic discharge curves. As presented in Figure 8b, the Zn-Air battery with Fe/NS/C-g-C$_3$N$_4$/TPTZ-1000 catalyst only presents a voltage loss of 4.03% after 20,000 s at a current density of 20 mA cm^{-2}. For the Pt/C, the output voltage loss reaches 8.17% under the same conditions, demonstrating the improved durability of the Fe/NS/C-g-C$_3$N$_4$/TPTZ-1000. The slight fluctuation may due to the disturbance of testing circumstance such as humidity, which cannot be completely avoided. These results suggest that the as-prepared Fe/NS/C-g-C$_3$N$_4$/TPTZ-1000 catalyst has great potential to replace the precious metal catalysts in practical application of Zn-air battery.

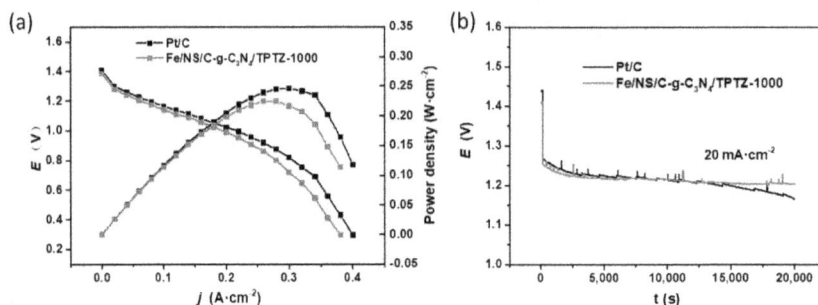

Figure 8. (a) Polarization curves and (b) discharge curves of Zn-air batteries with Fe/NS/C-g-C$_3$N$_4$/TPTZ-1000, and 20 wt % Pt/C as cathode catalysts at 298 K.

3. Experimental Section

3.1. Preparation of g-C$_3$N$_4$ Nanosheets

Bulk g-C$_3$N$_4$ powder was synthesized according to a procedure described in a previous paper [41]. Typically, dicyandiamide (Aldrich, Milwaukee, WI, USA, 99%) powder was placed in an alumina crucible with cover and heated at 550 °C for 4 h in air with a ramp rate of 2.3 °C/min. The obtained yellow agglomerates were grinded into powders. The light yellow g-C$_3$N$_4$ nanosheets were prepared by thermal etching of bulk g-C$_3$N$_4$ in air at 500 °C for 2 h with a ramp rate of 5 °C/min.

3.2. Catalyst Synthesis

Commercial Ketjenblack EC 600J (KJ 600) carbon black was first pretreated in 6.0 M HCl solution for 12 h to remove metal impurities and collected by filtration. The obtained carbon black was then treated in concentrated HNO$_3$ solution at 80 °C for 8 h to introduce carboxyl groups [42].

In a typical synthesis, 125 mg of 2,4,6-tri(2-pyridyl)-1,3,5-triazine (TPTZ, Adamas, Shanghai, China, 97%), 125 mg of g-C$_3$N$_4$ nanosheets, and 50 mg of acid-treated KJ600 were dispersed in 50 mL of alcohol under vigorous stirring for 30 min, then Fe(SCN)$_3$ solution, prepared by mixing FeCl$_3$ (0.2 M, 1.2 mL) and potassium thiocyanate (KSCN, 0.2 M, 3.6 mL) in 50 mL of alcohol, was added into the suspension and stirred for another 30 min. The solvent was then removed by rotary evaporation and vacuum drying at 80 °C for 3 h. The resulting powder was pyrolyzed at 800 °C in a N$_2$ atmosphere for 1 h with a ramp rate of 10 °C/min. The pyrolyzed sample was subjected to acid leaching in 0.5 M H$_2$SO$_4$ solution at 80 °C for 8 h to remove unstable and inactive species followed by filtration and thoroughly washed with ultrapure water. The sample was finally vacuum dried to obtain the Fe/NS/C-g-C$_3$N$_4$/TPTZ catalyst. As a comparison, Fe/NS/C-g-C$_3$N$_4$ and Fe/NS/C-TPTZ catalysts were prepared by the same procedure except using 250 mg of g-C$_3$N$_4$ or 250 mg of TPTZ as the single nitrogen precursor, respectively.

Moreover, the Fe/NS/C-g-C$_3$N$_4$/TPTZ sample was heat-treated again at 800 °C, 900 °C, and 1000 °C in a N$_2$ atmosphere for 3 h to obtain the Fe/NS/C-g-C$_3$N$_4$/TPTZ-T (T = 800, 900, 1000) catalysts. The synthesis approaches for Fe/N/C-g-C_3N_4/TPTZ-1000 and NS/C-g-C_3N_4/TPTZ-1000 are the same as that for Fe/NS/C-g-C_3N_4/TPTZ-1000 without adding KSCN and $FeCl_3$, respectively.

3.3. Characterizations

The morphology and elemental mapping of the samples were analyzed using field-emission transmission electron microscopy (FE-TEM, JEM-2100F, JEOL Ltd., Tokyo, Japan) with a working voltage of 200 kV. XRD analysis was performed using a D8 advance powder X-ray diffractometer (Bruker, Karlsruhe, Germany) with a Cu Ka (λ = 1.5418 Å) at 0.2° s^{-1}. The elemental composition and chemical states were measured by XPS (K-Alpha, Thermo Scientific, Waltham, MA, USA) with an Al Ka X-ray source. The surface areas and pore structures were characterized using a Micromeritics ASAP 2020 instrument (Micromeritics Instrument Corp., Norcross, GA, USA).

3.4. Electrochemical Measurement

All electrochemical measurements were conducted on a bipotentiostat (CHI-730E, Shanghai Chenhua, Shanghai, China) equipped with a rotating disk electrode (RDE) or rotating ring-disk electrode (RRDE) system (PINE Inc., Durham, NC, USA) at room temperature. An Ag/AgCl electrode and a graphite plate were used as the reference and counter electrode, respectively. The potential has been experimentally corrected to the range of RHE by the following equation: E(RHE) = E(Ag/AgCl) + 0.956 V. The electrolyte was O_2-saturated 0.1 M KOH solution.

For all Fe-NS/PC catalysts, 12 mg of catalyst was dispersed in the mixture of 0.5 mL of de-ionized water, 0.45 mL of isopropanol and 0.05 mL of 5 wt % Nafion solution (Aldrich) by sonication for 1 h to produce the ink. A certain volume of the catalyst ink was pipetted onto the pre-polished glassy carbon disk (0.196 cm^2 for RDE and 0.2475 cm^2 for RRDE) resulting a loading of 0.4 mg cm^{-2}. For commercial Pt/C catalyst (20 wt %, ElectroChem, Inc., Woburn, MA, USA), the ink was prepared by dispersing 10 mg catalyst in 1.0 mL of de-ionized water, 0.95 mL of isopropanol, and 0.05 mL of 5 wt % Nafion solution. The loading of the Pt/C catalyst was 0.1 mg cm^{-2}.

In RDE tests, ORR polarization curves were measured at a scan rate of 10 mV s^{-1}. The electrode rotational speed was 1600 rpm. The background current was determined by recording the voltammogram in N_2-saturated electrolyte. The accelerated durability tests (ADT) were carried out by cycling the potential in the range from 0.6 to 1.1 V in O_2-saturated electrolyte with a scan rate of 50 mV s^{-1}.

Hydrogen peroxide yield and the electron transfer number (n) can be calculated by the following equations with the potential of ring electrode fixed at 1.4 V (RHE) in RRDE:

$$H_2O_2 \% = 200\% \times \frac{\frac{I_r}{N}}{|I_d| + \frac{I_r}{N}} \tag{1}$$

$$n = 4 \times \frac{|I_d|}{|I_d| + \frac{I_r}{N}} \tag{2}$$

where I_d is the disk current, I_r is the ring current, N = 0.37 is the collection efficiency of the Pt ring.

3.5. Primary Zn-Air Battery Test

The cathode of the Zn-air battery was prepared by loading the Fe/NS/C-g-C_3N_4/TPTZ-1000 or 20 wt % Pt/C catalyst onto carbon fiber paper (3.0 × 4.0 cm) with a catalyst loading of 1.0 mg cm^{-2}. Electrolytic zinc powder was used as the anode. The electrolyte was 6.0 M KOH solution. Polarization curves and galvanostatic discharge curves were measured on Arbin battery testing system.

4. Conclusions

In summary, a Fe, N, S co-doped porous carbon as ORR electrocatalyst was developed based on the interaction of binary nitrogen precursors during the pyrolysis process. After secondary heat treatment, the Fe/NS/C-g-C$_3$N$_4$/TPTZ-1000 catalyst displays superior ORR activity and durability in alkaline media, in comparison with the commercial Pt/C. Enhanced ORR activity and durability can be attributed to its good porous structure, high surface area, high contents of pyridinic N and graphitic N, and the synergy of N and S co-doping. Moreover, the Zn-air battery assembled with Fe/NS/C-g-C$_3$N$_4$/TPTZ-1000 as a cathode exhibits comparable power density and better stability than that of the Pt/C, demonstrating its potential for substituting precious metal catalysts in practical energy devices.

Supplementary Materials: The following are available online at http://www.mdpi.com/2073-4344/8/4/158/s1, Figure S1: Thermogravimetric analysis (TGA) of (a) g-C$_3$N$_4$ and (b)TPTZ under N$_2$ atmosphere, Figure S2: ORR polarization curves of Fe/N/C-g-C$_3$N$_4$/TPTZ-1000, NS/C-g-C$_3$N$_4$/TPTZ-1000 and Fe/NS/C-g-C$_3$N$_4$/TPTZ-1000 in O$_2$-saturated 0.1 M KOH solution with a rotational speed of 1600 rpm and a scan rate of 10 mV/s, Figure S3: The polarization curves of Fe/NS/C-g-C$_3$N$_4$/TPTZ-1000 catalyst before and after adding SCN$^-$ and after rinsing and replacing fresh O$_2$-saturated 0.1 M KOH solution, Table S1: The element contents of Fe/NS/C-g-C$_3$N$_4$, Fe/NS/C-TPTZ and Fe/NS/C-g-C$_3$N$_4$/TPTZ obtained by XPS, Table S2: The element contents of Fe/NS/C-g-C$_3$N$_4$/TPTZ-T (T = 800, 900, 1000) obtained by XPS, Table S3: Comparison of ORR activity of Fe/NS/C-g-C$_3$N$_4$/TPTZ, Fe/NS/C-g-C$_3$N$_4$/TPTZ-T (T = 800, 900, 1000) and Pt/C catalysts.

Acknowledgments: The financial supports from the National Key Research and Development Program of China (grant no. 2017YFA0206500) and the National Natural Science Foundation of China (grant no. 21673275, 21533005) are greatly appreciated.

Author Contributions: Xiao Liu, Chi Chen, Zhiqing Zou, and Hui Yang conceived and designed the experiments; Xiao Liu performed the experiments; Xiao Liu, Chi Chen, Qingqing Cheng, and Liangliang Zou analyzed the data; Xiao Liu and Chi Chen wrote the paper; Zhiqing Zou and Hui Yang managed all the experiments and the writing process as the corresponding authors.

Conflicts of Interest: The authors declare no conflict of interest.

References

1. Cui, X.; Yang, S.; Yan, X.; Leng, J.; Shuang, S.; Ajayan, P.M.; Zhang, Z. Pyridinic-Nitrogen-Dominated Graphene Aerogels with Fe-N-C Coordination for Highly Efficient Oxygen Reduction Reaction. *Adv. Funct. Mater.* **2016**, *26*, 5708–5717. [CrossRef]

2. Zhu, Z.; Yang, Y.; Guan, Y.; Xue, J.; Cui, L. Constructing of cobalt-embedded in nitrogen-doped carbon material with desired porosity derived from MOFs confined growth within graphene aerogel as a superior catalyst towards HER and ORR. *J. Mater. Chem. A* **2016**, *4*, 15536–15545. [CrossRef]

3. Yang, J.; Liu, D.J.; Kariuki, N.N.; Chen, L.X. Aligned carbon nanotubes with built-in FeN$_4$ active sites for electrocatalytic reduction of oxygen. *Chem. Commun.* **2008**, *36*, 329–331. [CrossRef]

4. Zeng, L.; Cui, X.; Chen, L.; Ye, T.; Huang, W.; Ma, R.; Zhang, X.; Shi, J. Non-noble bimetallic alloy encased in nitrogen-doped nanotubes as a highly active and durable electrocatalyst for oxygen reduction reaction. *Carbon* **2017**, *114*, 347–355. [CrossRef]

5. Liu, H.; Shi, Z.; Zhang, J.; Zhang, L.; Zhang, J. Ultrasonic spray pyrolyzed iron-polypyrrole mesoporous spheres for fuel cell oxygen reduction electrocatalysts. *J. Mater. Chem.* **2009**, *19*, 468–470. [CrossRef]

6. Liang, H.W.; Wei, W.; Wu, Z.S.; Feng, X.; Müllen, K. Mesoporous Metal-Nitrogen-Doped Carbon Electrocatalysts for Highly Efficient Oxygen Reduction Reaction. *J. Am. Chem. Soc.* **2013**, *135*, 16002–16005. [CrossRef] [PubMed]

7. Wu, G.; Zelenay, P. Nanostructured nonprecious metal catalysts for oxygen reduction reaction. *Acc. Chem. Res.* **2013**, *46*, 1878–1889. [CrossRef] [PubMed]

8. Zhang, J.; Dai, L. Heteroatom-doped graphitic carbon catalysts for efficient electrocatalysts of oxygen reduction reaction. *ACS Catal.* **2015**, *5*, 7244–7253. [CrossRef]

9. Jia, Q.; Ramaswamy, N.; Hafiz, H.; Tylus, U.; Strickland, K.; Wu, G.; Barbiellini, B.; Bansil, A.; Holby, E.F.; Zelenay, P. Experimental observation of redox-induced Fe–N switching behavior as a determinant role for oxygen reduction activity. *ACS Nano* **2015**, *9*, 12496–12505. [CrossRef] [PubMed]

10. Kone, I.; Xie, A.; Tang, Y.; Chen, Y.; Liu, J.; Chen, Y.; Sun, Y.; Yang, X.; Wan, P. Hierarchical Porous Carbon Doped with Iron-Nitrogen-Sulfur for Efficient Oxygen Reduction Reaction. *ACS Appl. Mater. Interfaces* **2017**, *9*, 20963–20973. [CrossRef] [PubMed]

11. Yu, H.; Fisher, A.; Cheng, D.; Cao, D. Cu, N-codoped Hierarchical Porous Carbons as Electrocatalysts for Oxygen Reduction Reaction. *ACS Appl. Mater. Interfaces* **2016**, *8*, 21431–21439. [CrossRef] [PubMed]
12. Qiao, M.; Tang, C.; He, G.; Qiu, K.; Binions, R.; Parkin, I.; Zhang, Q.; Guo, Z.; Titirici, M. Graphene/ nitrogen-doped porous carbon sandwiches for the metal-free oxygen reduction reaction: Conductivity versus active sites. *J. Mater. Chem. A* **2016**, *4*, 12658–12666. [CrossRef]
13. Lefèvre, M.; Proietti, E.; Jaouen, F.; Dodelet, J.P. Iron-based catalysts with improved oxygen reduction activity in polymer electrolyte fuel cells. *Science* **2009**, *324*, 71–74. [CrossRef] [PubMed]
14. Li, Z.; Sun, H.; Wei, L.; Jiang, W.J.; Wu, M.; Hu, J.S. Lamellar Metal Organic Framework-Derived Fe–N–C Non-Noble Electrocatalysts with Bimodal Porosity for Efficient Oxygen Reduction. *ACS Appl. Mater. Interfaces* **2017**, *9*, 5272–5278. [CrossRef] [PubMed]
15. Gupta, S.; Zhao, S.; Ogoke, O.; Lin, Y.; Xu, H.; Wu, G. Engineering Favorable Morphology and Structure of Fe-N-C Oxygen-Reduction Catalysts through Tuning of Nitrogen/Carbon Precursors. *ChemSusChem* **2017**, *10*, 774–785. [CrossRef] [PubMed]
16. Fu, X.; Zamani, P.; Choi, J.Y.; Hassan, F.M.; Jiang, G.; Higgins, D.C.; Zhang, Y.; Hoque, M.A.; Chen, Z. In Situ Polymer Graphenization Ingrained with Nanoporosity in a Nitrogenous Electrocatalyst Boosting the Performance of Polymer-Electrolyte-Membrane Fuel Cells. *Adv. Mater.* **2016**, *29*, 1604456. [CrossRef] [PubMed]
17. Chung, H.T.; Cullen, D.A.; Higgins, D.; Sneed, B.T.; Holby, E.F.; More, K.L.; Zelenay, P. Direct atomic-level insight into the active sites of a high-performance PGM-free ORR catalyst. *Science* **2017**, *357*, 479–484. [CrossRef] [PubMed]
18. Wang, Y.C.; Lai, Y.J.; Song, L.; Zhou, Z.Y.; Liu, J.G.; Wang, Q.; Yang, X.D.; Chen, C.; Shi, W.; Zheng, Y.P.; et al. S-Doping of an Fe/N/C ORR Catalyst for Polymer Electrolyte Membrane Fuel Cells with High Power Density. *Angew. Chem. Int. Ed.* **2015**, *127*, 10045–10048. [CrossRef]
19. Chen, P.; Zhou, T.; Xing, L.; Xu, K.; Tong, Y.; Xie, H.; Zhang, L.; Yan, W.; Chu, W.; Wu, C. Atomically Dispersed Iron-Nitrogen Species as Electrocatalysts for Bifunctional Oxygen Evolution and Reduction Reactions. *Angew. Chem. Int. Ed.* **2017**, *56*, 610–614. [CrossRef] [PubMed]
20. Tian, J.; Morozan, A.; Sougrati, M.T.; Chenitz, R.; Dodelet, J.P.; Jones, D.; Jaouen, F. Optimized synthesis of Fe/N/C cathode catalysts for PEM fuel cells: A matter of iron-ligand coordination strength. *Angew. Chem. Int. Ed.* **2013**, *52*, 6867–6870. [CrossRef] [PubMed]
21. Yang, L.; Kong, J.; Zhou, D.; Ang, J.M.; Phua, S.L.; Yee, W.A.; Liu, H.; Huang, Y.; Lu, X. Transition-Metal-Ion-Mediated Polymerization of Dopamine: Mussel-Inspired Approach for the Facile Synthesis of Robust Transition-Metal Nanoparticle–Graphene Hybrids. *Chem. Eur. J.* **2014**, *20*, 7776–7783. [CrossRef] [PubMed]
22. Zhou, D.; Yang, L.; Yu, L.; Kong, J.; Yao, X.; Liu, W.; Xu, Z.; Lu, X. Fe/N/C hollow nanospheres by Fe(III)-dopamine complexation-assisted one-pot doping as nonprecious-metal electrocatalysts for oxygen reduction. *Nanoscale* **2015**, *7*, 1501–1509. [CrossRef] [PubMed]
23. Jaouen, F.; Herranz, J.; Lefèvre, M.; Dodelet, J.P.; Kramm, U.I.; Herrmann, I.; Bogdanoff, P.; Maruyama, J.; Nagaoka, T.; Garsuch, A. Cross-laboratory experimental study of non-noble-metal electrocatalysts for the oxygen reduction reaction. *ACS Appl. Mater. Interfaces* **2009**, *1*, 1623–1639. [CrossRef] [PubMed]
24. Jaouen, F.; Lefèvre, M.; Dodelet, J.P.; Cai, M. Heat-Treated Fe/N/C Catalysts for O_2 Electroreduction: Are Active Sites Hosted in Micropores? *J. Phys. Chem. B* **2006**, *110*, 5553–5558. [CrossRef] [PubMed]
25. Yasuda, S.; Furuya, A.; Uchibori, Y.; Kim, J.; Murakoshi, K. Iron–Nitrogen-Doped Vertically Aligned Carbon Nanotube Electrocatalyst for the Oxygen Reduction Reaction. *Adv. Funct. Mater.* **2016**, *26*, 738–744. [CrossRef]
26. Choi, I.A.; Kwak, D.H.; Han, S.B.; Park, J.Y.; Park, H.S.; Ma, K.B.; Kim, D.H.; Won, J.E.; Park, K.W. Doped porous carbon nanostructures as non-precious metal catalysts prepared by amino acid glycine for oxygen reduction reaction. *Appl. Catal. B Environ.* **2017**, *211*, 235–244. [CrossRef]
27. Liang, W.; Chen, J.; Liu, Y.; Chen, S. Density-functional-theory calculation analysis of active sites for four-electron reduction of O_2 on Fe/N-doped graphene. *ACS Catal.* **2014**, *4*, 4170–4177. [CrossRef]
28. Xiao, H.; Shao, Z.G.; Zhang, G.; Gao, Y.; Lu, W.; Yi, B. Fe–N–carbon black for the oxygen reduction reaction in sulfuric acid. *Carbon* **2013**, *57*, 443–451. [CrossRef]
29. Saidi, W.A. Oxygen reduction electrocatalysis using N-doped graphene quantum-dots. *J. Phys. Chem. Lett.* **2013**, *4*, 4160–4165. [CrossRef]

30. Liang, J.; Jiao, Y.; Jaroniec, M.; Qiao, S.Z. Sulfur and nitrogen dual-doped mesoporous graphene electrocatalyst for oxygen reduction with synergistically enhanced performance. *Angew. Chem. Int. Ed.* **2012**, *51*, 11496–11500. [CrossRef] [PubMed]

31. Jiang, T.; Wang, Y.; Wang, K.; Liang, Y.; Wu, D.; Tsiakaras, P.; Song, S. A novel sulfur-nitrogen dual doped ordered mesoporous carbon electrocatalyst for efficient oxygen reduction reaction. *Appl. Catal. B Environ.* **2016**, *189*, 1–11. [CrossRef]

32. Ai, W.; Luo, Z.; Jiang, J.; Zhu, J.; Du, Z.; Fan, Z.; Xie, L.; Zhang, H.; Huang, W.; Yu, T. Nitrogen and sulfur codoped graphene: Multifunctional electrode materials for high-performance Li-ion batteries and oxygen reduction reaction. *Adv. Mater.* **2014**, *26*, 6186–6192. [CrossRef] [PubMed]

33. Chen, C.; Yang, X.D.; Zhou, Z.Y.; Lai, Y.J.; Rauf, M.; Wang, Y.; Pan, J.; Zhuang, L.; Wang, Q.; Wang, Y.C. Aminothiazole-derived N, S, Fe-doped graphene nanosheets as high performance electrocatalysts for oxygen reduction. *Chem. Commun.* **2015**, *51*, 17092–17095. [CrossRef] [PubMed]

34. Shen, H.; Gracia-Espino, E.; Ma, J.; Zang, K.; Luo, J.; Wang, L.; Gao, S.; Mamat, X.; Hu, G.; Wagberg, T. Synergistic Effects between Atomically Dispersed Fe–N–C and C–S–C for the Oxygen Reduction Reaction in Acidic Media. *Angew. Chem. Int. Ed.* **2017**, *56*, 13800–13804. [CrossRef] [PubMed]

35. Zeng, S.; Lyu, F.; Nie, H.; Zhan, Y.; Bian, H.; Tian, Y.; Li, Z.; Wang, A.; Lu, J.; Li, Y.Y. Facile fabrication of N/S-doped carbon nanotubes with Fe_3O_4 nanocrystals enchased for lasting synergy as efficient oxygen reduction catalysts. *J. Mater. Chem. A* **2017**, *5*, 13189–13195. [CrossRef]

36. Ren, G.; Lu, X.; Li, Y.; Zhu, Y.; Dai, L.; Jiang, L. Porous Core–Shell Fe_3C Embedded N-doped Carbon Nanofibers as an Effective Electrocatalysts for Oxygen Reduction Reaction. *ACS Appl. Mater. Interfaces* **2016**, *8*, 4118–4125. [CrossRef] [PubMed]

37. Cao, L.; Li, Z.; Gu, Y.; Li, D.; Su, K.; Yang, D.; Cheng, B. Rational design of N-doped carbon nanobox-supported $Fe/Fe_2N/Fe_3C$ nanoparticles as efficient oxygen reduction catalysts for Zn–air batteries. *J. Mater. Chem. A* **2017**, *5*, 11340–11347. [CrossRef]

38. Zhao, Y.; Lai, Q.; Wang, Y.; Zhu, J.; Liang, Y.Y. Interconnected Hierarchically Porous Fe, N-Codoped Carbon Nanofibers as Efficient Oxygen Reduction Catalysts for Zn-Air Batteries. *ACS Appl. Mater. Interfaces* **2017**, *9*, 16178–16186. [CrossRef] [PubMed]

39. Yang, J.; Toshimitsu, F.; Yang, Z.; Fujigaya, T.; Nakashima, N. Pristine carbon nanotube/iron phthalocyanine hybrids with a well-defined nanostructure show excellent efficiency and durability for oxygen reduction reaction. *J. Mater. Chem. A* **2016**, *5*, 1184–1191. [CrossRef]

40. Cai, P.; Hong, Y.; Ci, S.; Wen, Z. In situ integration of CoFe alloy nanoparticles with nitrogen-doped carbon nanotubes as advanced bifunctional cathode catalysts for Zn-air batteries. *Nanoscale* **2016**, *8*, 20048–20055. [CrossRef] [PubMed]

41. Niu, P.; Zhang, L.; Liu, G.; Cheng, H.M. Graphene-like carbon nitride nanosheets for improved photocatalytic activities. *Adv. Funct. Mater.* **2012**, *22*, 4763–4770. [CrossRef]

42. Choi, J.Y.; Hsu, R.S.; Chen, Z. Highly active porous carbon-supported nonprecious metal-N electrocatalyst for oxygen reduction reaction in PEM fuel cells. *J. Phys. Chem. C* **2010**, *114*, 8048–8053. [CrossRef]

catalysts

MDPI

Article

Engineering Mesoporous NiO with Enriched Electrophilic Ni³⁺ and O⁻ toward Efficient Oxygen Evolution

Xiu Liu [1,2,3], Zhi-Yuan Zhai [1], Zhou Chen [1], Li-Zhong Zhang [2,*], Xiu-Feng Zhao [2], Feng-Zhan Si [3,*] and Jian-Hui Li [1,*]

[1] National Engineering Laboratory for Green Chemical Productions of Alcohols-Ethers-Esters, College of Chemistry and Chemical Engineering, Xiamen University, Xiamen 361005, China; guoguo11266@163.com (X.L.); jz@xmu.edu.cn (Z.-Y.Z.); zc7@ualberta.ca (Z.C.)

[2] Department of Chemistry and Applied Chemistry, Changji University, Changji 831100, China; zhaoxiufeng19670@126.com

[3] College of Materials Science and Engineering, Shenzhen University, Shenzhen 518060, China

* Correspondence: lzzhang@cjc.edu.cn (L.-Z.Z.); sifengzhan@szu.edu.cn (F.-Z.S.); jhli@xmu.edu.cn (J.-H.L.); Tel.: +86-592-2184591 (J.-H.L.)

Received: 29 June 2018; Accepted: 26 July 2018; Published: 30 July 2018

Abstract: Tremendous efforts have been devoted to develop low-cost and highly active electrocatalysts for oxygen evolution reaction (OER). Here, we report the synthesis of mesoporous nickel oxide by the template method and its application in the title reaction. The as-prepared mesoporous NiO possesses a large surface area, uniform mesopores, and rich surface electrophilic Ni³⁺ and O⁻ species. The overpotential of *meso*-NiO in alkaline medium is 132 mV at 10 mA cm⁻¹ and 410 mV at 50 mA cm⁻¹, which is much smaller than that of the other types of NiO samples. The improvement in the OER activity can be ascribed to the synergy of the large surface area and uniform mesopores for better mass transfer and high density of Ni³⁺ and O⁻ species favoring the nucleophilic attack by OH⁻ to form a NiOOH intermediate. The reaction process and the role of electrophilic Ni³⁺ and O⁻ were discussed in detail. This results are more conducive to the electrochemical decomposition of water to produce hydrogen fuel as a clean and renewable energy.

Keywords: oxygen evolution reaction; mesoporous NiO; water splitting; nucleophilic attack; electrophilic Ni³⁺ and O⁻

1. Introduction

Nowadays, since clean energy is playing an increasingly important role in environmental protection, hydrogen production by water decomposition has been a hot research topic all along [1]. The research and development of catalysts with high activity and stability has important theoretical significance [2] and application value in the fields of renewable resources for hydrogen production by electrolysis. Hydrogen evolution reaction (HER) and oxygen evolution reaction (OER) are the two half-reactions in electrochemical water that split to produce hydrogen. The OER process has slow reaction dynamics, which consume much more energy in practical applications, and has attracted a great amount of attention from researchers. Currently, the most widely used and efficient catalysts for OER are RuO₂ and IrO₂. However, these catalysts are scarce in resources and are also expensive, which impeded the large-scale application for commercial purpose. Therefore, in order to overcome these difficulties, it is our goal to develop advanced catalysts with high-stability and low-cost. The studies on first-transition metals, especially Mn, Fe, Co, and Ni-based materials have a growth spurt owing to the good OER performance approaching that of noble metals.

It has been found that in addition to Pt, Pd, and other precious metals, nickel-based electrocatalysts have emerged as outstanding OER catalysts owing to their low price, stability in alkaline solution, relatively low overpotential, and ease of being doped or modified with guest elements. For example, Ni-Fe-CNTs [3] and $Co_xNi_yS_z$ [4] octahedral nanocages showed the overpotential of 247 mV and 362 mV at 10 mA cm^{-2}, respectively. Among them, NiO is easily available, and possesses a high specific capacitance that is suitable in electrochemical reaction. It was reported that a NiO/Ni heterojunction [5] showed highlighted performance, with the overpotential reaching 1.5 V (versus Reversible Hydrogen Electrode, RHE) at 20 mA cm^{-2}. It was reported that the Ni-Co oxide hierarchical nanosheets with a Ni^{3+} rich surface benefits the formation of NiOOH, which promotes the OER as the main active site [6].

Since electrochemical processes with electron gain or loss are interfacial phenomena, the electrode materials should possess a high specific surface area with a suitable pore size distribution to enhance the charge–storage capability. The activity of a catalyst depends in part on its surface area [7]. Thus, mesoporous morphology is one of the most important design parameters for making an electrode material. Recently, the synthesis of metal oxide porous electrodes was greatly reported. Compared with micro or macroporous materials, mesoporous materials have unique advantages, such as the larger mesopores increase the internal transmission efficiency of electrons, facilitate contact with active sites, are stable in alkaline solution, have a pore size that is adjustable to control better selectivity, and have a regular pore structure for particles in the rapid proliferation of channels [8]. It is well known that mesopores with sizes in the range of 2 nm to 50 nm are highly desirable for electrochemical reactions, since the mesopores facilitate ionic motion easily. There have been a lot of reports using mesoporous first-transition metal oxides in electrochemical processes, especially in supercapacitors [9–12] and rechargeable batteries [13]. Mesoporous NiO [14] possessed a high surface area, good surface redox reactivity, and consequently a high capacitance of 124 F g^{-1}. The mosoporous structure provides OH$^-$ accessible pores and a very high surface area (477.7 m^2 g^{-1} when calcined at 250 °C) for charge storage. However, few attempts have been made to employ mesoporous NiO in OER reactions.

The oxygen species and lattice defects for transitional metal oxide may play an important role in electrochemical reactions. However, the role of Ni^{3+} in the title reaction has been well recognized, but the effect of oxygen species, especially the electrophilic O$^-$ active species, has been rarely studied. In our previous work [15], we have reported that the mesoporous NiO prepared by a surfactant-assisted route allowed efficient molecular transport by diffusion in the channels during the oxidative dehydrogenation of propane reaction. The density of Ni^{3+} and O$^-$ species are much higher for this mesoporous NiO than for other kinds of NiO. In this work, the same mesoporous nickel oxide with rich Ni^{3+} and O$^-$ species was synthesized and tested in the oxygen evolution reaction. For comparison, the conventional nano-sized NiO sample was synthesized by a modified sol–gel method. The role of electrophilic Ni^{3+} and O$^-$ species in the formation of active NiOOH intermediate was fully discussed.

2. Results and Discussion

2.1. Texture Characterizations

The mesoporous nickel oxide was prepared by a surfactant-templated method via urea hydrolysis of metal nitrates followed by calcination in air, and denoted as *meso*-NiO. For comparison, the traditional nano-sized NiO without the mesoporous structure was prepared by a modified sol–gel method [16] using citric acid as the ligand, and denoted as *bulk*-NiO. The mesoporous structure and surface properties of nickel oxide were characterized by XRD, N_2 physical adsorption, O_2 temperature programmed desorption (O_2-TPD), X-ray photoelectron spectroscopy (XPS) techniques, and laser Raman spectroscopy. From the XRD patterns shown in Figure 1a, their corresponding XRD patterns indicated the information of pure NiO phase for both samples. The characteristic peaks (111), (200), and (220) can be observed at approximately 2θ of 27°, 43°, and 63°. The crystal diffraction for *meso*-NiO is much wider, indicating that the crystallite size for this mesoporous sample is smaller. By the Scherrer formula ($L = 0.89\lambda/\beta\cos\theta$, where β is the half height of the diffraction line, λ is the wavelength, and k is the diffraction angle), the calculated crystallite size for *meso*-NiO is only 2.1 nm, which is much smaller than the *bulk*-NiO of 17.4 nm. The homogeneous precipitation with surfactant-assisted urea hydrolysis can ensure the smaller crystal size of the NiO catalyst [15].

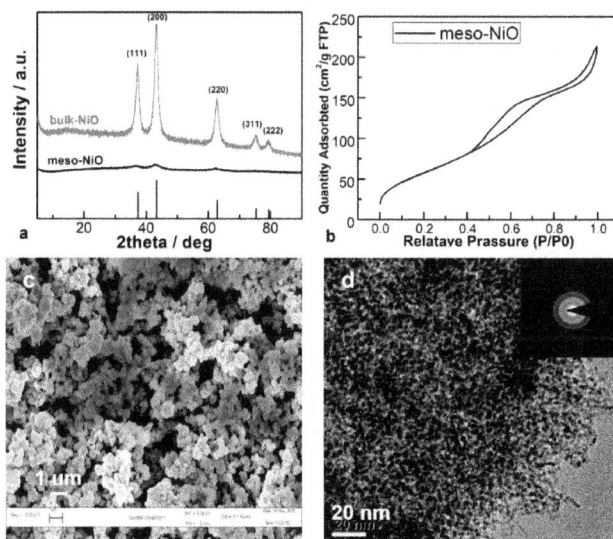

Figure 1. (**a**) XRD pattern of mesoporous nickel oxide and nano-NiO catalysts. (**b**) Pore size distribution plot of the *meso*-NiO; (**c**) SEM and (**d**) TEM images of the *meso*-NiO.

N_2 adsorption tests were used to analyze the surface area and pore size distribution. As shown in Figure 1b, the N_2 sorption isotherm shows a type-IV adsorption–desorption behavior with a well-defined pore-filling step at 0.45 P/P0, indicating the presence of mesopores for the *meso*-NiO sample. The mesoporous size distributions based on the Barrett–Joyner–Halenda (BJH) method are confirmed by the corresponding pore size distribution curves. The Brunauer–Emmett–Teller (BET) specific area is 220.8 m^2/g and 45.3 m^2/g for *meso*-NiO and *bulk*-NiO samples, respectively. The pore size and volume of *meso*-NiO are much larger than that of the *bulk*-NiO. In electrochemistry, a smaller particle size distribution and larger specific surface area will be more beneficial to the improvement of the catalytic efficiency, thereby enhancing the electrochemical activity [17].

The macroscopic pore and mesopore for the meso-NiO can be seen from SEM and TEM images, as shown in Figure 1c,d. The macroscopic pore was formed by the aggregations of large particulates, which processes the average size of about 1 µm (SEM, Figure 1c). The mesopores were wormhole-like and randomly arranging (TEM, Figure 1d), but the pore sizes were distributed homogeneously within the whole NiO nanoparticles. These wormhole-like pores were formed between NiO nanoparticles due to their aggregation. Unlike SiO_2, NiO could not form a space-stable ordered structure, because it has a low framework cross-linking and a very strong Ni–O–Ni bond angle, so the stability of the nickel oxide-ordered mesoporous structure is relatively difficult to obtain. Hence, during the electrochemical process, the formation of a wormhole structure allows the electrolyte solution to penetrate and release gases very quickly [18], which can improve the mass transfer and enhance the electrocatalytic activity.

2.2. OER Activity

The OER catalytic activities of *meso*-NiO and *bulk*-NiO were tested using a typical three-electrode system in 1.0 M of KOH electrolyte. The samples were deposited onto glassy carbon electrodes to prepare the working electrodes. Figure 2a,b displays the linear scanning voltammetry (LSV) polarization and Tafel slope curves of *meso*-NiO and *bulk*-NiO with IR-compensation. Obvious oxidation peaks are obtained for *meso*-NiO near 380 mV (versus Ag/AgCl), which represents the oxidation of Ni (II) to Ni (III) [19,20]. Obviously, the specific current density of *meso*-NiO is indeed higher than that of *bulk*-NiO under a curtain applied voltage. *Meso*-NiO also presents a slightly earlier onset potential. The overpotentials at 50 mA cm^{-2} are 410 mV and 494 mV for *meso*-NiO and *bulk*-NiO, respectively. It is noted that the overpotential at 10 mA cm^{-2} is only 132 mV for *meso*-NiO, but it is obtained from the anodic oxidation process, which leads to an unstable potential with running time. Moreover, the corresponding Tafel plots (η versus log(j)) were drawn to study the OER kinetics (Figure 2b). The linear region of the plots is fitted to the equation (η = a + blog $|j|$) yielding the Tafel plots of 192 mV dec^{-1} and 230 mV dec^{-1} for *meso*-NiO and *bulk*-NiO, respectively. Both Tafel slope values are comparable with the values reported in Zhou's work, where the values are between 159–503 mV dec^{-1}, depending on the different catalysts [21]. Since the Tafel slope is directly associated with the reaction kinetics of an electrocatalyst, the reduced Tafel slope for *meso*-NiO means smaller overpotential changes [22,23], resulting in a faster reaction rate constant, which implies better electrocatalytic kinetics for OER as compared with *bulk*-NiO.

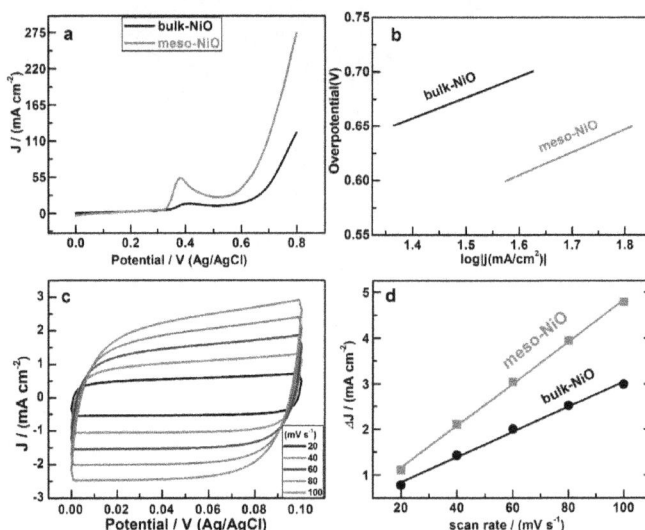

Figure 2. (**a**) Linear scanning voltammetry (LSV) curves over *meso*-NiO and *bulk*-NiO; (**b**) The corresponding Tafel curves; (**c**) Cyclic voltammetry scans over *meso*-NiO with different scan rates; (**d**) Electrochemically active surface area estimated by double-layer capacitance measurements.

The roughness factor (RF) is used to determine the ratio of the electrochemically active surface areas (ECSA) to the geometric surface area of the electrode. To understand the origin of the better OER activity of *meso*-NiO in comparison with *bulk*-NiO and estimate the ECSA and RF, we measured the double-layer capacitance (C_{dl}) via a simple cyclic voltammetry (CV) method in a non-Faradaic region at different scan rates [24], as can be seen from the CV curves recorded at different scan rates in Figure 2c. *Meso*-NiO has a larger current density when the scan rate was the same. As demonstrated in Figure 2d, the C_{dl} of *meso*-NiO (23.0 mF/cm^2) was higher than that of *bulk*-NiO (8.9 mF/cm^2), especially when the scan rate was high. This indicated that *meso*-NiO had a greatly increased active site and higher efficient mass and charge transport capability due to the high density of mesopores in the bulk of *meso*-NiO. The larger ECSA values and more active sites can contribute to the better OER electrocatalytic performance. It is important to note that the different catalytic activity has relevance to the morphology. It seems that the lower overpotential of *meso*-NiO than that of *bulk*-NiO results from the better intrinsic activity of *meso*-NiO than *bulk*-NiO for OER under alkaline condition.

2.3. Characterizations

The composition and chemical valence states of the samples were probed by X-ray photoelectron spectroscopy (XPS). Figure 3a,b show the XPS characterization of O 1s and Ni $2p_{3/2}$ for different nickel oxide samples. As shown in Figure 3a, O 1s can be deconvoluted into two spectral peaks, one with a lower binding energy peak (peak 1) at near 530 eV, and another with a higher binding energy peak (peak 2) at about 532 eV, respectively. Peak 1 can be assigned to the surface lattice oxygen species O^{2-}, while peak 2 can be assigned to the oxygen species O^-, which can be attributed to the formation of an oxygen–Ni^{3+} bond [25]. It was noted that the ratio of peak 2 to peak 1 in the *meso*-NiO sample is significantly greater than that of *bulk*-NiO, showing that the O^- species of the former is significantly higher than that of the latter.

As shown in Figure 3b, Ni $2p_{3/2}$ can be deconvoluted into Ni^{2+} (855 eV) and Ni^{3+} (856 eV) and shake-up satellite (~862 eV) for all of the samples. In the case of *meso*-NiO, the intensity of the Ni^{3+} $2p_{3/2}$ peak was much stronger than that of the Ni^{3+} $2p_{3/2}$ peak. This may be related to the particle size and the surface effect of the catalyst. *Meso*-NiO has a small particle size and a strong surface effect resulting in a relatively high peak of the binding energy spectrum. In addition, the valence and composition of Ni elements are also factors that determine the strength of the binding energy. The non-stoichiometric *meso*-NiO contains more Ni^{2+} cation vacancies; part of Ni^{2+} is oxidized to higher valence states to maintain electrical neutrality. Therefore, the presence of more high-valence Ni cations indicates a higher density of non-stoichiometric nickel oxide.

Figure 3c is the Raman spectra of the two samples. A Ni–O stretching vibration peak at near 500 cm^{-1} appeared in the spectra [26]. The shoulder peak at about 375 cm^{-1} is related to the vacancy concentration of Ni cations [27]. It can be seen from Figure 3c that the shoulder of *meso*-NiO is more obvious than that of *bulk*-NiO. Therefore, it can be concluded that the concentration of Ni^{3+} cations or Ni vacancy in the former is greater. This result is consistent with the results of XPS characterization.

The distinct difference of active oxygen species between these two NiO can also be proved from the O_2-TPD tests, as shown in Figure 3d. The desorption oxygen species below 700 °C are non-stoichiometric oxygen, which is attributed mainly to the O^{2-} and O^- species. As can be seen from Figure 3d, the amount of O^- species in the *meso*-NiO sample is significantly higher than that of the *bulk*-NiO sample. This result is consistent with the results of XPS.

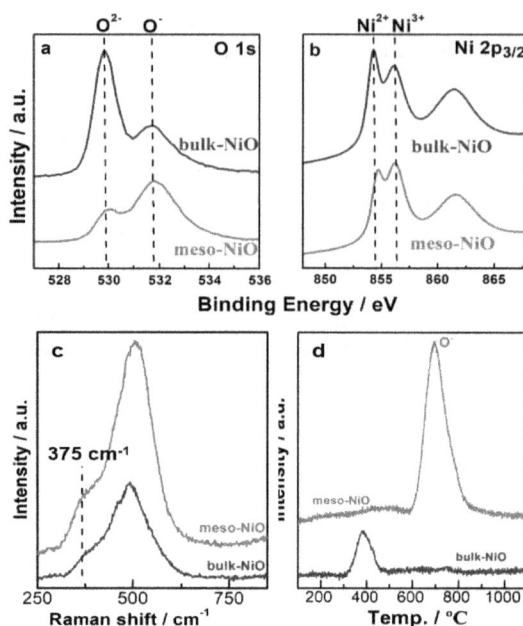

Figure 3. (a) O 1s for XRD, (b) Ni $2p_{3/2}$ for X-ray photoelectron spectroscopy (XPS), (c) Raman, and (d) O_2 temperature programmed desorption (O_2-TPD) characterizations of the *meso*-NiO and *bulk*-NiO catalysts.

2.4. Discussion

From the above results, it was concluded that the mesoporous nickel oxide p exhibits a superior electrocatalytic performance compared to regular nickel oxide in the electrochemical OER reaction. These enhanced activity was beneficial from several aspects, such as geometric and electron structure effect and surface properties.

Firstly, the geometric structure, such as particle size, specific surface area, and mesopores of the catalyst, have a certain influence on the OER performance of nickel oxide. Electrocatalytic reactions commonly take place on the surface of a catalyst, which means that electrochemical performance can be strongly influenced by the geometric structure. The pore structure of mesoporous materials plays an essential role to improve the catalytic performance of the catalyst. The features of a larger surface area and pore volume formed by the aggregation of nanocrystalline NiO in *meso*-NiO is beneficial for efficient electron transfer, therefore enhancing the conductivity and improving the catalytic efficiency. Also, a large aperture can enlarge the range of reaction by promoting the full contact of the active center and reaction species and intermediates.

Secondly, the OER activity has a relationship with the type of metal oxide catalysts. An electrical double layer is formed when a metal oxide electrode is inserted into an electrolyte. Metal oxides provide a much lower charge carrier concentration compared with metals. Therefore, a large space charge layer is formed at the surface of the electrode. In general, due to the accumulation of surface pores, the potential drop in the space charge layer is negligible for the *p*-type metal oxide at anodic potentials, which is typical for OERs. In contrast, the space charge layer of the interface can cause additional obstacles to the charge carrier for an *n*-type oxide. Therefore, as a catalyst material, a *p*-type semiconducting oxide should be more suitable for OER than an *n*-type oxide [28]. NiO is a typical *p*-type oxide, which has the advantage as being an OER catalyst. For *p*-type semiconductors, the oxygen species O^{2-}, O^-, and holes (h^+) will be balanced by the conversion: $O^{2-} + h^+ \leftrightarrow O^-$, and the number of holes determines the number of O^-. Meanwhile, the as-prepared *meso*-NiO in this work is rich in nickel vacancies (holes), which will lead to the increasing of Ni^{3+} for electric neutrality. The holes derived from the vacancies of Ni cations, which depend on Ni^{3+} and other high-valent Ni ions, play an important role in the conductance. Therefore, the greater the Ni cation gap concentration, the higher density of holes, and then the better the electrical conductivity.

Thirdly, the surface properties of the NiO catalyst are also important factors affecting the electrocatalytic performance. As evidenced from the XPS, O_2-TPD, and Raman results, the *meso*-NiO is rich in Ni^{3+} and O^- species; both are electrophilic agents. It was reported that the enriched Ni^{3+} on the surface can initiate the formation of NiOOH, and was responsible for most of the redox sites acting for OH^- adsorption in alkaline solution, which was critical for enhancing OER [5]. Pfeifer proposed that the reactive species in OER over IrO_2 is not solely a property of the metal; it is also intimately tied to the electronic structure of oxygen [29]. The surface electrophilic oxygen species is extremely susceptible to nucleophilic attack by water or hydroxide, which as the nucleophile donate electrons to the reactive oxygen to facilitate O–O bond formation on transition metal oxide during the OER.

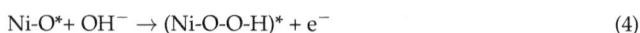

$$2MO \rightarrow 2M + O_2 \text{ (g)} \tag{1}$$

$$MO + OH^- \rightarrow MOOH + e^- \tag{2}$$

$$MOOH + OH^- \rightarrow M + O_2 \text{ (g)} + H_2O \text{ (l)} \tag{3}$$

$$Ni\text{-}O^* + OH^- \rightarrow (Ni\text{-}O\text{-}O\text{-}H)^* + e^- \tag{4}$$

Finally, the reaction mechanism could be concluded based on understanding the relationship between geometric/electron structure properties and the reaction process of the electrochemical OER process (Figure 4). In the process of OER over the NiO catalyst, there are two different paths that separate out the oxygen in an alkaline solution. One is the direct combination of Ni–O to produce oxygen (Equation (1)), and the other is to produce the intermediates (NiOOH) and then decompose to produce oxygen (Equations (2) and (3)) [26]. When a nickel metal electrode is immersed in an alkaline solution (the solution of 1 M KOH in this article), a hydrous layer can be aged in base or in vacuum to give the hydrous β-Ni(OH)$_2$, which will oxidize to γ-NiOOH and β-NiOOH when the potential reaches a certain value. The latter (β-NiOOH) has better water oxidation activity. The electrochemical oxidation of nickel–oxide in electrochemical conditions is increased by the in situ formation of Ni(OH)$_2$/NiOOH hydroxide/oxyhydroxide species [30,31]. The hydroperoxy (-OOH) species are the key intermediates in OER. In our case, the electrophilic Ni^{3+} and O^- species is prone to be nucleophilic attacked by OH^- in the KOH solution (Equation (4), where Ni-O* means the electrophilic Ni^{3+} and O^- species). The formed (Ni-O-O-H)* is also facilitated to further react with OH^- to form the product O_2. For mesoporous NiO, the high density of electrophilic Ni^{3+} and O^- species result in the enhanced OER activity.

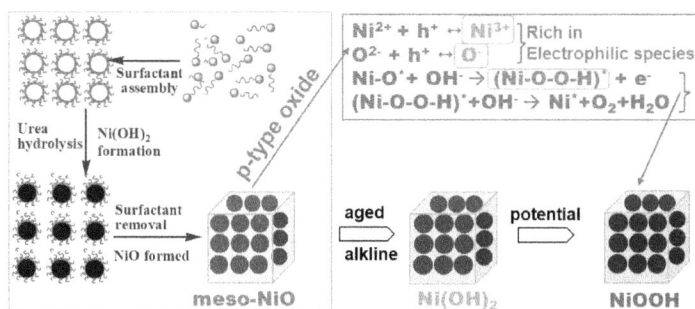

Figure 4. Representation of the fabrication process and reaction mechanism of *meso*-NiO.

3. Materials and Methods

3.1. Materials

Materials were purchased in the grade indicated and used as received: nickel hexahydrate ($NiCl_2 \cdot 6H_2O$, AR, Sinopharm Chemical Reagent Co. Ltd., Shanghai, China), lauryl sodium sulfate (SDS, 95%, Energy Chemical Reagent Co. Ltd, Shanghai, China), urea (CH_4N_2O, AR, Sinopharm Chemical Reagent Co. Ltd., Shanghai, China), nickel nitrate ($Ni(NO_3)_2$, AR, Sinopharm Chemical Reagent Co. Ltd., Shanghai, China), citric acid ($C_6H_8O_7H_2O$, AR, Sinopharm Chemical Reagent Co. Ltd., Shanghai, China), caustic potash (KOH, 95%, Energy Chemical Reagent Co. Ltd., Shanghai, China), and muriate (KCl, 99.5%, Sinopharm Chemical Reagent Co. Ltd., Shanghai, China).

3.2. Analytical Equipment

The equilibrium potential, stability, current density, etc. of the catalyst were determined by the electrochemical workstation (CHI660E, A17104 Chenhua Co. Ltd., Shanghai, China). All of the testing was done in a three-electrode system at 25 °C.

3.3. Synthesis of Nano-Nickel Oxide

Nano-nickel oxide was synthesized by the sol–gel method [15,16]. 0.03 M $Ni(NO_3)_2$, and 0.03 M of citric acid was added to 50 mL of ethanol and then stirred with a glass rod until completely dissolved. The solution was placed in a water bath and heated to 80 °C to allow it to evaporate. When one-third of the solution remained, it was put in an oven and dried at 100 °C for 24 h, and then put in a muffle furnace, heated to 300 °C, and calcined for 4 h to obtain *bulk*-NiO.

3.4. Synthesis of Meso-Nickel Oxide

The template method to synthesize mesoporous nickel oxide [14] takes the anionic surfactant sodium dodecyl sulfate (SDS) as a guide agent structure. Urea slowly hydrolyzes to form NH_3, which reacts with Ni^{2+} to form a $Ni(OH)_2$ precursor. In a typical process, $NiCl_2$, SDS, urea, and H_2O were mixed at quality ratio of 1: 2: 30: 60 and stirred at 40 °C for 1 h to obtain a clear solution. The clear solution was poured into a reaction kettle and treated at 80 °C for 20 h. The resulting material was washed several times with water and ethanol and dried overnight at 60 °C, thus obtaining a precursor $Ni(OH)_2$. The precursor was calcined at 300 °C for 4 h to give *meso*-NiO.

3.5. Characterizations

The prepared sample was characterized by XRD (Model d/Max-C, Rigaku Corporation, Tokyo Japan) [32]. The morphology of the sample was observed by SEM (LEO-1530, LEO Electron Microscope Com., Berlin, Germany) and TEM (Phillps FEI, Tecnai 30, Thermo Fisher Scientific Com., Oregan, OR, USA). The BET (AutoChem2920, Micromeritics Co. California, CA, USA) was used to test the particle size and pore volume of the sample. Raman was used to test the relationship between the Ni cation and vacancy concentration. O_2-TPD measurements were performed on a mass spectrometer and used to test the presence of oxygen species. The catalyst sample (0.2 g) was pretreated in pure oxygen flowing at 15 mL/min at the sample's calcining temperature for 1 h, and then cooled to room temperature under the oxygen flow. After the system was subsequently flushed with He to achieve a smooth baseline, the sample was heated to 900 °C at a rate of 15 °C/min in He flow (30 mL/min). The reactor exit was monitored online by a mass spectrometer. XPS was used to test the surface composition of the catalyst and the change in valence of the surface element.

3.6. Electrode Preparation

The catalyst was made in a certain ratio of ink: 0.1 g of catalyst was mixed with 0.1 g of Ketjen black; then, 800 μL of isopropanol and 200 μL of Nafion were added and ultrasound mixed to disperse evenly. The geometric surface area of glassy carbon electrode is $0.196\ cm^2$. Before the electrochemical test, the working electrode was smoothed with α-AI_2O_3 with different particle sizes in a three-electrode system for testing at 25 degrees. The three-electrode system consisted of the glassy carbon electrode as a working electrode, the Pt wire as an auxiliary electrode, and Ag/AgCl as a reference electrode. The electrolyte solution was 1M KOH. We took 8 μL of ink drops on the working electrode, dried it at room temperature, and maintained the test temperature at 25 degrees. Then, we passed O_2 in the 1 M of KOH for half an hour prior to testing, and then placed the oxygen tube above the liquid level to isolate the air from creating a saturated oxygen condition.

3.7. Electrochemical Characterization

The electrochemical capacitance measurements were determined by cyclic voltammetry (CV). The potential range was selected as the 0.1-V potential window, which had the system's open-point as the center. The test was first activated at a sweep rate of $50\ mV\ s^{-1}$ for 5 min, aiming to clean the surface of electrodes and enhance the number of oxygen-containing functional groups in the catalysts [33]. The CV test was carried out in static solution. We set five different scanning rates for scanning: $0.02\ V\ s^{-1}$, $0.04\ V\ s^{-1}$, $0.06\ V\ s^{-1}$, $0.08\ V\ s^{-1}$, and $0.1\ V\ s^{-1}$.

The electrochemical activity for OER was assessed by linear sweep voltammetry (LSV) in the potential range of 0–0.8 V at a scan rate of $50\ mV\ s^{-1}$. During the test, the reaction process should be continuously stirred in order to prevent the working electrode reaction bubbles from affecting the test results. The scan rate was $0.05\ V\ s^{-1}$.

4. Conclusions

In summary, the *meso*-NiO composed of uniform mesopores were synthesized via the surfactant-assisted urea hydrolysis method. The as-prepared *meso*-NiO shows high-rich Ni^{3+} cations and O^- species on the surface, as well as a large surface area, which concurrently benefited the OER reaction. An overpotential of 410 mV to generate a current density of $50\ mA/cm^2$ were obtained for *meso*-NiO compared with the value of 494 mV for regular *bulk*-NiO. The outstanding electrocatalytic performance of *meso*-NiO can be attributed to the synergy of the large surface area offered by the mesoporous structure to facilitate the mass transport and surface properties with a higher density of Ni^{3+} and O^- as the main active sites on the surface. The surface electrophilic Ni^{3+} and O^- species are prone to be nucleophilic attacked by OH^- in the alkaline solution to form the reactive intermediate, which is also facilitated to further react with OH^- to generate oxygen. Although the performance of mesoporous nickel oxide in electrocatalysis is superior to that of nano-nickel oxide, there is still room for improvement in its performance.

Author Contributions: Conceptualization, L.-Z.Z.; Data curation, Z.-Y.Z., Z.C.; Funding acquisition, F.-Z.S., J.-H.L.; Investigation, X.L.; Methodology, X.-F.Z.; Supervision, J.-H.L.; Writing–original draft, X.L.; Writing—review & editing, F.-Z.S., J.-H.L.

Funding: This research was funded by the National Natural Science Foundation of China (No. 21773195), the Fundamental Research Funds for the Central Universities (No. 20720170030) and State Key Laboratory of New Ceramic and Fine Processing Tsinghua University (No. KF201706).

Acknowledgments: We thank all the graduate students for their assistance with the fieldwork.

Conflicts of Interest: The authors declare no conflict of interest.

References

1. Kim, J.; Jin, H.; Oh, A.; Baik, H.; Joo, S.H.; Lee, K. Synthesis of compositionally tunable, hollow mixed metal sulphide coxniysz octahedral nanocages and their composition-dependent electrocatalytic activities for oxygen evolution reaction. *Nanoscale* **2017**, *13*, 3437–3444. [CrossRef] [PubMed]
2. Michalska-Domanska, M.; Norek, M.; Jóźwik, P.; Jankiewicz, B.; Stępniowski, W. Catalytic stability and surface analysis of microcrystalline Ni3Al thin foils in methanol decomposition. *Appl. Surf. Sci.* **2014**, *293*, 169–176. [CrossRef]

3. Tang, C.; Asiri, A.M.; Sun, X. Highly-active oxygen evolution electrocatalyzed by a Fe-doped NiSe nanoflake array electrode. *Chem. Commun.* **2016**, *52*, 4529–4532. [CrossRef] [PubMed]
4. Wang, J.; Cui, W.; Liu, Q.; Xing, Z.; Asiri, A.M.; Sun, X. Recent progress in cobalt-based heterogeneous catalysts for electrochemical water splitting. *Adv. Mater.* **2016**, *28*, 215–230. [CrossRef] [PubMed]
5. Trotochaud, L.; Ranney, J.K.; Williams, K.N.; Boettcher, S.W. Solution-cast metal oxide thin film electrocatalysts for oxygen evolution. *J. Am. Chem. Soc.* **2012**, *134*, 17253–17261. [CrossRef] [PubMed]
6. Wang, H.Y.; Hsu, Y.Y.; Chen, R.; Chan, T.S.; Chen, H.M.; Liu, B. Ni^{3+}-induced formation of active NiOOH on the spinel Ni-Co oxide surface for efficient oxygen evolution reaction. *Adv. Energy Mater.* **2015**, *5*, 395–417. [CrossRef]
7. Matsumura, Y.; Tanaka, K.; Tode, N.; Yazawa, T.; Haruta, M. Catalytic methanol decomposition to carbon monoxide and hydrogen over nickel supported on silica. *J. Mol. Catal. A Chem.* **2000**, *152*, 157–165. [CrossRef]
8. Taguchi, A.; Schüth, F. Ordered mesoporous materials in catalysis. *Micropor. Mesopor. Mater.* **2005**, *77*, 1–45. [CrossRef]
9. Chang, J.; Park, M.; Ham, D.; Ogale, S.B.; Mane, R.S.; Han, S.H. Liquid-phase synthesized mesoporous electrochemical supercapacitors of nickel hydroxide. *Electrochim. Acta* **2008**, *53*, 5016–5021. [CrossRef]
10. Yuan, C.; Gao, B.; Su, L.; Zhang, X. Interface synthesis of mesoporous MnO_2 and its electrochemical capacitive behaviors. *J. Colloid Interface Sci.* **2008**, *322*, 545–550. [CrossRef] [PubMed]
11. Wang, Y.G.; Xia, Y.Y. Electrochemical capacitance characterization of NiO with ordered mesoporous structure synthesized by template SBA-15. *Electrochim. Acta* **2006**, *51*, 3223–3227. [CrossRef]
12. Xing, W.; Li, F.; Yan, Z.F.; Lu, G.Q. Synthesis and electrochemical properties of mesoporous nickel oxide. *J. Power Sources* **2004**, *134*, 324–330. [CrossRef]
13. Shaju, K.M.; Jiao, F.; Débart, A.; Bruce, P.G. Mesoporous and nanowire Co_3O_4 as negative electrodes for rechargeable lithium batteries. *Phys. Chem. Chem. Phys.* **2007**, *9*, 1837–1842. [CrossRef] [PubMed]
14. Li, J.-H.; Wang, C.-C.; Huang, C.-J.; Sun, Y.-F.; Weng, W.-Z.; Wan, H.-L. Mesoporous nickel oxides as effective catalysts for oxidative dehydrogenation of propane to propene. *Appl. Catal. A Gen.* **2010**, *382*, 99–105. [CrossRef]
15. He, Y.; Wu, Y.; Chen, T.; Weng, W.; Wan, H. Low-temperature catalytic performance for oxidative dehydrogenation of propane on nanosized Ti(Zr)-Ni-O prepared by modified sol-gel method. *Catal. Commun.* **2006**, *7*, 268–271. [CrossRef]
16. Alberto, N.; Vigario, C.; Duarte, D.; Almeida, D.; Gonçalves, G. Characterization of graphene oxide coatings onto optical fibers for sensing applications. *Mater. Today Proc.* **2015**, *2*, 171–177. [CrossRef]
17. Schenk, A.S.; Eiben, S.; Goll, M.; Reith, L.; Kulak, A.N.; Meldrum, F.C.; Jeske, H.; Wege, C.; Ludwigs, S. Virus-directed formation of electrocatalytically active nanoparticle-based Co_3O_4 tubes. *Nanoscale* **2017**, *9*, 6334–6345. [CrossRef] [PubMed]
18. Han, L.; Dong, C.; Zhang, C.; Gao, Y.; Zhang, J.; Gao, H.; Wang, Y.; Zhang, Z. Dealloying-directed synthesis of efficient mesoporous CoFe-based catalysts towards the oxygen evolution reaction and overall water splitting. *Nanoscale* **2017**, *9*, 16467–16475. [CrossRef] [PubMed]
19. Lyons, M.E.; Brandon, M.P. The oxygen evolution reaction on passive oxide covered transition metal electrodes in aqueous alkaline solution. Part 1-nickel. *Int. J. Electrochem. Sci.* **2008**, *3*, 1386–1424.
20. Li, W.; Gao, X.; Wang, X.; Xiong, D.; Huang, P.P.; Song, W.G.; Bao, X.; Liu, L. From water reduction to oxidation: Janus Co-Ni-P nanowires as high-efficiency and ultrastable electrocatalysts for over 3000h water splitting. *J. Power Sources* **2016**, *330*, 156–166. [CrossRef]
21. Zhou, W.; Wu, X.; Cao, X.; Huang, X.; Tan, C.; Tian, J.; Liu, H.; Wang, J.; Zhang, H. Ni_3S_2 nanorods/Ni foam composite electrode with low overpotential for electrocatalytic oxygen evolution. *Energ. Environ. Sci.* **2013**, *6*, 2921–2924. [CrossRef]
22. Sun, H.; Zhao, Y.; Mølhave, K.; Zhang, M.; Zhang, J. Simultaneous modulation of surface composition, oxygen vacancies and assembly in hierarchical Co_3O_4 mesoporous nanostructures for lithium storage and electrocatalytic oxygen evolution. *Nanoscale* **2017**, *9*, 14431–14441. [CrossRef] [PubMed]
23. Suen, N.T.; Hung, S.F.; Quan, Q.; Zhang, N.; Xu, Y.J.; Chen, H.M. Electrocatalysis for the oxygen evolution reaction: Recent development and future perspectives. *Chem. Soc. Rev.* **2017**, *46*, 337–365. [CrossRef] [PubMed]
24. Zhang, B.; Yu, H.L.; Ni, H.; Hu, S. Bimetallic $(Fe_xNi_{1-x})2P$ nanoarrays as exceptionally efficient electrocatalysts for oxygen evolution in alkaline and neutral media. *Nano Energy* **2017**, *38*, 553–560. [CrossRef]

25. Stoyanova, M.; Konova, P.; Nikolov, P.; Naydenov, A.; Christoskova, S.; Mehandjiev, D. Alumina-supported nickel oxide for ozone decomposition and catalytic ozonation of CO and VOCs. *Chem. Eng. J.* **2006**, *122*, 41–46. [CrossRef]
26. Cordoba-Torresi, S.I. Electrochromic behavior of nickel oxide electrodes. *J. Electrochem. Soc.* **1991**, *138*, 1554–1559. [CrossRef]
27. Heracleous, E.; Lemonidou, A.A. Ni-nb-o mixed oxides as highly active and selective catalysts for ethene production via ethane oxidative dehydrogenation. Part i: Characterization and catalytic performance. *J. Catal.* **2006**, *237*, 162–174. [CrossRef]
28. Fabbri, E.; Habereder, A.; Waltar, K.; Kötz, R.; Schmidt, T.J. Developments and perspectives of oxide-based catalysts for the oxygen evolution reaction. *Catal. Sci. Technol.* **2014**, *4*, 3800–3821. [CrossRef]
29. Pfeifer, V.; Jones, T.E.; Wrabetz, S.; Massue, C.; Velasco Velez, J.J.; Arrigo, R.; Scherzer, M.; Piccinin, S.; Havecker, M.; Knop-Gericke, A. Reactive oxygen species in iridium-based oer catalysts. *Chem. Sci.* **2016**, *7*, 6791–6795. [CrossRef] [PubMed]
30. Corrigan, D.A.; Bendert, R.M. Effect of coprecipitated metal ions on the electrochemistry of nickel hydroxide thin films: Cyclic voltammetry in 1 M KOH. *J. Electrochem. Soc.* **1989**, *20*, 723–728. [CrossRef]
31. Wehrens-Dijksma, M.; Notten, P.H.L. Electrochemical quartz microbalance characterization of $Ni(OH)_2$-based thin film electrodes. *Electrochim. Acta* **2006**, *51*, 3609–3621. [CrossRef]
32. Michalska-Domańska, M.; Jóźwik, P.; Jankiewicz, B.J.; Bartosewicz, B.; Siemiaszko, D.; Stępniowski, W.J.; Bojar, Z. Study of Cyclic Ni_3Al Catalyst Pretreatment Process for Uniform Carbon Nanotubes Formation and Improved Hydrogen Yield in Methanol Decomposition. *Mater. Today Proc.* **2016**, *3S*, S171–S177. [CrossRef]
33. Chen, G.F.; Ma, T.Y.; Liu, Z.Q.; Li, N.; Su, Y.Z.; Davey, K.; Qiao, S.Z. Efficient and stable bifunctional electrocatalysts Ni/NixMy (M = P, S) for overall water splitting. *Adv. Funct. Mater.* **2016**, *26*, 3314–3323. [CrossRef]

![catalysts logo] *catalysts*

MDPI

Article

Cobalt and Nitrogen Co-Doped Graphene-Carbon Nanotube Aerogel as an Efficient Bifunctional Electrocatalyst for Oxygen Reduction and Evolution Reactions

Xiaochang Qiao [1,2], Jutao Jin [2], Hongbo Fan [2,*], Lifeng Cui [2], Shan Ji [3], Yingwei Li [1] and Shijun Liao [1,*]

[1] The Key Laboratory of Fuel Cell Technology of Guangdong Province, School of Chemistry and Chemical Engineering, South China University of Technology, Guangzhou 510641, China; qiaoxc@dgut.edu.cn (X.Q.); liyw@scut.edu.cn (Y.L.)

[2] School of Environment and Civil Engineering, Dongguan University of Technology, Dongguan 523808, China; 2016809@dgut.edu.cn (J.J.); lifeng.cui@gmail.com (L.C.)

[3] College of Biological, Chemical Science and Chemical Engineering, Jiaxing University, Jiaxing 314001, China; jishan@mail.zjxu.edu.cn

* Correspondence: fhb@dgut.edu.cn (H.F.); chsjliao@scut.edu.cn (S.L.); Tel.: +86-20-8711-3586 (S.L.)

Received: 13 June 2018; Accepted: 3 July 2018; Published: 7 July 2018

Abstract: In this study, a low-cost and environmentally friendly method is developed to synthesize cobalt and nitrogen co-doped graphene-carbon nanotube aerogel (Co-N-GCA) as a bifunctional electrocatalyst for the oxygen reduction reaction (ORR) and the oxygen evolution reaction (OER). The as-prepared Co-N-GCA has a hierarchical meso- and macroporous structure with a high N doping level (8.92 at. %) and a large specific surface area (456 m^2 g^{-1}). In an alkaline medium, the catalyst exhibits superior ORR electrocatalytic activity with an onset potential 15 mV more positive than Pt/C, and its diffusion-limiting current density is 29% higher than that of commercial Pt/C. The obtained Co-N-GCA is also highly active toward the OER, with a small overpotential of 408 mV at a current density of 10 mA cm^{-2}. Its overall oxygen electrode activity parameter (ΔE) is 0.821 V, which is comparable to most of the best nonprecious-metal catalysts reported previously. Furthermore, Co-N-GCA demonstrates superior durability in both the ORR and the OER, making it a promising noble-metal-free bifunctional catalyst in practical applications for energy conversion and storage.

Keywords: graphene-carbon nanotube aerogel; cobalt and nitrogen co-doped; oxygen reduction reaction; oxygen evolution reaction

1. Introduction

Global energy and environmental issues have stimulated tremendous ongoing research into developing sustainable and environmentally friendly energy conversion and storage systems [1]. Creating highly active electrocatalysts for the oxygen reduction reaction (ORR) and the oxygen evolution reaction (OER) is crucial for the practical application of fuel cells, metal-air batteries, and electrolyzers. Currently, Pt-based catalysts are regarded as the most active ORR catalysts but are poor for the OER [2,3], while IrO_2 and RuO_2 are the most efficient catalysts for the OER but are poor for the ORR [4]. However, their commercialization has been greatly hindered by the scarcity of their materials, their consequent high cost, and their poor durability. As a result, extensive research efforts have been devoted to searching for low-cost alternatives with comparable catalytic activity to noble metal-based catalysts [5,6]. Due to the sluggish kinetics of the ORR and OER [7–9], it is highly challenging but imperative to develop a cheap and effective bifunctional electrocatalyst for the ORR and OER.

Extensive efforts have been made to develop non-noble-metal bifunctional electrocatalysts, such as transition metal oxides/sulfides [10–15], carbon-based materials [9,16–19], and perovskites [20]. Among these promising bifunctional ORR/OER catalysts, metal-nitrogen-doped carbon materials have attracted much attention due to their remarkable activity, tunable surface chemistry, fast electron transfer capacity, and economic viability [21–23]. However, substantially increasing the active sites of metal-nitrogen-doped carbon materials remains a great challenge. Recently, it was found that using nanosized carbon materials as supports, such as one-dimensional carbon nanotubes (CNTs) and two-dimensional graphene, can further improve electrocatalytic performance [24–27]. However, methods for the synthesis of metal-nitrogen-doped graphene/CNTs often requires the high temperature (600–1200 °C) which limits the practical application. And both CNTs and graphene nanosheets are inclined to aggregate with each other due to strong Van der Waals interactions, and this greatly hinders the full utilization of the active sites for catalytic reactions [21,28,29].

Because of their three-dimensional network structure, extremely low density, high porosity, and large specific surface area, carbon-based aerogels have emerged as promising nanomaterials and are providing fascinating options for preparing new functional electrode materials [30]. The incorporation of CNTs into graphene to produce a graphene–CNTs hybrid aerogel not only could favor the dispersion of graphene and CNTs while maintaining the original properties of both graphene and CNTs, but also could yield a 3D, interconnected, conductive network structure. This aerogel network structure, for ORR, favors the O_2 transformation to the active sites; for OER, facilitates timely transfer evolved O_2 molecules. It is expected that a graphene-carbon nanotube aerogel could result in better electrocatalytic performance. However, to the best of our knowledge, there is no report about M-N-C graphene-CNTs aerogel as an efficient oxygen electrode catalyst.

Herein, we report a one-pot hydrothermal method to prepare a cobalt and nitrogen co-doped graphene-CNTs aerogel (Co-N-GCA) as a bifunctional electrocatalyst for the ORR and OER. Benefiting from abundant catalytic active sites and a unique hierarchical structure that increases the exposure of active sites and favors electron transfer and the diffusion of O_2 molecules, the obtained electrocatalyst exhibits superior electrocatalytic activity towards both the ORR and the OER in an alkaline medium.

2. Results and Discussions

The inset in Figure 1a is a photo of the as-prepared Co-N-GCA aerogel after freeze drying, showing a black, low-weight, sponge-like material. The SEM images (Figure 1a,b) reveal that the Co-N-GCA exhibits a well-defined and interconnected 3D porous network structure. The CNTs are randomly and uniformly distributed between the graphene sheets. The pore walls consist of thin layers of a network, which are cross-linked with graphene sheets and CNTs. TEM images (Figure 1c,d) further confirm that the CNTs adhere tightly on the graphene substrates. No obvious CNT bundles or graphene agglomerates are observable. Few scattered metal nanoparticles could also be found by TEM inspection. The high-resolution TEM image (Figure S1) reveals the lattice fringe space of 0.47 nm is consistent with the (111) plane of cubic Co_3O_4 spinel-phase. Photos of the prepared Co-GCA and N-GCA aerogels and the corresponding SEM images are also provided in Figure S2.

The N_2 adsorption–desorption isotherms and corresponding pore-size distribution curve of Co-N-GCA are shown in Figure 2. According to the IUPAC classification, the N_2 adsorption–desorption isotherms are of type IV, with the amount of absorbed N_2 monotonically increasing/decreasing at high relative pressure ($P/P_0 > 0.9$); this is typically associated with capillary condensation, indicating the existence of mesopores (2–50 nm). Notably, the capillary condensation phenomenon occurs at a high relative pressure, indicating a large pore-size distribution. The hysteresis loop resembles type H_3, suggesting open slit-shaped capillaries between the parallel layers of graphene. The surface area of Co-N-GCA was calculated to be 456 m^2 g^{-1}. The Barrett-Joyner-Halenda (BJH) method indicated that a hierarchical, meso- and macroporous material with a pore volume of 1.64 cm^3 g^{-1} was formed by this method. This result is consistent with the SEM observations. The high surface area and hierarchically porous structure of Co-N-GCA would provide plenty of active sites and favor the mass transport of reactants and products, resulting in enhanced ORR/OER electrocatalytic activity.

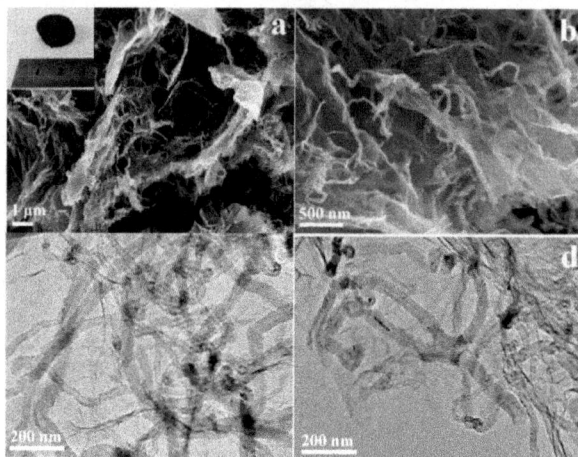

Figure 1. SEM (**a,b**) and TEM (**c,d**) images of Co-N-GCA (inset in (**a**) shows a digital photograph of the prepared aerogel).

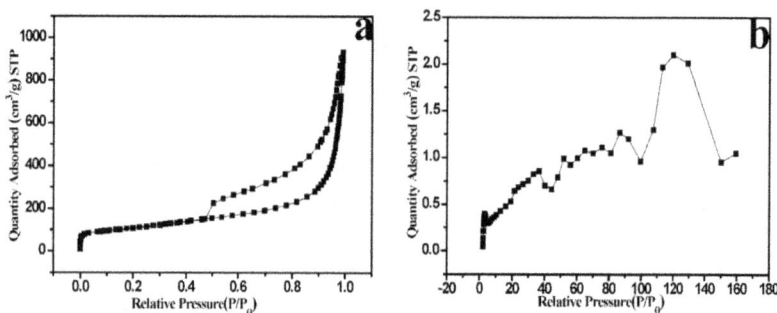

Figure 2. Nitrogen adsorption–desorption isotherms (**a**) and the corresponding pore-size distribution curve (**b**) of Co-N-GCA.

The elemental compositions of Co-N-GCA and N-GCA were investigated by X-ray photoelectron spectroscopy (XPS) analysis. As shown in Figure S2, the XPS survey spectrum of Co-N-GCA exhibited the signals of a C 1s peak (~284.5 eV), a N 1s peak (~398.1 eV), an O 1s peak (~531.1 eV), and a Co 2p peak (~780.1 eV) [22]. The XPS results confirmed that Co and N elements were successfully doped into the carbon matrix, and the amounts of doped Co and N in Co-N-GCA were 0.44 and 8.92 at. %, respectively. In the case of N-GCA, the doped N content was 8.68 at. %, which was similar to that of Co-N-GCA (Table S1). XPS analysis shows that a high amount of doped N can be achieved by this method.

The deconvoluted high-resolution N 1s spectrum (Figure 3a) revealed four types of N species in Co-N-GCA: pyridinic N/Co–N_x (398.9 eV), pyrrolic N (400.0 eV), graphitic N (401.2 eV), and oxidated N (405.1 eV) [31]. Their corresponding contents are 47.5, 34.8, 13.5, and 4.2 at. %, respectively (Table S2) [32]. Notably, most of the doped N is pyridinic (47.5 at. %). It has been reported that the ORR active sites in N-doped carbon materials are carbon atoms with Lewis basicity next to pyridinic N [33].

We expected the as-prepared Co-N-GCA to exhibit outstanding electrocatalytic performance due to the high amount of pyridinic N. Meanwhile, M–N$_x$ structure have also been reported to contribute to ORR/OER active sites apart from the N–C active sites [34]. The high-resolution N 1s spectrum and distribution of each N species of N-GCA are presented in Figure S3 and Table S2, respectively.

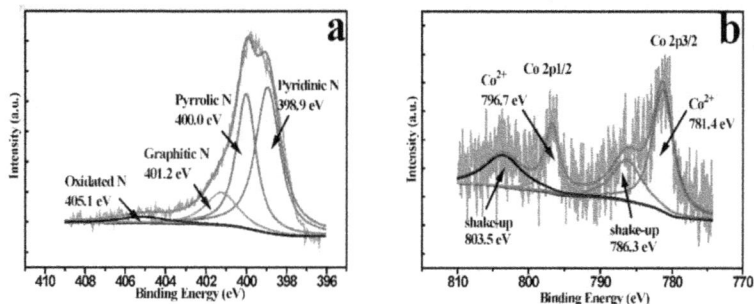

Figure 3. The X-ray photoelectron spectroscopy (XPS) high-resolution N 1s (**a**) and Co 2p (**b**) spectra of Co-N-GCA.

Figure 3b presents a high-resolution XPS spectrum of Co 2p. The peaks situated at 781.4 eV and 796.7 eV with two weak shake-up (satellite) peaks are assigned to the Co 2p 3/2 and Co 2p 1/2 atomic orbitals.

We have also performed inductively coupled plasma mass spectroscopic measurements to analyze the content of Co, the result showed a relatively low cobalt content of ~0.91 wt. % in the Co-N-GCA sample. Meanwhile, we also performed element analysis test, the result showed the N content in the Co-N-GCA is 17.5 wt. %.

Figure 4a presents the XRD patterns of Co-N-GCA, N-GCA, and GCA. All three samples show two broad, weak diffraction peaks at $2\theta \approx 24°$ and $43.7°$, which correspond to the (002) and (100) reflections of the graphitic peak (PDF#41-1487), respectively, confirming the graphitic crystal structure [35]. For Co-N-GCA, no metal peaks or other than carbon are observed, which may be due to the low Co content.

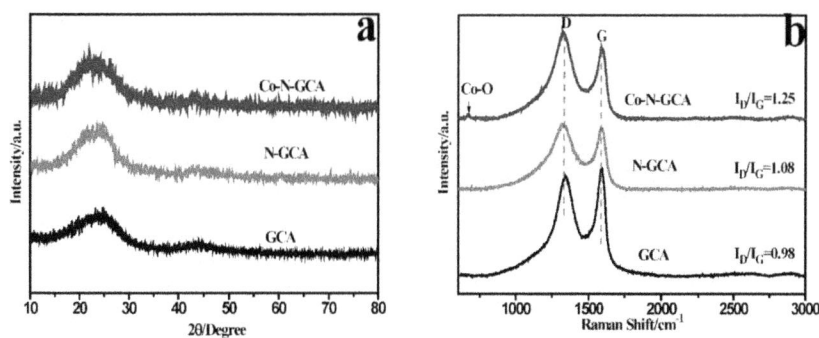

Figure 4. XRD patterns (**a**) and Raman spectra (**b**) of Co-N-GCA, N-GCA, and GCA.

Figure 4b shows the Raman spectra of Co-N-GCA, N-GCA, and GCA. All of the samples exhibit two prominent peaks at ca. 1580 cm^{-1} and 1335 cm^{-1}, corresponding to the characteristic D and G bands, respectively. The D band belongs to the breathing modes of sp^2 atoms in rings, whereas the G band is assigned to stretching in all pairs of sp^2 atoms in both rings and chains [36]. Generally, the intensity ratio of the D and G bands (I_D/I_G) is used to quantify the extent of defects in carbon materials. The I_D/I_G ratios for Co-N-GCA, N-GCA, and GCA were 1.25, 1.08, and 0.98, respectively, indicating that Co-N-GCA had more structural defects. In the Raman spectra of Co-N-GCA, there is a small peak at 672 cm^{-1}, which can be indexed to Co-O.

The electrocatalytic performance of Co-N-GCA for the ORR was initially evaluated separately by cyclic voltammetry (CV) in N$_2$- and O$_2$-saturated 0.1 M KOH solutions. As shown in Figure 5a, the CV curve of Co-N-GCA was virtually featureless in the N$_2$-saturated electrolyte, while displayed well-defined oxygen reduction cathodic peaks in the O$_2$-saturated electrolyte. Notably, Co-N-GCA had the more positive ORR peak potential (0.837 V vs. RHE) than commercial 20 wt. % Pt/C (0.841 V vs. RHE) (Figure S4), indicating the superior ORR catalytic activity of Co-N-GCA.

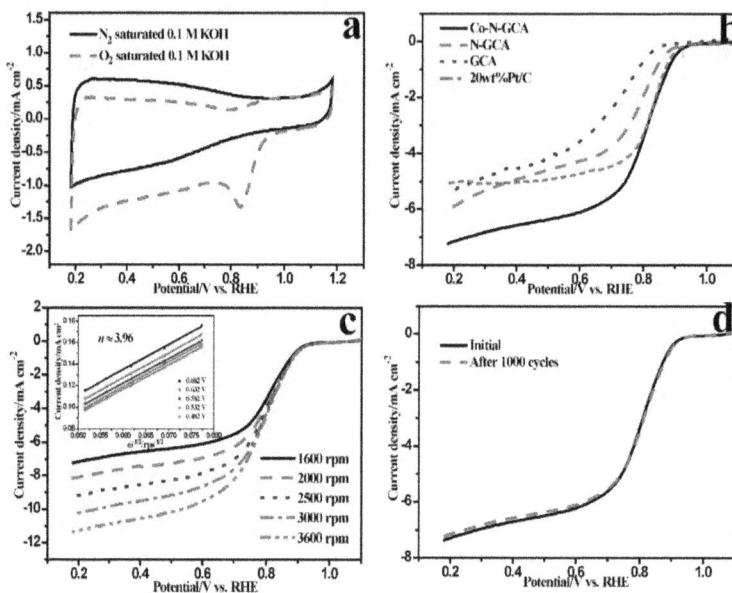

Figure 5. (a) Cyclic voltammetry (CV) curves of Co-N-GCA in N$_2$- and O$_2$-saturated 0.1 M KOH solutions; (b) linear sweep voltammetry (LSV) curves of Co-N-GCA, N-GCA, GCA, and commercial 20 wt. % Pt/C at a rotation speed of 1600 rpm; (c) LSV curves of Co-N-GCA at 1600–3600 rpm rotation speeds (inset shows the corresponding K–L plots); (d) LSV plots of Co-N-GCA at a rotation speed of 1600 rpm before and after 1000 CV (50 mV s^{-1}) cycles.

To further investigate the ORR activity, the linear sweep voltammetry (LSV) curve of Co-N-GCA was recorded at 1600 rpm in O$_2$-saturated 0.1 M KOH solution. For comparison, we present the LSV curves of Co-N-GCA, N-GCA, GCA, and commercial 20 wt. % Pt/C in Figure 5b. In terms of the onset potentials and diffusion-limiting current densities of the ORR, Figure 5b show that the ORR activity follows the order GCA < N-GCA < Co-N-GCA. The Co-N-GCA is more electrocatalytically active toward the ORR than Pt/C, with an onset potential of 0.975 V (vs. RHE), which is 15 mV more positive than that of commercial Pt/C. The diffusion-limiting current density of Co-N-GCA (at 0.6 V) is about 29% higher than that of commercial Pt/C, which further indicates that the mass transport on Co-N-GCA is more efficient than on Pt/C.

To gain more information on the ORR kinetics of the Co-N-GCA catalyst, LSV curves were recorded in an O$_2$-saturated 0.1 M KOH solution at various rotation rates, increasing from 1600 to 3600 rpm (Figure 5c). The diffusion current density of oxygen reduction increased with the rotation rate, owing to enhanced mass transport. In addition, the K–L plots at different electrode potentials displayed good linearity. The K–L equation was adopted to calculate the electron transfer number

(n) of Co-N-GCA in the potential range from 0.482 V to 0.682 V, and an average n value of 3.96 was obtained, indicating that the ORR proceeded via a four-electron pathway.

The stability of the Co-N-GCA catalyst was further evaluated by LSV in O_2-saturated 0.1 M KOH at 1600 rpm for 1000 continuous cycles. As shown in Figure 5d, there wasn't an obvious change in the half-wave potential of Co-N-GCA after 1000 continuous cycles, suggesting that Co-N-GCA has excellent ORR stability in an alkaline medium.

The electrocatalytic OER activity of Co-N-GCA was investigated by LSV measurements, which were conducted in O_2-saturated 0.1 M KOH at 1600 rpm. For comparison, the OER with N-GCA, GCA, and IrO_2/C was also performed under the same conditions (Figure 6a). OER catalytic activities are commonly judged by the potential required to oxidize water at a current density of 10 mA cm^{-2}. Significantly, compared with the standard reaction potential (1.23 V), the Co-N-GCA composite can reach 10 mA cm^{-2} with a small overpotential (η) of 408 mV, while N-GCA, GCA, and IrO_2/C reach the same current density with overpotentials (η) of 515, 685, and 377 mV, respectively.

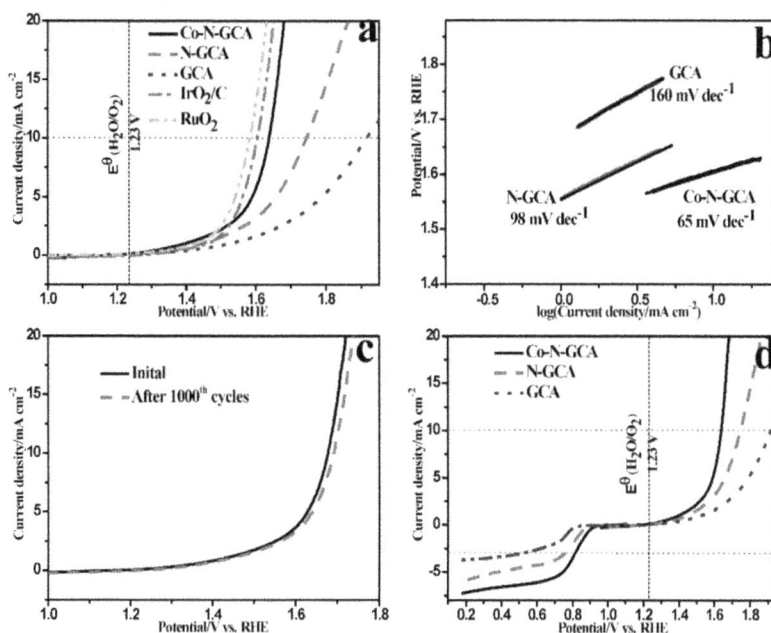

Figure 6. Oxygen evolution reaction (OER) LSV curves of Co-N-GCA, N-GCA, GCA, IrO_2/C, and RuO_2 at 1600 rpm in O_2-saturated 0.1 M KOH solution (**a**); OER Tafel plots of Co-N-GCA, N-GCA, and GCA (**b**); OER LSV plots of Co-N-GCA in the beginning and after 1000 cycles (**c**); and the overall LSV curves of Co-N-GCA, N-GCA, and GCA at 1600 rpm in O_2-saturated 0.1 M KOH solution (**d**).

The Tafel plots were also generated to study the OER kinetics of these catalysts. As shown in Figure 6b, the resulting Tafel slopes of Co-N-GCA, N-GCA, and GCA were found to be ~65, ~98, and ~160 mV dec^{-1}, respectively. The Co-N-GCA composite exhibited the smallest Tafel slope, suggesting its extremely favorable reaction kinetics.

Accelerated stability tests in O_2-saturated 0.1 M KOH at room temperature for Co-N-GCA were also carried out to investigate its durability for the OER. As shown in Figure 6c, after 1000 continuous potential cycles, the overpotential of Co-N-GCA increased by only 20 mV at a current density of 10 mA cm^{-2}, indicating Co-N-GCA has superior electrocatalytic stability for the OER.

The overall electrocatalytic activity of a bifunctional electrocatalyst as an oxygen electrode can be evaluated by taking the difference in potential between the OER current density at 10 mA cm^{-2} and the ORR current density at $-3\,\text{mA cm}^{-2}$ (ΔE). If the difference ΔE is smaller, this usually indicates the material is more suitable for practical applications. The overall electrocatalytic activities for the above catalysts are shown in Figure 6d. The ΔE values for Co-N-GCA, N-GCA, and GCA are 0.821, 0.983, and 1.389 V, respectively, showing that the bifunctional catalytic activities of the samples followed the order Co-N-GCA > N-GCA > GCA. More importantly, the ΔE value for Co-N-GCA is comparable to or even much smaller than many nonprecious metal-based bifunctional oxygen electrode catalysts reported previously (Table S3), suggesting the as-prepared Co-N-GCA is an effective bifunctional catalyst for the ORR and OER.

3. Materials and Methods

3.1. Materials

Graphite powder was purchased from Sigma Aldrich (Saint Louis, MO, USA). Multiwall carbon nanotubes were bought from Tsinghua University (Beijing, China). Urea and cobalt(II) acetate tetrahydrate were obtained from Sinopharm Chemical Reagent Co., Ltd. (Shanghai, China). Nafion (5 wt. %) was obtained from DuPont (Wilmington, DE, USA). Commercial Pt/C (20 wt. %) catalyst was purchased from Johnson Matthey (London, UK). RuO_2 was obtained from Sigma-Aldrich (Saint Louis, MO, USA). IrO_2/C (20 wt. %) catalyst was prepared by a method reported in the literature [37]. Carbon nanotubes were acid-treated with concentrated HNO_3 at 110 °C for 3 h to remove metal impurities and enhance wettability. Other reagents were of analytical grade and were used as received without further purification. Graphite oxide (GO) was synthesized from natural graphite flakes by a modified Hummers method [38].

3.2. Preparation of the Catalysts

GO (40 mg) was dispersed in 40 mL deionized water. Acid-treated carbon nanotubes (4 mg), urea (6 g), and cobalt(II) acetate tetrahydrate (30 mg) were added to the GO solution. The obtained mixture was ultrasonically treated for 2 h to form a suspension. Subsequently, the stable suspension was transferred into a 50 mL Teflon-lined autoclave and heated at 170 °C for 10 h. After the autoclave cooled to room temperature, the resulting hydrogel was washed with deionized water several times, followed by freeze drying. The resulting material is denoted as Co-N-GCA. For comparison, Co-GCA, N-GCA, and GCA were also prepared via similar procedures without the addition of urea, cobalt(II) acetate tetrahydrate, or both, respectively.

3.3. Physical Characterization

Scanning electron microscopy (SEM) images were collected on a Nova Nano 430 field emission scanning electron microscope (FEI, Hillsboro, OR, USA). Transmission electron microscopy (TEM) images were recorded on a JEM-2100HR transmission electron microscope (JEOL, Tokyo, Japan). Powder X-ray diffraction (XRD) patterns were obtained on a TD-3500 powder diffractometer (Tongda, Liaoning, China). Raman spectroscopy measurements were conducted on a Lab RAM Aramis Raman spectrometer (HORIBA Jobin Yvon, Edison, NJ, USA) with a laser wavelength of 632.8 nm. Surface area and pore structure characteristics were analyzed by Brunauer-Emmett-Teller (BET) and Barrett-Joyner-Halenda (BJH) methods using the adsorption branch of the isotherms obtained on a Tristar II 3020 gas adsorption analyzer (Micromeritics, Norcross, GA, USA). X-ray photoelectron spectroscopy (XPS) measurements were carried out on an ESCALAB 250 X-ray photoelectron spectrometer (Thermo-VG Scientific, Waltham, MA, USA).

3.4. Electrochemical Measurements

Electrochemical measurements were performed on an electrochemical workstation (Ivium, Eindhoven, the Netherlands) in a three-electrode configuration. An Ag/AgCl electrode (3 M NaCl) and a Pt wire were used as the reference electrode and the counter electrode, respectively. 0.1 M

KOH solution was selected as the electrolyte. The catalyst ink was prepared by dispersing 5 mg of the corresponding catalyst in 1 mL Nafion ethanol solution (0.25 wt. %), then dropping 20 μL of the ink on a polished glassy carbon electrode with a diameter of 5 mm, followed by drying under an infrared lamp.

Before the measurements, the electrolyte solution was purged with high-purity N_2 or O_2 gas for at least 30 min. Cyclic voltammetry (CV) and linear sweep voltammetry (LSV) tests were performed with a sweep rate of 10 mV s^{-1}. The Ag/AgCl (3 M NaCl) electrode was calibrated with respect to reversible hydrogen electrode (RHE), in 0.1 M KOH, E (RHE) = E (Ag/AgCl) + 0.982 V.

For the ORR, the electron transfer number (n) per oxygen molecule was calculated with the Koutecky–Levich (K–L) equation as follows [11]:

$$j^{-1} = j_k^{-1} + (0.62nFCD^{2/3}r^{-1/6}\omega^{1/2})^{-1} \tag{1}$$

where j and j_k are the measured current density and the kinetic current density, respectively; n is the transferred electron number per oxygen molecule; F is the Faraday constant (96,485 C mol^{-1}); C is the bulk concentration of O_2 in 0.1 M KOH (1.2×10^{-3} mol L^{-1}); D is the O_2 diffusion coefficient of in 0.1 M KOH (1.9×10^{-5} cm^2 s^{-1}); γ is the kinetic viscosity of the electrolyte (0.01 cm^2 s^{-1}); and ω is the angular velocity.

The Tafel plots were obtained by correcting the measured current density against the diffusion-limiting density to give the kinetic current density, calculated from:

$$j_{kin} = \frac{jj_{diff}}{j_{diff} - j}. \tag{2}$$

For the OER, the OER potential was IR corrected to compensate for the electrolyte's ohmic resistance by using the E–iR relation, where i is the current and R is the uncompensated for ohmic electrolyte resistance (~45 Ω) measured via high-frequency AC impedance.

4. Conclusions

In summary, a highly efficient bifunctional cobalt and nitrogen co-doped graphene-carbon nanotube aerogel (Co-N-GCA) has been designed and prepared using a simple one-pot hydrothermal method. By combining the merits of: (1) a synergistic effect between the abundant highly active sites, including cobalt oxide species, nitrogen dopant, and possibly Co-N$_x$ species and (2) a 3D conductive, interconnected, hierarchical porous structure for the exposure of active sites and efficient electron and mass transport, the Co-N-GCA exhibited excellent bifunctional catalytic performance and long-term durability in the ORR and OER. These results reveal its promising potential as a low-cost and highly active ORR and OER catalyst for applications in regenerative fuel cells and rechargeable metal–air batteries.

Supplementary Materials: The following are available online at http://www.mdpi.com/2073-4344/8/7/275/s1, Figure S1. HRTEM image of Co-N-GCA. Figure S2: Representative SEM images of N-GCA (a,b) and GCA (c,d) (inset in (a and c) shows the digital photograph of the corresponding prepared aerogels); Figure S3: XPS survey spectrum of Co-N-GCA and N-GCA (a); high-resolution N 1s spectra of N-GCA (b); Table S1: Surface composition of Co-N-GCA and N-GCA, calculated from XPS results; Table S2: Distribution of each N species, obtained from the fitting results of N1s XPS spectra (normalized to the surface N atoms of each material); Figure S4: CV curves of 20 wt. % Pt/C in O$_2$-saturated 0.1 M KOH solution; Table S3: Comparison of ORR and OER activity parameters with other recently reported highly active non-noble metal bifunctional electrocatalysts.

Author Contributions: Methodology, X.Q. and Y.L.; Date curation, X.Q. and J.J.; Resources, L.C.; Validation, S.J.; Supervision, H.F.; Writing-original draft, X.Q.; Writing-Review & Editing, S.L.

Acknowledgments: The authors would like to acknowledge the financial support of the China Postdoctoral Science Foundation (Project No. 2017M612664), Natural Science Foundation of Guangdong Province (Project No. 2017A030310645).

Conflicts of Interest: The authors declare no conflict of interest.

References

1. Chu, S.; Majumdar, A. Opportunities and challenges for a sustainable energy future. *Nature* **2012**, *488*, 294–303. [CrossRef] [PubMed]
2. Stamenković, V.R.; Fowler, B.; Mun, B.S.; Wang, G.; Ross, P.N.; Lucas, C.A.; Marković, N.M. Improved oxygen reduction activity on $Pt_3Ni(111)$ via increased surface site availability. *Science* **2007**, *315*, 493–497. [CrossRef] [PubMed]
3. Winter, M.; Brodd, R.J. What are batteries, fuel cells, and supercapacitors? *Chem. Rev.* **2004**, *104*, 4245–4270. [CrossRef] [PubMed]
4. McCrory, C.C.L.; Jung, S.; Peters, J.C.; Jaramillo, T.F. Benchmarking Heterogeneous Electrocatalysts for the Oxygen Evolution Reaction. *J. Am. Chem. Soc.* **2013**, *135*, 16977–16987. [CrossRef] [PubMed]
5. Chen, Z.; Higgins, D.; Yu, A.; Zhang, L.; Zhang, J. a review on non-precious metal electrocatalysts for PEM fuel cells. *Energy Environ. Sci.* **2011**, *4*, 3167–3192. [CrossRef]
6. Tahir, M.; Pan, L.; Idrees, F.; Zhang, X.W.; Wang, L.; Zou, J.J.; Wang, Z.L. Electrocatalytic oxygen evolution reaction for energy conversion and storage: a comprehensive review. *Nano Energy* **2017**, *37*, 136–157. [CrossRef]
7. Yeo, B.S.; Bell, A.T. Enhanced Activity of Gold-Supported Cobalt Oxide for the Electrochemical Evolution of Oxygen. *J. Am. Chem. Soc.* **2011**, *133*, 5587–5593. [CrossRef] [PubMed]
8. Duan, Z.; Wang, G. a first principles study of oxygen reduction reaction on a Pt(111) surface modified by a subsurface transition metal M (M = Ni, Co, or Fe). *Phys. Chem. Chem. Phys.* **2011**, *13*, 20178–20187. [CrossRef] [PubMed]
9. Zhao, Y.; Kamiya, K.; Hashimoto, K.; Nakanishi, S. Efficient Bifunctional Fe/C/N Electrocatalysts for Oxygen Reduction and Evolution Reaction. *J. Phys. Chem. C* **2015**, *119*, 2583–2588. [CrossRef]
10. Zhao, Y.; Chen, S.; Sun, B.; Su, D.; Huang, X.; Liu, H.; Yan, Y.; Sun, K.; Wang, G. Graphene-Co_3O_4 nanocomposite as electrocatalyst with high performance for oxygen evolution reaction. *Sci. Rep.* **2015**, *5*, 7629. [CrossRef] [PubMed]
11. Liang, Y.; Li, Y.; Wang, H.; Zhou, J.; Wang, J.; Regier, T.; Dai, H. Co_3O_4 nanocrystals on graphene as a synergistic catalyst for oxygen reduction reaction. *Nat. Mater.* **2011**, *10*, 780–786. [CrossRef] [PubMed]
12. Zhang, Z.; Chen, Y.; Bao, J.; Xie, Z.; Wei, J.; Zhou, Z. Co_3O_4 Hollow Nanoparticles and Co Organic Complexes Highly Dispersed on N-Doped Graphene: An Efficient Cathode Catalyst for $Li-O_2$ Batteries. *Part. Part. Syst. Charact.* **2015**, *32*, 680–685. [CrossRef]
13. Cao, X.; Zheng, X.; Tian, J.; Jin, C.; Ke, K.; Yang, R. Cobalt Sulfide Embedded in Porous Nitrogen-doped Carbon as a Bifunctional Electrocatalyst for Oxygen Reduction and Evolution Reactions. *Electrochim. Acta* **2016**, *191*, 776–783. [CrossRef]
14. Chen, B.; Li, R.; Ma, G.; Gou, X.; Zhu, Y.; Xia, Y. Cobalt sulfide/N,S codoped porous carbon core-shell nanocomposites as superior bifunctional electrocatalysts for oxygen reduction and evolution reactions. *Nanoscale* **2015**, *7*, 20674–20684. [CrossRef] [PubMed]
15. Liu, Q.; Jin, J.; Zhang, J. $NiCo_2S_4$@graphene as a bifunctional electrocatalyst for oxygen reduction and evolution reactions. *ACS Appl Mater. Interfaces* **2013**, *5*, 5002–5008. [CrossRef] [PubMed]
16. Zhang, J.; Qu, L.; Shi, G.; Liu, J.; Chen, J.; Dai, L. N,P-Codoped Carbon Networks as Efficient Metal-free Bifunctional Catalysts for Oxygen Reduction and Hydrogen Evolution Reactions. *Angew. Chem. Int. Ed. Engl.* **2016**, *55*, 2230–2234. [CrossRef] [PubMed]
17. Qu, K.G.; Zheng, Y.; Dai, S.; Qiao, S.Z. Graphene oxide-polydopamine derived N, S-codoped carbon nanosheets as superior bifunctional electrocatalysts for oxygen reduction and evolution. *Nano Energy* **2016**, *19*, 373–381. [CrossRef]
18. Qiao, X.; Liao, S.; Zheng, R.; Deng, Y.; Song, H.; Du, L. Cobalt and Nitrogen Codoped Graphene with Inserted Carbon Nanospheres as an Efficient Bifunctional Electrocatalyst for Oxygen Reduction and Evolution. *ACS Sustain. Chem. Eng.* **2016**, *4*, 4131–4136. [CrossRef]
19. Zhang, J.; Zhao, Z.; Xia, Z.; Dai, L. a metal-free bifunctional electrocatalyst for oxygen reduction and oxygen evolution reactions. *Nat. Nanotechnol.* **2015**, *10*, 444–452. [CrossRef] [PubMed]

20. Petrie, J.R.; Cooper, V.R.; Freeland, J.W.; Meyer, T.L.; Zhang, Z.; Lutterman, D.A.; Lee, H.N. Enhanced Bifunctional Oxygen Catalysis in Strained LaNiO$_3$ Perovskites. *J. Am. Chem. Soc.* **2016**, *138*, 2488–2491. [CrossRef] [PubMed]

21. He, D.; Xiong, Y.; Yang, J.; Chen, X.; Deng, Z.; Pan, M.; Li, Y.; Mu, S. Nanocarbon-intercalated and Fe-N-codoped graphene as a highly active noble-metal-free bifunctional electrocatalyst for oxygen reduction and evolution. *J. Mater. Chem. A* **2017**, *5*, 1930–1934. [CrossRef]

22. Su, Y.; Zhu, Y.; Jiang, H.; Shen, J.; Yang, X.; Zou, W.; Chen, J.; Li, C. Cobalt nanoparticles embedded in N-doped carbon as an efficient bifunctional electrocatalyst for oxygen reduction and evolution reactions. *Nanoscale* **2014**, *6*, 15080–15089. [CrossRef] [PubMed]

23. Wu, Y.; Zang, J.; Dong, L.; Zhang, Y.; Wang, Y. High performance and bifunctional cobalt-embedded nitrogen doped carbon/nanodiamond electrocatalysts for oxygen reduction and oxygen evolution reactions in alkaline media. *J. Power Sources* **2016**, *305*, 64–71. [CrossRef]

24. Qiao, X.; You, C.; Shu, T.; Fu, Z.; Zheng, R.; Zeng, X.; Li, X.; Liao, S. a one-pot method to synthesize high performance multielement co-doped reduced graphene oxide catalysts for oxygen reduction. *Electrochem. Commun.* **2014**, *47*, 49–53. [CrossRef]

25. Liu, X.; Amiinu, I.S.; Liu, S.; Cheng, K.; Mu, S. Transition Metal/Nitrogen dual-doped Mesoporous Graphene-like Carbon Nanosheets for Oxygen Reduction and Evolution Reactions. *Nanoscale* **2016**, *8*, 13311–13320. [CrossRef] [PubMed]

26. Mao, S.; Wen, Z.; Huang, T.; Hou, Y.; Chen, J. High-performance bi-functional electrocatalysts of 3D crumpled graphene–cobalt oxide nanohybrids for oxygen reduction and evolution reactions. *Energy Environ. Sci.* **2014**, *7*, 609–616. [CrossRef]

27. Yasuda, S.; Furuya, A.; Uchibori, Y.; Kim, J.; Murakoshi, K. Iron-Nitrogen-Doped Vertically Aligned Carbon Nanotube Electrocatalyst for the Oxygen Reduction Reaction. *Adv. Funct. Mater.* **2016**, *26*, 738–744. [CrossRef]

28. Zhao, J.; Liu, Y.; Quan, X.; Chen, S.; Zhao, H.; Yu, H. Nitrogen and sulfur co-doped graphene/carbon nanotube as metal-free electrocatalyst for oxygen evolution reaction: The enhanced performance by sulfur doping. *Electrochim. Acta* **2016**, *204*, 169–175. [CrossRef]

29. Tang, C.; Zhang, Q.; Zhao, M.Q.; Huang, J.Q.; Cheng, X.B.; Tian, G.L.; Peng, H.J.; Wei, F. Nitrogen-doped aligned carbon nanotube/graphene sandwiches: Facile catalytic growth on bifunctional natural catalysts and their applications as scaffolds for high-rate lithium-sulfur batteries. *Adv. Mater.* **2014**, *26*, 6100–6105. [CrossRef] [PubMed]

30. Zhu, C.; Fu, S.; Song, J.; Shi, Q.; Su, D.; Engelhard, M.H.; Li, X.; Xiao, D.; Li, D.; Estevez, L.; et al. Self-Assembled Fe-N-Doped Carbon Nanotube Aerogels with Single-Atom Catalyst Feature as High-Efficiency Oxygen Reduction Electrocatalysts. *Small* **2017**, *13*, 1603407. [CrossRef] [PubMed]

31. Sun, T.T.; Xu, L.B.; Li, S.Y.; Chai, W.X.; Huang, Y.; Yan, Y.S.; Chen, J.F. Cobalt-nitrogen-doped ordered macro-/mesoporous carbon for highly efficient oxygen reduction reaction. *Appl. Catal. B Environ.* **2016**, *193*, 1–8. [CrossRef]

32. Wohlgemuth, S.-A.; White, R.J.; Willinger, M.-G.; Titirici, M.-M.; Antonietti, M. a one-pot hydrothermal synthesis of sulfur and nitrogen doped carbon aerogels with enhanced electrocatalytic activity in the oxygen reduction reaction. *Green Chem.* **2012**, *14*, 1515–1523. [CrossRef]

33. Guo, D.; Shibuya, R.; Akiba, C.; Saji, S.; Kondo, T.; Nakamura, J. Active sites of nitrogen-doped carbon materials for oxygen reduction reaction clarified using model catalysts. *Science* **2016**, *351*, 361–365. [CrossRef] [PubMed]

34. Cui, X.Y.; Yang, S.B.; Yan, X.X.; Leng, J.G.; Shuang, S.; Ajayan, P.M.; Zhang, Z.J. Pyridinic-Nitrogen-Dominated Graphene Aerogels with Fe-N-C Coordination for Highly Efficient Oxygen Reduction Reaction. *Adv. Funct. Mater.* **2016**, *26*, 5708–5717. [CrossRef]

35. Liu, X.; Zhou, W.; Yang, L.; Li, L.; Zhang, Z.; Ke, Y.; Chen, S. Nitrogen and sulfur co-doped porous carbon derived from human hair as highly efficient metal-free electrocatalysts for hydrogen evolution reactions. *J. Mater. Chem. A* **2015**, *3*, 8840–8846. [CrossRef]

36. Ferrari, A.C.; Robertson, J. Raman spectroscopy of amorphous, nanostructured, diamond–like carbon, and nanodiamond. *Philos. Trans. R. Soc. Lond. a Math. Phys. Eng. Sci.* **2004**, *362*, 2477–2512. [CrossRef] [PubMed]

37. Prabu, M.; Ramakrishnan, P.; Shanmugam, S. CoMn$_2$O$_4$ nanoparticles anchored on nitrogen-doped graphene nanosheets as bifunctional electrocatalyst for rechargeable zinc–air battery. *Electrochem. Commun.* **2014**, *41*, 59–63. [CrossRef]
38. Wang, H.; Yang, Y.; Liang, Y.; Cui, L.F.; Sanchez Casalongue, H.; Li, Y.; Hong, G.; Cui, Y.; Dai, H. LiMn$_{1-x}$Fe$_x$PO$_4$ Nanorods Grown on Graphene Sheets for Ultrahigh-Rate-Performance Lithium Ion Batteries. *Angew. Chem.* **2011**, *123*, 7502–7506. [CrossRef]

![catalysts logo] *catalysts*

MDPI

Article

Influence of the Structure-Forming Agent on the Performance of Fe-N-C Catalysts

Sven Schardt [1], Natascha Weidler [2], W. David Z. Wallace [1], Ioanna Martinaiou [2,3], Robert W. Stark [4] and Ulrike I. Kramm [1,2,3,*]

[1] TU Darmstadt, Department of Chemistry, Catalysts and Electrocatalysts Group (EKAT), 64287 Darmstadt, Germany; s-schardt@online.de (S.S.); wallace@ese.tu-darmstadt.de (W.D.Z.W.)
[2] TU Darmstadt, Department of Materials and Earth Sciences, EKAT Group, 64287 Darmstadt, Germany; weidler@ese.tu-darmstadt.de (N.W.); Martinaiou@ese.tu-darmstadt.de (I.M.)
[3] TU Darmstadt, Graduate School of Excellence Energy Science and Engineering, 64287 Darmstadt; Germany
[4] TU Darmstadt, Department of Materials and Earth Sciences, Physics of Surfaces Group, 64287 Darmstadt, Germany; stark@csi.tu-darmstadt.de
* Correspondence: kramm@ese.tu-darmstadt.de; Tel.: +49-6151-1620356

Received: 6 June 2018; Accepted: 25 June 2018; Published: 28 June 2018

Abstract: In this work, the influence of the structure-forming agent on the composition, morphology and oxygen reduction reaction (ORR) activity of Fe-N-C catalysts was investigated. As structure-forming agents (SFAs), dicyandiamide (DCDA) (nitrogen source) or oxalic acid (oxygen source) or mixtures thereof were used. For characterization, cyclic voltammetry and rotating disc electrode (RDE) experiments were performed in 0.1 M H_2SO_4. In addition to this, N_2 sorption measurements and Raman spectroscopy were performed for the structural, and elemental analysis for chemical characterization. The role of metal, nitrogen and carbon sources within the synthesis of Fe-N-C catalysts has been pointed out before. Here, we show that the optimum in terms of ORR activity is achieved if both *N*- and *O*-containing SFAs are used in almost similar fractions. All catalysts display a redox couple, where its position depends on the fractions of SFAs. The SFA has also a strong impact on the morphology: Catalysts that were prepared with a larger fraction of N-containing SFA revealed a higher order in graphitization, indicated by bands in the 2nd order range of the Raman spectra. Nevertheless, the optimum in terms of ORR activity is obtained for the catalyst with highest D/G band ratio. Therefore, the results indicate that the presence of an additional oxygen-containing SFA is beneficial within the preparation.

Keywords: Fe-N-C catalyst; oxygen reduction reaction

1. Introduction

Me-N-C catalysts play an emerging role as possible catalyst materials for the oxygen reduction reaction in fuel cells. It is clear that they can be prepared from any kind of metal, nitrogen and carbon precursor, but that the overall achievable activity will be strongly related to the optimization of the mixtures. In the first approaches, macrocycles were supported on a carbon black and then pyrolysed at temperatures above 600 °C to form significantly more active catalysts in comparison to the non-pyrolysed macrocycles [1,2]. To enable cheaper preparation and independent optimization of metal and nitrogen content in the synthesis, alternative preparation approaches had to be established. In 1989, Gupta et al. showed that the combination of metal acetate with polyacrylonitrile and carbon black leads to the formation of active materials when pyrolysed at temperatures >600 °C [3]. Main efforts to find suitable nitrogen and metal precursors were made by Dodelet's group and others [4–7]. These early approaches were strongly limited in the density of active sites, as the carbon black always displays a relative mass in the final catalyst that is not contributing in terms of active sites [8–11].

To overcome this limitation, the first method that enabled the preparation of Me-N-C catalysts without additional carbon support, is the oxalate-supported pyrolysis of porphyrins [12]. In this approach oxalate works as a structure-forming agent (SFA) as the final catalyst resembles its morphology [13–15]. In addition to this, the use of sulfur strongly affected the morphology and performance of the catalysts [16,17]. Sulfur usually enables a higher porosity and lower graphitization [16,18], but more important it was found to prevent carbide formation that was assigned to active site destruction by works of us and of others [17,19,20]. That indeed excess iron was at the origin of active site destruction was further confirmed through an alternative strategy to prevent active site destruction [21]. Namely, instead of following the state-of-the art preparation protocol (pyrolysis plus acid leaching), the pyrolysis was interrupted at 550 °C (a temperature below the temperature limit for iron carbide formation), excess iron was removed by an acid leaching. Afterwards the synthesis was continued up to the usual end temperature followed by an additional acid-leaching. Indeed, such catalysts prepared with intermediate acid leaching (IAL) reach the same activities as the original sulfur-added catalysts.

Motivated by these findings, we started to develop a new synthesis route in our group that uses dicyandiamide (DCDA) as a structure-forming agent in combination with a nitrogen and iron precursor. In recent studies we varied the amount of sulfur within this synthesis [22]. The trends in kinetic current density as a function of S/Me ratio were similar to trends observed previously for the combination of iron oxalate with CoTMPP [23], with a maximum in ORR activity at S/Fe = 0.8. However, in contrast to these earlier finding, our new catalysts still contained considerable fractions of iron carbide in the optimum of ORR activity. Only for the highest S/Me ratios iron carbide formation was inhibited but iron sulfide species were formed [22]. This showed us, that while sulfur-addition strongly affects the achievable ORR activity, sulfur itself helps not to prevent iron carbide formation in our selected system.

In order to get more insights on the missing parameter that is required for further optimization, we focused on a molar ratio of S/Fe = 0.8 in this work and stepwise exchanged the DCDA with oxalic acid as alternative a structure-forming agent with strong similarities to the metal oxalates used within the oxalate-supported pyrolysis of porphyrins [14]. The results show that by tuning the composition of the SFAs the activity and morphology is changing considerably.

2. Results and Discussion

In the following chapter, the composition of the precursor is expressed as the relative oxalate content which is the molar fraction of oxalic acid in the overall sum of structure-forming agent (SFA = oxalic acid and DCDA). In Figure 1a,b the yield and the BET surface area; respectively, are given as a function of the relative oxalate content within the SFA.

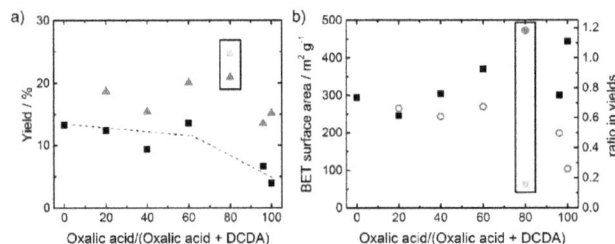

Figure 1. (**a**) Change of the synthesis yield as a function of relative part of oxalic acid in the overall SFA (considering molar fractions), both the yield after heat-treatment (▲) and after heat-treatment plus acid-leaching (■) are shown. In (**b**) the relation between BET surface area (left axis, ■); respectively, ratio of the yields (right axis, ○) and the amount of oxalic acid in the structure-forming agent (SFA) are given.

The values plotted as squares in Figure 1a were all obtained after pyrolysis and acid leaching; the yields after the pyrolysis only are shown as triangles. Due to strong gas evolution, the pyrolysis product of the catalyst with only DCDA as SFA was transferred out of the quartz boat. Therefore, no yield without acid leaching is given for this catalyst (and for all other syntheses it was taken care that sufficiently large boats were used). In general, the yields of final catalysts are larger when DCDA is used as SFA in comparison to oxalic acid. One important aspect can be noted from the comparison of the yields after pyrolysis and after pyrolysis plus acid-leaching: In all cases, the as-obtained yield after pyrolysis was larger compared to the yield after pyrolysis plus acid leaching. This is expected, as it is supposed that during the acid-leaching, inorganic iron species which were formed during the pyrolysis are leached [24,25]. However, for the sample prepared with 80% oxalic acid in the SFA the yield increases during acid leaching. This illustrates, that beside a possible removal of some species, the implementation of additional ones must have appeared so that the yield could increase. From the ratio of the yield after pyrolysis and acid leaching in relation to the yield just after pyrolysis the values as plotted in Figure 1b are obtained. The observed trend for this curve, is almost opposite with respect to the trend observed for the BET surface area. Hence, it can be concluded, that a not successful leaching (or change from an anhydrous iron species to a hydrated one) is at the origin of the low surface area observed for this catalyst.

The data for the catalyst at 80% oxalic acid as SFA is highlighted in both graphs of Figure 1. Initially, we assumed that these out-of-trend values of yield and BET surface area were due to failures during the synthesis of this catalyst. However, repeating the synthesis confirmed the reproducibility in all respects. This catalyst (80% oxalic acid in SFA) will therefore not be included in detail in the following discussion, as it seems that the precursor combination leads to undesired reactions, most probably already in the precursor stage. This is shortly addressed in Section 3.1.

The results of N_2 sorption measurements are summarized in Table 1. As indicated, only some of the catalysts contain surface areas within the micropore regime (<20 nm); most of the samples are basically mesoporous.

Table 1. Summary of the results obtained from N_2 sorption measurements. SA is used as surface area and DCDA as dicyandiamide.

Oxalic Acid/(Oxalic Acid + DCDA)	100	96	80	60	40	20	0
BET SA/$m^2 g^{-1}$	444	300	63	371	305	245	295
Meso SA/$m^2 g^{-1}$	441	300	35	147	277	245	295
Micro SA/$m^2 g^{-1}$	3	<1	28	224	28	0	0
Micro/BET/%	1	<1	44	60	9	0	0

It becomes clear that in order to obtain a significant fraction of micropores, both nitrogen and oxygen-containing SFAs should be used at the same time. In order to be able to verify to what extent it is the microporosity that is of importance for the high activity, further experiments are required.

In order to see the impact of the synthesis on the electrochemical performance, cyclic voltammetry and RDE experiments were performed in N_2 saturated and O_2 saturated electrolytes, respectively. The results are given in Figure 2. It becomes clear that there is a strong impact of the SFA-ratio on the capacitive current density. Interestingly, all catalysts exhibit a redox peak at about 0.7 V. Furthermore, for some catalysts a peak in the anodic sweep is visible between 0.3 and 0.4 V. This peak was also previously observed for phenanthroline-based catalysts but usually vanishes upon cycling. So far, we were not able to make a final assignment. For the intense redox peak at ~0.7 V it might be assigned either to Fe^{2+}/Fe^{3+} couple or quinone/hydroquine [11]. Considering the double layer capacity, it roughly seems to increase with an increasing ratio of oxalic acid (maximum for the sample with a ratio of 60%) and then to drop down (see discussion below).

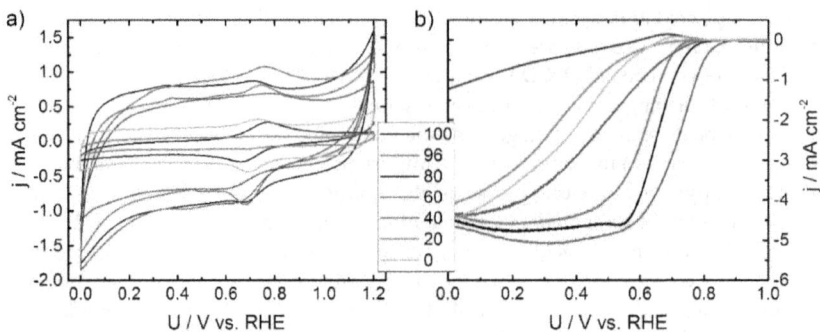

Figure 2. Cyclic voltammetry (100 mV s^{-1}) in N$_2$ saturated 0.1 M H$_2$SO$_4$ (**a**) and rotating disc electrode experiments (rpm 1500, 10 mV s^{-1}) (**b**) of the variation series.

During RDE experiments it is again the 60% sample that gives the highest onset potential and largest kinetic current density in comparison to all other samples. Most of the samples give the usual shape of RDE curves, but at lower fractions of oxalic acid, the curves indicate some kinetic hindrance. It should be noted that none of the catalysts reach the expected diffusion limiting current density for a four electron reduction (expected value 5.9 mA cm^{-2}) [26]. There are two different explanations possible: First, the selectivity is changing according to the observed value of maximum current density. To further verify this, it would be important to perform experiments with the Rotating Ring Disc Electrode (RRDE) and to have a more detailed analysis of the composition of the catalysts by ideally X-ray induced photoelectron spectroscopy and Mößbauer spectroscopy. Unfortunately, this is beyond the scope of this publication but might be addressed in a future work. Second, it is well known for Me-N-C catalysts that the observable diffusion limiting current density is depending on the loading. This might be due to a 2 × 2 electron mechanism, where the probability of re-adsorbing and further reducing any of the formed hydrogen peroxide increases with catalyst film thickness on the electrode [27,28]. Also here, RRDE experiments at different loadings would assist in answering this question. Both, the cyclic voltammogram (CV) and RDE curves of the outlier related to 80% look pretty much like iron oxide species.

In Figure 3 the Tafel plots of all investigated catalysts are shown. It becomes apparent that the ORR activity in terms of kinetic current density changes drastically with the ratio of oxalic acid in the SFA. The Tafel slopes of the more-active catalysts show values of about 70 mV dec^{-1}. This value is in good agreement with the values that were observed for other catalysts prepared by the (oxalate supported) pyrolysis of porphyrins [15,24,29] as well as alternatively prepared catalysts (without the use of a reactive gas heat-treatment) [19]. When the fraction of oxalic acid goes down the Tafel slope increases, getting closer to 100 mV dec^{-1}. While not intended, this might be related to an interaction of the catalyst with ammonia (from DCDA decomposition) during the heat-treatment. A larger Tafel slope value for catalysts that involve ammonia in their preparation was also previously observed [19]. The lowest value of 64 mV dec^{-1} is obtained for the catalyst with exclusively oxalic acid as SFA. It might illustrate slightly better kinetics that might result in a higher turn-over frequency (TOF). If a change in the TOF is also at the origin of the observed changes in Tafel slope, in this work, it might be concluded that the catalyst with 60% oxalic acid in the SFA need to have a significant higher number of ORR active sites. This will be discussed below with respect to the data from combustion analysis.

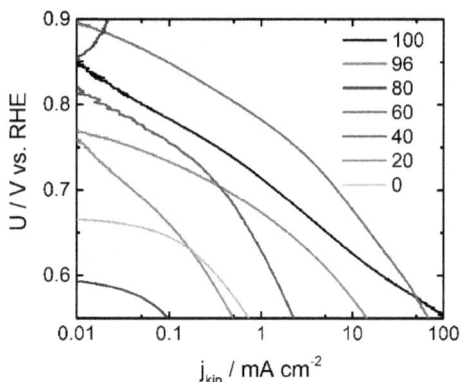

Figure 3. Tafel plots of all investigated catalysts (calculated from RDE data provided in Figure 2b).

As mentioned above, the capacity of the different catalysts seems to be affected by the ratio of oxalic acid in the SFA. A plot of the related current density in the double layer region of the CVs (named as j(DL capacity)) vs ratio of oxalic acid in SFA shows that there is a maximum in capacity current for 60% oxalic acid in the SFA and then the capacitive current remains constant or slightly decreases, this is shown in Figure 4a. In contrast to this a clear maximum is visible for the ORR activity in Figure 4b.

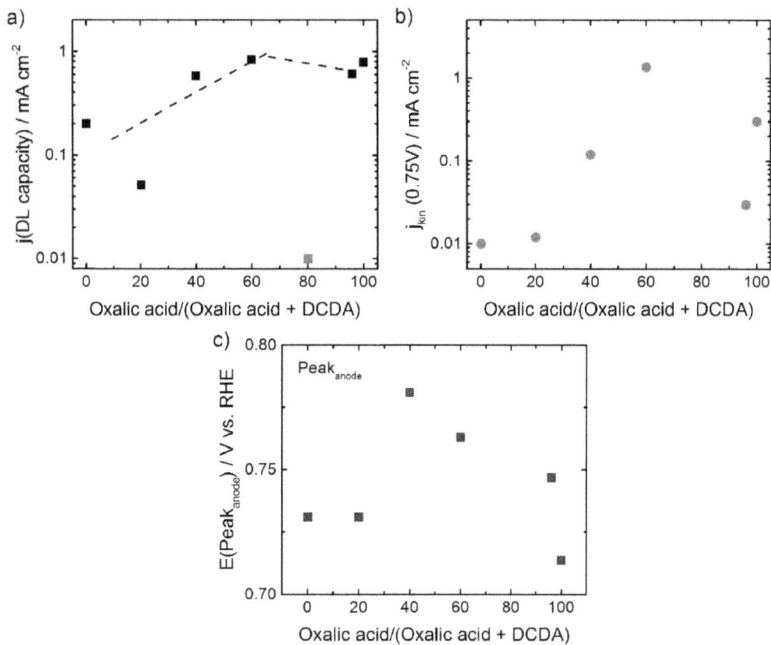

Figure 4. Analysis of electrochemical data. In (**a**) the current density related to the double layer capacity ($100 \, mV \, s^{-1}$) (■), outlier at 80% (■), in (**b**) the kinetic current density j_{kin} (•) and in (**c**) the potential positions of the peak maxima in the anodic sweep of the CVs are given as a function of the ratio of oxalic acid in SFA (■).

In Figure 4c the potentials related to the peak maxima in the anodic sweep of the cyclic voltammograms are given as a function of the ratio of oxalic acid in the SFA. Roughly, the trend is similar as for the kinetic current density: The resulting kinetic current density was getting as higher as more positive the anodic peak position was, what is in agreement with previous reports [30–32]. Nevertheless, it becomes clear, that E (Peak$_{anode}$) is only a weak activity descriptor for this group of catalysts. While the peak position might be an indicator for the turn-over frequency (TOF) on active sites, the charge under the redox peak might be related to the number of active sites (if assigned to Fe^{2+}/Fe^{3+}), or number of sites promoting the ORR (evtl. quinone/hydroquinone). As both (TOF and site density) will contribute to the kinetic current density, this might explain the only weak correlation. Furthermore, it should be noted, that the difference between the anodic and cathodic peak is not constant but increases from about 40 to 220 mV and then decreases to about 50 mV (not displayed). Further characterization will be required to fully understand this trend.

It is well known that the nitrogen content is an important parameter as it is required for the formation of active sites. In order to get further insights to what extend the SFA composition affects the composition of the final catalyst combustion analysis for CHN was performed. The nitrogen and hydrogen contents as a function of SFA composition are shown in Figure 5a. Also, the amount of residuals was determined as difference between 100% minus nitrogen, carbon and hydrogen contents. The data are shown in Figure 5b. The expected residuals were basically iron, sulfur and oxygen, and maybe some chlorine from the acid-leaching procedure.

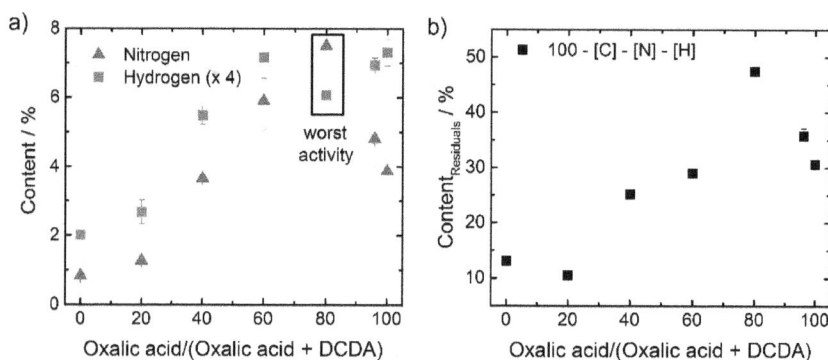

Figure 5. Results of elemental analysis for nitrogen (▲) and hydrogen (▦) (values multiplied by factor of four) are shown (**a**). In (**b**) the content assigned to residuals is given (■).

Rather unexpected, the outlier at 80% with the worst activity yields the highest nitrogen content. A similar trend was observed for the amount of residuals. The very high value of residuals of almost 50% in this catalyst could be assigned (at least in parts) to iron oxide as indicated in the electrochemical measurements, but also in the Raman spectra (below).

If one has to identify a distinct difference between the outlier versus the others, it was the ratio between [H] and [N] that was changing, as visible from Figure 5a.

In general, there was a good correlation between the nitrogen content and the onset potential. This is visible from Figure 6a. Also here, the worst performing catalyst gave an exception. It could have its origin in the very low surface area observed for this catalyst (right axis). Figure 6b correlates the content of nitrogen with the intensity ratio of the D$_3$ band and G band, we will refer to it, below after discussing the Raman spectra.

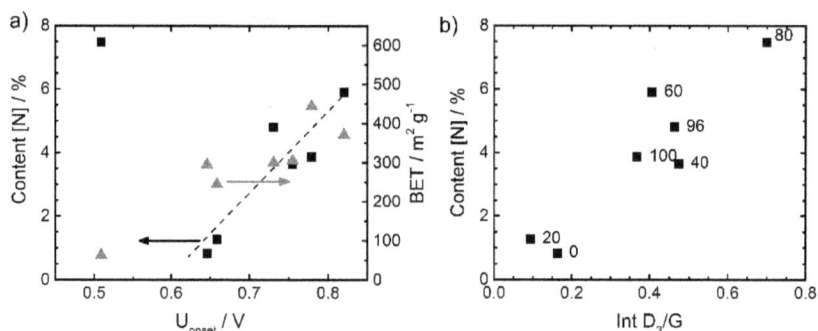

Figure 6. Correlation between the nitrogen content from elemental analysis (left axis, ■) as well as BET surface area (right axis, ▲) and the onset potential (**a**) (defined as potential at -0.1 mA cm^{-2}) and correlation between the nitrogen content and the intensity ratio D_3/G from Raman spectroscopy (**b**).

For characterization of the carbon morphology, Raman spectra were recorded for all catalysts (Figure 7a,b). The analysis of the Raman spectra was made in relation to literature on carbon blacks [33,34]. All samples reveal the typical shape of carbon blacks with distinct D-band and G-band intensities. While the latter is related to in-plane vibration modes, the former is related to vibrations "at the edges" of the graphene layer [35]. Hence, the intensity ratio D/G is inverse proportional to the graphene layer extension.

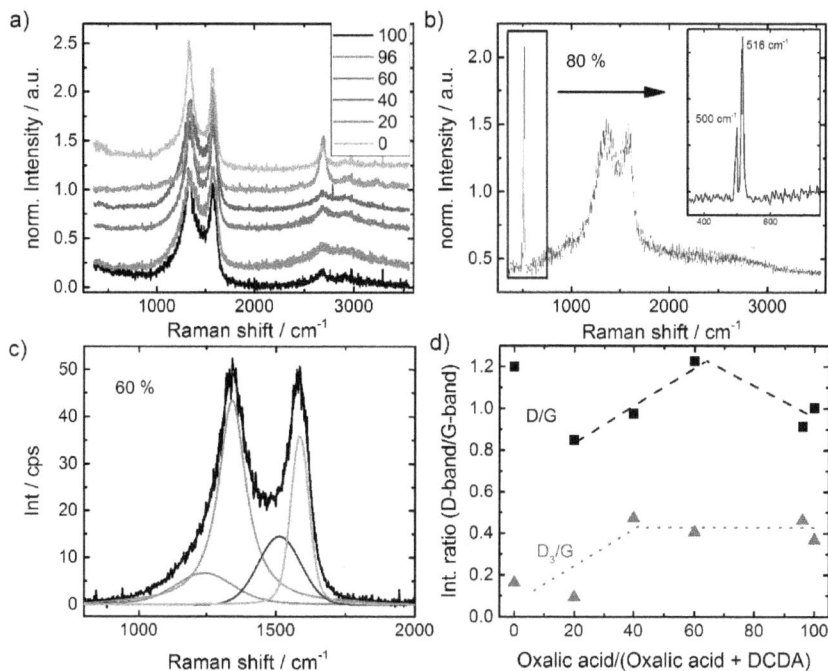

Figure 7. Raman spectra of the variation series (**a**), spectrum for 80% oxalic acid in SFAs (**b**) and deconvoluted Raman spectrum (1st order region) of the best performing catalyst with 60% oxalic acid in SFAs (**c**). Influence of the ratio of oxalic acid in the SFA on the D/G (■) and D_3/G band ratios (▲) (**d**).

It is interesting to note, that for those samples with dominant ratios of DCDA in the SFA additional bands in the 2nd order region become pronounced. This band is related to the 2D band. In a recent work by Larouche et al. [36] it was discussed that the ratio of 2D/D areas can be used to calculate a graphene layer extension labelled L_{eq} that also considers the curvature within graphene flakes [36]. That is why in the paragraph above "at the edges" was written in quotation marks, because the D band will be formed when the vibrations occur at the edges, but also if the symmetry (e.g., by implementation of defects, five-ring groups, etc.) is broken.

Also, the 1st order region was analyzed in more detail by fitting the spectra assuming the presence of four bands: The aforementioned D and G bands and in addition a D_3 and a D_4 band. These are assigned to heteroatoms in carbon and polyaromatic hydrocarbons, respectively [33,34,36]. As an example, the fit of the most active catalyst is provided in Figure 7c.

In Figure 7d the intensity ratios of D/G band and D_3/G band are shown as a function of oxalic acid in the SFA. Except for the sample prepared exclusively with DCDA, first an increase in D/G ratio and then a decrease (above 60% oxalic acid) is observed. Comparing the trends in Figures 4a and 7d it seems that the intensity ratio D/G follows the trend in capacity. There is also a strong relation between the capacity and the BET surface area (not shown) underlining that in most of the cases indeed edges of graphene layers might contribute to the D-band intensity (rather than curvatures).

In a previous work, we were able to show by analyzing the Raman spectra of a group of Me-N-C catalysts before and after accelerated stress tests (AST), that a decrease in the D_3/G ratio correlated with the decrease in ORR activity observed during AST [37]. As this loss in activity was also correlating with the displacement of metal out of the N_4 plane, we attributed a contribution of MeN_4 related vibrations to be located in the D_3 band region [37]. As given in Figure 7d, up to a ratio of 40% oxalic acid in the SFA the D_3/G band ratio is increasing and then does not change significantly (It should be noted that the worst performing catalyst reached a value of 0.7). On the basis of our previous work, this might be an indicator for a change of the relative amount of FeN_4 sites. The correlation between nitrogen content and the D_3/G band ratio in Figure 6b pointed into this direction. However, it must be noted that a similar good correlation would be given for the content of residuals with Int D_3/G.

Nevertheless, as discussed above, the D_3 band is assigned to vibrations at defects/ heteroatoms in the graphene sheets. Hence, most possibly all these heteroatoms contributed to some extend to the D_3 band intensity.

Even though, it was already discussed above, the sample with 80% is out of trend, its Raman spectrum as displayed in Figure 7b should be addressed, shortly. Distinct bands can be identified at 500 and 516 cm^{-1}. On first view, these bands were assigned to iron oxide vibrations [38]. This is in agreement with the band at 500 cm^{-1}, but so far, we were not able to assign the band at 516 cm^{-1}. On the basis of the observations made so far for this sample, we believe that the formation of these two bands is in strong relation to its worse composition and activity.

Why does the activity increase so much by changing the fraction of oxalic acid in the SFA?

In our previous publication, iron acetate was the only source of oxygen within the synthesis. In this study we replaced the iron acetate by iron chloride which allowed us to systematically vary the content of oxygen in the overall precursor from zero to a maximum of 70 at % (hydrogen not considered). It is interesting to note, that the catalyst with the maximum activity in this work has a relative content of oxygen (44 at %) very similar to the one calculated for the precursor of the oxalate support pyrolysis of porphyrins (43 at %). This might be an indicator for its very good performance. Furthermore, the results indicate that only by using relative fractions of oxalic acid $\geq$40% in the SFA, a sufficiently amorphous carbon is formed that enables high kinetic current densities.

Nevertheless, it must be pointed out that further characterization including Mößbauer spectroscopy and X-ray induced photoelectron spectroscopy will be required to enable a conclusive interpretation on the role of oxygen in the precursors to obtain highly active Fe-N-C catalysts.

On the basis of the results provided in this work, we can conclude that both a high surface area and sufficient nitrogen content are required to achieve a good performing Fe-N-C catalyst.

Catalysts **2018**, *8*, 260

3. Materials and Methods

3.1. Catalyst Preparation

In order to prepare a precursor, first sulfur (0.78 mol eq) is grounded in a mortar together with iron chloride hexahydrate (1 mol eq). Then, 1,10-phenanthroline (0.48 mol eq) is added and mixed until a homogeneous mixture is obtained. Finally, the structure-forming agents (SFAs) (6.5 mol eq) are added and the mixture is grounded until homogeneous. The preparation is given as Scheme in Figure 8, below.

As mentioned in the scheme, the color of the precursor changed upon addition of DCDA and/or oxalic acid. At the same time the initially solid precursor mixture turned into a viscous paste. While DCDA addition gave a distinct red color to this pulp (might be indicative of $[Fe(CN)_6]^{3+}$ formation), the addition of oxalic acid turned the color more orange/light brown. In addition, gas evolution took then place. If the precursor was then left in air, finally the precursor appeared like a meringue as visible in the picture. Regarding the precursor to obtain the 80% catalyst of this variation series, it contains the minimum amount of DCDA that had to be present to enable this gas evolution. Gas evolution and formation of this meringue-like morphology was not observed at higher oxalic acid contents. Except for this; however, the precursor of the 80% sample behaved rather similar during preparation compared to the other samples.

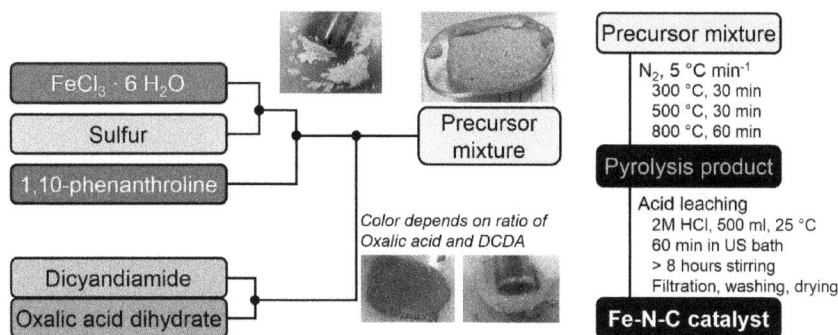

Figure 8. Scheme of the preparation procedure to obtain Fe-N-C catalysts by the use of two different type of SFA.

The as-obtained precursor mixture is filled in quartz boats and then subjected to a heat-treatment. For the pyrolysis a heating ramp of 5 °C min^{-1} is chosen. To allow the system to balance in-between there are two dwell times at 300 and 500 °C, each for 30 min, before heated to the final end temperature of 800 °C with a dwell time of 60 min.

After cooling down (<80 °C) the quartz boat with the main part of the pyrolysis product is transferred to 2 M HCl. The composition is first acid-leached for 1 h in an ultrasonic bath and then further stirred for additional 8 h (usually over-night). After filtration, washing with H_2O and drying the final mass is obtained.

Note, for determining the yields the final catalyst mass (yield after pyrolysis and acid leaching) or the mass after pyrolysis was divided by the related precursor masses of iron chloride hexahydrate, 1,10-phenanthroline and sulfur, without considering the masses of DCDA and/or oxalic acid. This was applied, as otherwise, due to the different molar masses of DCDA and oxalic acid also the absolute mass of the precursor mixture would have changed.

3.2. Electrochemical Measurements

Electrochemistry is performed in 0.1 M H_2SO_4 with a conventional 3-electrode setup. The working electrode is a glassy carbon disc (0.196 cm^2) coated with the catalyst ink. As counter and reference electrode; respectively, a glassy carbon rod and an Ag/AgCl were used. All given potentials refer to the reversible hydrogen electrode (RHE). In order to prepare the catalyst ink, 5 mg of catalyst powder were dispersed in a mixture of 25 μL Nafion (5 wt %), 83.3 μL H_2O and 142 μL ethanol, hence the Nafion to catalyst ratio is 0.25. After 30 min of sonication the ink is homogenized with an ultrasound homogenizer. Then 5 μL of the ink are dropped on the GC disc and left to dry (catalyst loading: 0.5 mg cm^{-2}).

First measurements are performed in N_2 saturated electrolyte with a conditioning of the electrode (20 scans with 300 mV s^{-1}, 0.0 to 1.2 V). Then a CV is measured with 100 mV s^{-1}, this CV is used for evaluating the capacitive current density of our catalysts. In addition, one Linear Scan Voltammogram (LSV) is measured from 1.2 to 0.0 V for later background correction of the RDE data with 10 mV s^{-1}.

After saturating the electrolyte for 15–20 min with oxygen, three LSVs are recorded in the same potential range with 0, 900 and 1500 rpm as rotation speed. After background correction, the kinetic current density j_{kin} is determined by Equation (1).

$$j_{kin}(U) = [j_{Diff,lim} \times j(U)] \times [j_{Diff,lim} - j(U)]^{-1} \tag{1}$$

In this equation, $j_{Diff,lim}$ is the as-measured diffusion limiting current density. For those catalysts that do not display a pronounced diffusion plateau the current density at 0.0 V was chosen as $j_{Diff,lim}$. The parameter $j(U)$ is the as-measured current density.

3.3. Raman Spectroscopy

Raman spectroscopy was performed with an "alpha300 R" confocal Raman microscope (WITEC, Ulm, Germany) with a laser wavelength of 532.2 nm. Spectra were recorded in a range of 0 to 4000 cm^{-1} as overlay of ten scans each with 10 s of integration time. For the measurements, catalyst ink was dropped on a silicon disc and dried. For each sample measurements are performed at two to three different locations and the average graphs are plotted in this work.

For the fitting of the 1st order range typically found for carbon blacks (800–2000 cm^{-1}) four bands were introduced with Voigt line shape.

3.4. N_2 Sorption Measurements

In order to determine the BET surface area and micropore surface area (from V-t plots) of all catalysts, N_2 sorption measurements were made with an Autosorb-3B (Quantachrome, Boynton Beach, FL, USA). Previous to the measurements, the samples (about 100 mg of the catalyst powders) were degassed at 200 °C over-night. Only for some of the catalysts a micropore surface area was obtained.

3.5. Combustion Analysis (CHN)

Combustion analysis was performed for the overall sample series with a VarioEL III instrument (Elementar Analysesysteme GmbH, Langenselbold, Germany). Each sample was measured two times, average values as well as the standard deviation were determined and are included in the plots. In all cases the sum of the three components did not count up to 100%, what shows that additional species are present within the catalysts. This might be iron, sulfur and oxygen and maybe chlorine from the acid leaching step.

Author Contributions: Conceptualization and Methodology, U.I.K.; Preparation and formal analysis, S.S.; Structural characterization, W.D.Z.W., S.S., I.M., R.W.S., N.W.; Writing: Original Draft Preparation, U.I.K., S.S., N.W.; Writing: Review & Editing, all authors; Supervision, N.W.; Project Administration, and Funding Acquisition, U.I.K.

Funding: This research was funded by the German ministry of education and research (BMBF) by grant number 03XP0092. In addition, I.M. and U.I.K. like to acknowledge further funding by the Graduate School of Excellence Energy Science and Engineering (GSC1070) funded by the German research foundation (DFG).

Conflicts of Interest: The authors declare no conflict of interest.

References

1. Jahnke, H.; Schönborn, M.; Zimmermann, G. Organic dyestuffs as catalysts for fuel cells. *Topics Curr. Chem.* **1976**, *61*, 133–182.

2. Bagotzky, V.S.; Tarasevich, M.R.; Radyushkina, K.A.; Levina, O.A.; Andrusyova, S.I. Electrocatalysis of the oxygen reduction process on metal chelates in acid electrolyte. *J.Power Sour.* **1977**, *2*, 233–240. [CrossRef]

3. Gupta, S.L.; Tryk, D.; Bae, I.; Aldred, W.; Yeager, E.B. Heat-treated polyacrylonitride-based catalysts for oxygen electroreduction. *J. Appl. Electrochem.* **1989**, *19*, 19–27. [CrossRef]

4. Lalande, G.; Côté, R.; Guay, D.; Dodelet, J.-P.; Weng, L.T.; Bertrand, P. Is nitrogen important in the formulation of Fe-based catalysts for oxygen reduction in solid polymer fuel cells? *Electrochim. Acta* **1997**, *42*, 1379–1388. [CrossRef]

5. Côté, R.; Lalande, G.; Guay, D.; Dodelet, J.-P. Influence of nitrogen-containing precursors on the electrocatalytic activtiy of heat-treated Fe(OH)$_2$ on carbon black for O$_2$ reduction. *J. Electrochem. Soc.* **1998**, *145*, 2411–2418. [CrossRef]

6. Faubert, G.; Côté, R.; Dodelet, J.-P.; Lefèvre, M.; Bertrand, P. Oxygen reduction catalysts for polymer electrolyte fuel cells from the pyrolysis of Fe^{II} acetate adsorbed on 3,4,9,10-perylenetetracarboxylic dianhydride. *Electrochim. Acta* **1999**, *44*, 2589–2603. [CrossRef]

7. He, P.; Lefèvre, M.; Faubert, G.; Dodelet, J.-P. Oxygen reduction catalysts for polymer electrolyte fuel cells from the pyrolysis of various transition metal acetates adsorbed on 3,4,9,10-perylenetetracarboxylic dianhydride. *J. New Mater. Electrochem. Syst.* **1999**, *2*, 243–251.

8. Herranz, J.; Lefèvre, M.; Larouche, N.; Stansfield, B.; Dodelet, J.-P. Step-by-Step synthesis of non-noble metal electrocatalysts for O$_2$ reduction under proton exchange membrane fuel cell conditions. *J. Phys. Chem. C* **2007**, *111*, 19033–19042. [CrossRef]

9. Jaouen, F.; Dodelet, J.-P. Turn-over frequency of O$_2$ electro-reduction for Fe/N/C and Co/N/C catalysts in PEFCs. *Electrochim. Acta* **2007**, *52*, 5975–5984. [CrossRef]

10. Herranz, J.; Lefevre, M.; Dodelet, J.-P. Metal-precursor adsorption effects on Fe-based catalysts for oxygen reduction in PEM fuel cells. *J. Electrochem. Soc.* **2009**, *156*, B593–B601. [CrossRef]

11. Jaouen, F.; Herranz, J.; Lefèvre, M.; Dodelet, J.-P.; Kramm, U.I.; Herrmann, I.; Bogdanoff, P.; Maruyama, J.; Nagaoka, T.; Garsuch, A. A Cross-laboratory experimental review of non-noble-metal catalysts for oxygen electro-reduction. *Appl. Mater. Interf.* **2009**, *1*, 1623–1639. [CrossRef] [PubMed]

12. Bogdanoff, P.; Herrmann, I.; Hilgendorff, M.; Dorbandt, I.; Fiechter, S.; Tributsch, H. Probing structural effects of pyrolysed CoTMPP-based electrocatalysts for oxygen reduction via new preparation strategies. *J. New. Mat. Electrochem. Syst.* **2004**, *7*, 85–92.

13. Herrmann, I.; Bogdanoff, P.; Schmithals, G.; Fiechter, S. Influence of the molecular and mesoscopic structure on the electrocatalytic activity of pyrolysed CoTMPP in the oxygen reduction. *ECS Trans.* **2006**, *3*, 211–219.

14. Herrmann, I.; Kramm, U.I.; Fiechter, S.; Bogdanoff, P. Oxalate supported pyrolysis of CoTMPP as electrocatalysts for the oxygen reduction reaction. *Electrochim. Acta* **2009**, *54*, 4275–4287. [CrossRef]

15. Koslowski, U.I.; Abs-Wurmbach, I.; Fiechter, S.; Bogdanoff, P. Nature of the catalytic centres of porphyrin based electrocatalysts for the ORR—A correlation of kinetic current density with the site density of Fe-N$_4$ centres. *J. Phys. Chem. C* **2008**, *112*, 15356–15366. [CrossRef]

16. Herrmann, I.; Kramm, U.I.; Radnik, J.; Bogdanoff, P.; Fiechter, S. Influence of sulphur on the pyrolysis of CoTMPP as electrocatalyst for the oxygen reduction reaction. *J. Electrochem. Soc.* **2009**, *156*, B1283–B1292. [CrossRef]

17. Kramm, U.I.; Herrmann-Geppert, I.; Fiechter, S.; Zehl, G.; Zizak, I.; Dorbandt, I.; Schmeißer, D.; Bogdanoff, P. Effect of iron-carbide formation on the number of active sites in Fe-N-C catalysts for the oxygen reduction reaction in acidic media. *J. Mater. Chem. A* **2014**, *2*, 2663–2670. [CrossRef]

18. Kiciński, W.; Dembinska, B.; Norek, M.; Budner, B.; Polański, M.; Kulesza, P.J.; Dyjak, S. Heterogeneous iron-containing carbon gels as catalysts for oxygen electroreduction: Multifunctional role of sulfur in the formation of efficient systems. *Carbon* **2017**, *116*, 655–669. [CrossRef]

19. Kramm, U.I.; Lefèvre, M.; Larouche, N.; Schmeisser, D.; Dodelet, J.-P. Correlations between mass activity and physicochemical properties of Fe/N/C catalysts for the ORR in PEM fuel cell via ^{57}Fe Mössbauer spectroscopy and other techniques. *J. Am. Chem. Soc.* **2014**, *136*, 978–985. [CrossRef] [PubMed]

20. Zhang, S.; Zhang, H.; Liu, Q.; Chen, S. Fe-N doped carbon nanotube/graphene composite: Facile synthesis and superior electrocatalytic activity. *J. Mater. Chem. A* **2013**, *1*, 3302–3308. [CrossRef]

21. Kramm, U.I.; Zana, A.; Vosch, T.; Fiechter, S.; Arenz, M.; Schmeißer, D. On the structural composition and stability of Fe-N-C catalysts prepared by an intermediate acid leaching. *J. Solid State Electrochem.* **2016**, *20*, 969–981. [CrossRef]

22. Janßen, A.; Martinaiou, I.; Wagner, S.; Weidler, N.; Shahraei, A.; Kramm, U.I. Influence of sulfur in the precursor mixture on the structural composition of Fe-N-C catalysts. *Hyperfine Interact.* **2018**, *239*, 7. [CrossRef]

23. Herrmann, I. *Innovative Elektrokatalyse: Platinfreie Kathodenkatalysatoren für Brennstoffzellen*; VDM (Verlag Dr. Müller): Saarbrücken, Germany, 2008; p. 248.

24. Kramm, U.I.; Abs-Wurmbach, I.; Herrmann-Geppert, I.; Radnik, J.; Fiechter, S.; Bogdanoff, P. Influence of the electron-density of FeN$_4$-centers towards the catalytic activity of pyrolysed FeTMPPCl-based ORR-electrocatalysts. *J. Electrochem. Soc.* **2011**, *158*, B69–B78. [CrossRef]

25. Bouwkamp-Wijnoltz, A.L.; Visscher, W.; van Veen, J.A.R.; Boellaard, E.; Kraan, A.M.v.d.; Tang, S.C. On active-site heterogeneity in pyrolysed carbon-supported iron porphyrin catalysts for the electrochemical reduction of oxygen: An in situ Mössbauer study. *J. Phys. Chem. B* **2002**, *106*, 12993–13001. [CrossRef]

26. Jaouen, F.; Dodelet, J.-P. O$_2$ Reduction mechanism on non-noble metal catalysts for PEM fuel cells. Part I: Experimental rates of O$_2$ electroreduction, H$_2$O$_2$ electroreduction, and H$_2$O$_2$ disproportionation. *J. Phys. Chem. C* **2009**, *113*, 15422–15432. [CrossRef]

27. Serov, A.; Tylus, U.; Artyushkova, K.; Mukerjee, S.; Atanassov, P. Mechanistic studies of oxygen reduction on Fe-PEI derived non-PGM electrocatalysts. *Appl. Catal. B: Environ.* **2014**, *150–151*, 179–186. [CrossRef]

28. Leonard, N.D.; Barton, S.C. Analysis of adsorption effects on a metal-nitrogen-carbon catalyst using a rotating ring-disk study. *J. Electrochem. Soc.* **2014**, *161*, H3100–H3105. [CrossRef]

29. Meng, H.; Larouche, N.; Lefèvre, M.; Jaouen, F.; Stansfield, B.; Dodelet, J.-P. Iron porphyrin-based cathode catalysts for polymer electrolyte membrane fuel cells: Effect of NH$_3$ and Ar mixtures as pyrolysis gases on catalytic activity and stability. *Electrochim. Acta* **2010**, *55*, 6450–6461. [CrossRef]

30. Li, J.; Ghoshal, S.; Liang, W.; Sougrati, M.-T.; Jaouen, F.; Halevi, B.; McKinney, S.; McCool, G.; Ma, C.; Yuan, X. Structural and mechanistic basis for the high activity of Fe-N-C catalysts toward oxygen reduction. *Energy Environ. Sci.* **2016**, *9*, 2418–2432. [CrossRef]

31. Zagal, J.H.; Páez, M.; Tanaka, A.A.; dos Santos, J.R., Jr.; Linkous, C.A. Electrocatalytic activity of metal phthalocyanines for oxygen reduction. *J. Electroanal. Chem.* **1992**, *339*, 13–30. [CrossRef]

32. Zagal, J.H.; Ponce, I.; Baez, D.; Venegas, R.; Pavez, J.; Paez, M.; Gulppi, M. A Possible interpretation for the high catalytic activity of heat-treated non-precious metal Nx/C catalysts for O$_2$ reduction in terms of their formal potentials. *Electrochem. Solid-State Lett.* **2012**, *15*, B90–B92. [CrossRef]

33. Sadezky, A.; Muckenhuber, H.; Grothe, H.; Niessner, R.; Pöschl, U. Raman microspectroscopy of soot and related carbonaceous materials: Spectral analysis and structural information. *Carbon* **2005**, *43*, 1731–1742. [CrossRef]

34. Robertson, J. Diamond-like amorphous carbon. *Mater. Sci. Eng. R* **2002**, *37*, 129–281. [CrossRef]

35. Tuinstra, F.; König, J.L. Raman spectrum of graphite. *J. Chem. Phys.* **1970**, *53*, 1126–1130. [CrossRef]

36. Larouche, N.; Stansfield, B.L. Classifying nanostructured carbons using graphitic indices derived from Raman spectra. *Carbon* **2010**, *48*, 620–629. [CrossRef]
37. Martinaiou, I.; Shahraei, A.; Grimm, F.; Zhang, H.; Wittich, C.; Klemenz, S.; Dolique, S.J.; Kleebe, H.-J.; Stark, R.W.; Kramm, U.I. Effect of metal species on the stability of Me-N-C catalysts during accelerated stress tests mimicking the start-up and shut-down conditions. *Electrochim. Acta* **2017**, *243*, 183–196. [CrossRef]
38. Jia, L.; Harbauer, K.; Bogdanoff, P.; Herrmann-Geppert, I.; Ramirez, A.; van de Krol, R.; Fiechter, S. α-Fe$_2$O$_3$ films for photoelectrochemical water oxidation—Insights of key performance parameters. *J. Mater. Chem. A* **2014**, *2*, 20196–20202. [CrossRef]

catalysts

MDPI

Review

Three-Dimensional Heteroatom-Doped Nanocarbon for Metal-Free Oxygen Reduction Electrocatalysis: A Review

Dongbin Xiong [1,2], Xifei Li [2,3,*], Linlin Fan [2] and Zhimin Bai [1,*]

[1] Beijing Key Laboratory of Materials Utilization of Nonmetallic Minerals and Solid Wastes,
    National Laboratory of Mineral Materials, School of Materials Science and Technology,
    China University of Geosciences, Beijing 100083, China; xiong_db@163.com
[2] Institute of Advanced Electrochemical Energy, Xi'an University of Technology, Xi'an 710048, China;
    tiantangzuoguai217@163.com
[3] Tianjin International Joint Research Centre of Surface Technology for Energy Storage Materials,
    College of Physics and Materials Science, Tianjin Normal University, Tianjin 300387, China
*   Correspondence: xfli@xaut.edu.cn (X.L.); zhimibai@cugb.edu.cn (Z.B.); Tel.: +86-180-0927-5876 (X.L.);
    +86-136-9111-5187 (Z.B.)

Received: 12 June 2018; Accepted: 27 June 2018; Published: 29 June 2018

Abstract: The oxygen reduction reaction (ORR) at the cathode is a fundamental process and functions a pivotal role in fuel cells and metal–air batteries. However, the electrochemical performance of these technologies has been still challenged by the high cost, scarcity, and insufficient durability of the traditional Pt-based ORR electrocatalysts. Heteroatom-doped nanocarbon electrocatalysts with competitive activity, enhanced durability, and acceptable cost, have recently attracted increasing interest and hold great promise as substitute for precious-metal catalysts (e.g., Pt and Pt-based materials). More importantly, three-dimensional (3D) porous architecture appears to be necessary for achieving high catalytic ORR activity by providing high specific surface areas with more exposed active sites and large pore volumes for efficient mass transport of reactants to the electrocatalysts. In this review, recent progress on the design, fabrication, and performance of 3D heteroatom-doped nanocarbon catalysts is summarized, aiming to elucidate the effects of heteroatom doping and 3D structure on the ORR performance of nanocarbon catalysts, thus promoting the design of highly active nanocarbon-based ORR electrocatalysts.

Keywords: oxygen reduction reaction; heteroatom doping; metal-free catalysts; nanocarbon; three-dimensional

---

1. Introduction

With the increasing energy consumption and environmental issues, there has been an urgent demand for the development of renewable and sustainable energy storage and conversion technologies [1–3]. Among the various technologies, rechargeable batteries, electrochemical capacitors, and fuel cells are recognized as the most efficient and feasible choices, particularly for electronic and transportation applications [4–6]. Compared with other batteries, such as nickel–metal hydride, lead–acid, and lithium–ion batteries, fuel cells and metal–air batteries have the higher theoretical energy density and higher efficiency with low emission of pollutants due to the direct conversion of chemical energy to electrical energy through chemical reaction [7–9]. For example, the typical fuel cell with H_2 as the fuel, O_2 as the oxidizing agent, and water as the end product with zero emissions and high efficiency, has drawn much attention in terms of fundamentals and applications [10,11]. However, the practical application of such fuel cells is largely restricted by the high activation barriers of electrochemical process, especially for the sluggish oxygen reduction reaction (ORR) kinetics at the cathode [12,13]. The cathodic ORR is much slower than the anodic oxidation reaction and therefore greatly limits the output performance of these promising technologies [14]. Fortunately, electrocatalysts can play an important role to lower the activation energy barriers of the sluggish ORR. Nowadays, platinum (Pt)-based materials are considered as the most efficient ORR electrocatalysts for their excellent catalytic performance with relatively high current density and low overpotential [15–18]. However, the large-scale application of Pt-based electrocatalysts is severely hindered by the drawbacks of high cost, fuel crossover effect, instability due to CO deactivation, and Pt dissolution. Thus, it is highly urgent to develop advanced non-precious metal or even metal-free catalysts to substitute Pt-based catalysts with high catalytic activity, enhanced durability, and satisfactory cost in the long term.

Recently, much effort has been devoted to developing a series of non-precious metal ORR catalysts, such as transition metal oxides, chalcogenides, and transition metal–nitrogen–carbon (M–N/C, M = Fe, Co, Ni, Mn, etc.) materials [19–22]. However, these non-precious metal-based catalysts have some disadvantages, such as low catalytic activity compared to Pt/C and poor durability caused by the metal leaching during usage. On the other hand, metal-free carbon-based catalysts have achieved great development due to their outstanding catalytic ORR performance, high chemical stability, relatively low cost, and environmental friendliness during the past decades [23–26]. Since the first accomplishment of vertically aligned nitrogen-doped carbon nanotubes (VA-NCNTs) as metal-free ORR catalysts reported by Dai's group [27], various heteroatom-doped nanocarbon materials—such as graphene, carbon nanotubes, porous carbon, and their hybrids—have been extensively exploited as substitutes for Pt-based catalysts [28–31]. Most of the heteroatom-doped nanocarbon materials exhibit comparable or even better catalytic activity to Pt-based catalysts due to the doping-induced charge redistribution, which can facilitate the chemisorption of O_2 and electron transfer for the ORR process [32]. More importantly, the metal-free catalysts of doped nanocarbon materials surprisingly exhibit efficient catalytic activity and long-term stability without CO deactivation and fuel crossover effects. Therefore, developing advanced and low-cost heteroatom-doped metal-free nanocarbon materials with superior ORR catalytic activities is highly desired.

Besides the heteroatom-doping (e.g., N, S, P, B, etc.), structure engineering of nanocarbon is also a crucial strategy to determine the catalytic ORR performance, considering that the high surface area and suitable pore structure of a superior electrode configuration are the prerequisites in ensuring accessible active sites and efficient transport of electrons and ions [33–35]. For instance, engineering 3D porous structure of nanocarbon materials is a feasible approach to improve the ORR performance by providing better electrolyte permeability, electron-transfer path, and mass transport/diffusion. In recent years, for example, the fast development of nanocarbon materials (e.g., graphene) enables them to play an important role in the improvement of metal-free ORR catalysts performance [36], but 2D graphene sheets are readily to restack, which would block the active sites of catalysts and increase the resistance for mass transfer, leading to poor catalytic properties. Hence, 3D structured nanocarbon materials (cross-linked CNTs, 3D graphene, porous carbon, etc.) with chemical doping assuredly hold great prospect as efficient ORR catalysts to replace noble-metal-based catalysts [37–39].

Several reviews have been published in the past few years regarding the design of heteroatom-doped carbon nanomaterials and their applications in ORR [31,40–42]. However, special attention has not been paid to these heteroatom-doped carbon nanomaterials from the standpoint of structural effects on ORR performance. As shown in Figure 1, in this review, the current widely accepted mechanisms for ORR is briefly introduced, the recent rational design of different 3D doped-nanocarbon materials, such as 3D CNTs nanostructures, 3D graphene, porous carbon, and their hybrids is discussed as well, and then we focus on the recent achievements of 3D doped-nanocarbon materials and their enhanced ORR performance. The structure-dependent ORR performance of 3D doped-nanocarbon are well discussed, which will be beneficial to future development of non-precious metal electrocatalysts with both exceptional activity and durability in the near future.

Figure 1. The relationships between ORR mechanism, catalyst design, and ORR performance.

2. The Mechanisms for ORR

The ORR is a fundamental reaction and a major limiting factor of performance for metal–air batteries and fuel cells [7,11]. In general, ORR, which involves multiple electrochemical reactions, can proceed either a four-electron path to directly produce H_2O (in acidic medium, $4H^+ + O_2 + 4e^- \rightarrow 2H_2O$) and OH^- (in alkaline medium, $2H_2O + O_2 + 4e^- \rightarrow 4HO^-$) as the final products or a less efficient two-step two-electron pathway with the formation of H_2O_2 (in acidic medium, $O_2 + 2H^+ + 2e^- \rightarrow H_2O_2$, $H_2O_2 + 2H^+ + 2e^- \rightarrow 2H_2O$) or HO_2^- (in alkaline media, $O_2 + H_2O + 2e^- \rightarrow HO_2^- + OH^-$, $HO_2^- + H_2O + 2e^- \rightarrow 3OH^-$) as the intermediate specie [41,43]. As schematically shown in Figure 2, in a typical proton exchange membrane fuel cell (PEMFC), hydrogen and oxygen/air continuously enter the anode and the cathode, respectively [44]. The fuel molecules (H_2) are oxidized at the anode ($H_2 \rightarrow 2H^+ + 2e^-$). In this process, the electrons flow out of the anode to provide electrical power, while protons diffuse across the electrolyte membrane towards the cathode and react with adsorbed oxygen to produce water (ORR, $4H^+ + O_2 + 4e^- \rightarrow 2H_2O$). The ORR occurs by efficient four-electron reduction.

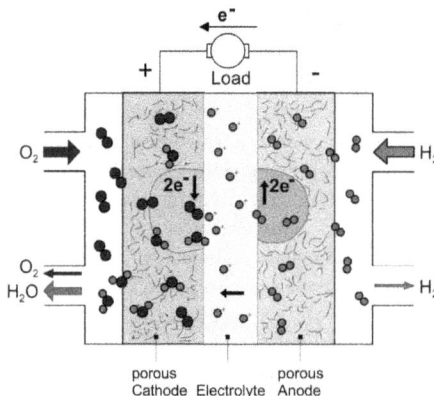

Figure 2. Schematic illustration of a hydrogen/oxygen fuel cell and its reactions based on the PEMFC. Reproduced from [44], Copyright 2001, Wiley.

Generally, the four-electron reduction mechanism is favorable for ORR owing to its efficiency advantage and the avoiding of hydrogen peroxide intermediate species that can damage the membrane and ionomer [45]. However, the ORR at the cathode suffers from complicated electron transfers, as illustrated in Figure 3, in a full reduction pathway (Figure 3a), the O–O bond of adsorbed O_2 breaks into two O* intermediates which could be reduced to OH^- and H_2O as the final products in the alkaline and acidic conditions, respectively. While for partial reduction to take place, as shown in Figure 3b, O_2 is first adsorbed on to the catalyst surface, then the adsorbed O_2 couples two protons to form HOOH* intermediates before the O-O bond is cleaved, leading to a high yield of H_2O_2 or HO_2^- via a two-electron pathway, which should be avoided during the ORR process [46,47].

Under acidic conditions, the ORR is inherently several orders of magnitude slower than the HOR, and catalysts are required to lower activation barriers of the sluggish ORR. Only a few types of materials, to date, have been found to provide suitable activity and stability towards the ORR in acidic media, such as heteroatom-doped nanocarbon materials and the transition metal-based NP functionalized carbon nanomaterials [21,48]. While the ORR kinetics are more favorable in alkaline mediums, providing an opportunity to use non-precious metal catalysts, such as metal-oxides and doped nanocarbon materials [45].

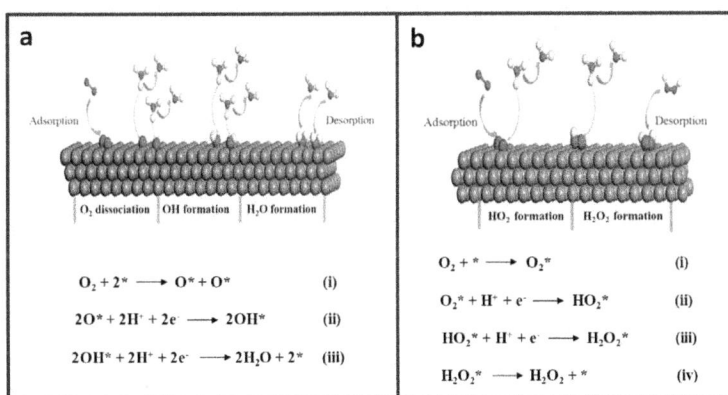

Figure 3. Proposed mechanism schematics of (**a**) full reduction and (**b**) partial reduction of oxygen. Reproduced from [47], Copyright 2016, Elsevier.

3. 3D Heteroatom-Doped Nanocarbon Electrocatalysts for ORR

Carbon materials, especially nanostructured carbons, has been widely used as electrode materials for energy storage and conversion devices originating from their general advantages, such as good conductivity for rapid electron transfer, super large active surface area for ion adsorption, and superior chemical stability for resistance of acid and alkaline corrosion [33,49–51]. Additionally, the tailorable surface chemistry, abundant structural variety, and low cost of nanocarbons together boost their harvesting as active catalysts for ORR in metal–air batteries and fuel cells [22,34]. Nevertheless, structure optimization of nanocarbon catalysts is one of the main strategies to sufficiently expose and/or activate the catalytic sites. It is noted that low-dimensional nanocarbons—including 0D fullerene, 1D carbon nanotubes, and 2D grapheme—usually have low utilization efficiency due to the embedded active sites on the limited surface, which is unfavorable for both mass transport and electron transfer during ORR process [33,52]. However, when fabricated to be a continuous 3D porous structure, the 3D nanocarbon can provide a high surface area with abundant exposed active sites, contributing to good electrocatalytic performance. More importantly, the 3D nanostructures play a critical role in greatly accommodating discharge products and providing channels for ion transfer and oxygen diffusion, further accelerating reaction kinetics [53].

Over the past decade, carbon-based metal-free electrocatalysts doped with heteroatom, such as N, S, B, P, and their mixtures, have emerged as front runners to replace Pt and other noble metals for highly efficient ORR [22,54]. Both experiments and theoretical calculations show that the doping of heteroatoms in the sp^2 lattice of graphitic carbon can alter the electronic arrangement of the carbon-based material and tailor their electron donor properties, as a result, breaking the electroneutrality of sp^2 carbon to create charged sites favorable for O_2 adsorption and enhancing effective utilization of carbon π electrons for O_2 reduction, thus leading to improved ORR electrocatalytic activity [24,29,43]. Despite the great improvements that have been achieved, most of the carbon-based metal-free catalysts still show inferior intrinsic ORR activities compared with Pt-based ones though with better stability and lower costs. Increasing evidences profile that efficient ORR nanocarbon catalysts should be favored to have abundant accessible active sites for implementing reaction, good conductivity for charge transfer, and suitable porous structure for mass transport. The presence of adequate reactive sites in combination with novel structural design makes them attractive metal-free catalysts. To achieve the aforementioned merits, various heteroatom-doped 3D nanocarbon materials, such as N-doped CNT aerogels [37], nanoporous N-doped graphene [55], N, P-doped porous carbon, and doped graphene/CNT hybrid materials have been developed as high-performance ORR catalysts (Figure 4) [56,57].

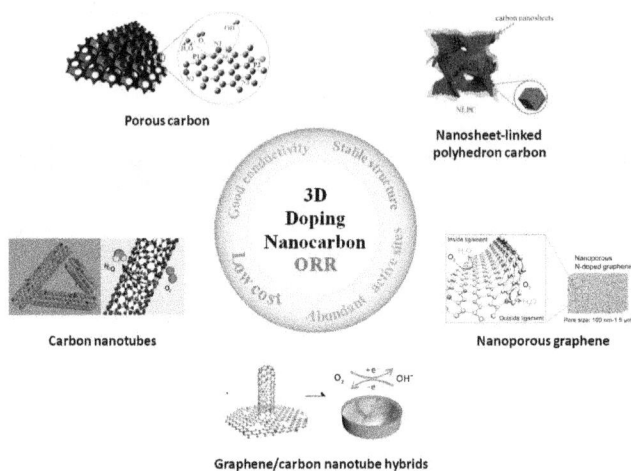

Figure 4. Advantages of metal-free 3D doped-nanocarbon materials and their applications as high-efficiency ORR electrocatalysts.

3.1. Heteroatom-Doped 3D CNTs for ORR

As typical 1D sp^2-hybridized carbon nanomaterials, CNTs can be considered as 2D graphene sheets rolled up into nanoscale tubes [8,58]. Because of their unique structural characteristics and outstanding physicochemical properties—such as large surface area, high mechanical property, good electrical conductivity, and excellent chemical stability—CNTs have stimulated continuous interest in the field of nanotechnology, especially in environmental and energy areas, in recent decades [52,59]. When heteroatoms are appropriately doped into the carbon matrix of CNTs, enhanced ORR performance can be achieved [8]. Doped-CNTs function in enhancing electrical conductivity, oxygen mass transfer, corrosion resistance, and water removal of catalysts, leading to improved catalytic activity and durability [60,61]. More importantly, putting the tiny cylindrical nanotubes together into an integral 3D framework with interconnection and rational distribution and controlling the porous structure can provide abundant exposed active sites and stable electron/mass transport skeleton, and hence ORR activity [37].

3.1.1. Single Heteroatom-Doped 3D CNTs

In 2009, Dai et al. synthesized N-doped VA-CNTs through pyrolyzing iron(II) phthalocyanine (FePc) in the presence of additional NH$_3$ gas (Figure 5a) [27]. The proposed 3D N-doped VA-CNTs were discovered to be superior to commercially Pt/C catalysts for the electrocatalysis of the ORR with a much higher catalytic activity, lower overpotential, smaller crossover sensitivity, and better long-term stability in alkaline electrolytes (Figure 5b). The catalytic mechanism of nitrogen-doped VA-CNTs for the ORR was investigated using quantum mechanical calculations based on the B3LYP hybrid density functional theory (DFT), results suggest that the introduce of nitrogen dopants changes the charge density of carbon atoms (Figure 5c). Based on the theoretical and experimental results, Dai and co-workers presented that the N-induced charge delocalization could change the chemisorption mode of O_2 from the end-on adsorption (Pauling model) at the non-doped CNT surface (top, Figure 5d) to a side-on adsorption (Yeager model) onto the N-doped CNT surface (bottom, Figure 5d). Interestingly, the parallel diatomic adsorption could effectively weaken the O–O bonding to facilitate ORR at the N-doped VA-CNTs electrodes. Besides the high surface area, good electrical and mechanical properties of 3D vertically aligned CNTs provide additional advantages for the nanotube electrode in fuel cells. It should be noted that a certain amount of residual Fe catalyst herein may exist in the N-doped VA-CNTs. The metal contaminants are believed to have certain effects on ORR performance. Therefore, it is urgently needed to develop a metal-free synthesis route to produce N-doped CNTs in order to confirm the actual electroactivity of N-doped CNTs for ORR without the disturbance of metal residue.

To reveal the intrinsic catalytical mechanism of N-doped CNTs, several technologies including metal-free catalysis growth and detonation-assisted chemical vapor deposition (CVD) have been developed [62,63]. By using an erasable-promoter-assisted hydrothermal reaction coupling with pyrolysis, Zhang and co-workers synthesized clean, high-specific surface area (~869 $m^2 \cdot g^{-1}$), and highly conductive (~10.9 $S \cdot m^{-1}$) N-doped CNT aerogels [37]. During the synthetic process, N doping was realized by pyrolysis of aerogels with pyrrole molecules as 'built-in' nitrogen sources (post-treatment), importantly, the electrical conductivity of CNT aerogels was restored and the π-π stacking between CNTs was enhanced through thermal treatment, which can repair the π-conjugated skeleton of CNTs, thereby maintaining the 3D gel network (Figure 5e,f). Owing to the unique structure with 3D frameworks constructed by randomly entangled CNTs, the obtained N-CNT aerogels (Figure 5g) exhibited superior activity and high stability for ORR catalysis in alkaline conditions. Moreover, the crossover test (Figure 5h) demonstrated that the N-CNT aerogels electrode efficiently inhibit the crossover effect of methanol. Zhu et al. prepared porous N-doped CNTs by activation and pyrolysis of polypyrrole nanotubes [64], which result in a hierarchical porous structure with in situ N doping. This material exhibited excellent catalytic activity with a four-electron pathway for ORR in alkaline condition. They believe that the favorable ORR activities of N-doped CNTs were attributed to its porous tube structure, high surface area, and uniform N distribution. Template method is considered as the most effective strategy for the preparation of ordered porous nanostructures. By using an anodic alumina oxide template, Yang et al. fabricated a N-doped macroporous carbonaceous nanotube array for ORR catalyst (Figure 5i,j) [65]. The nanotubes with macroporous feature can facilitate mass transfer within the electrodes. The doping of electron-rich N which can activate the p electrons of sp^2 carbon by conjugating with the lone-pair electrons from nitrogen dopants can provide more active sites to adsorb O_2. These merits together contribute to the high performance for ORR in terms of onset potential, half-wave potential values, reaction current, and durability.

There are several different nitrogen configurations including pyrrolic N, pyridinic N, and graphitic N in carbon matrix [54]. Pyrrolic N and pyridinic N are located at the edges and bonded to two carbon atoms, while graphitic N is incorporated in the core structure of the carbon materials by replacing the sp^2-hybridized carbon atom. Generally, different configurations with different N type would affect the electronic structure of neighboring carbon atoms, leading to different catalytic properties. Up to date, numerous N-doped CNT materials have been prepared via different synthetic strategies for ORR electrocatalyst [37,64,66,67]. However, it is still unclear whether the pyridinic or graphitic N is mainly responsible for the active sites for the ORR. Recent theoretical work and experiments suggest that pyridinic N improves the onset potential, whereas the graphitic N determines the limiting current density for the ORR [24]. Pyridinic N can provide one p electron to the aromatic π system with a lone electron pair in the plane of the carbon matrix to enhance the electron-donating capability of the catalyst. Therefore, pyridinic N can weaken the O–O bond via the bonding of O with N and/or the adjacent C atom to facilitate the reduction of O_2.

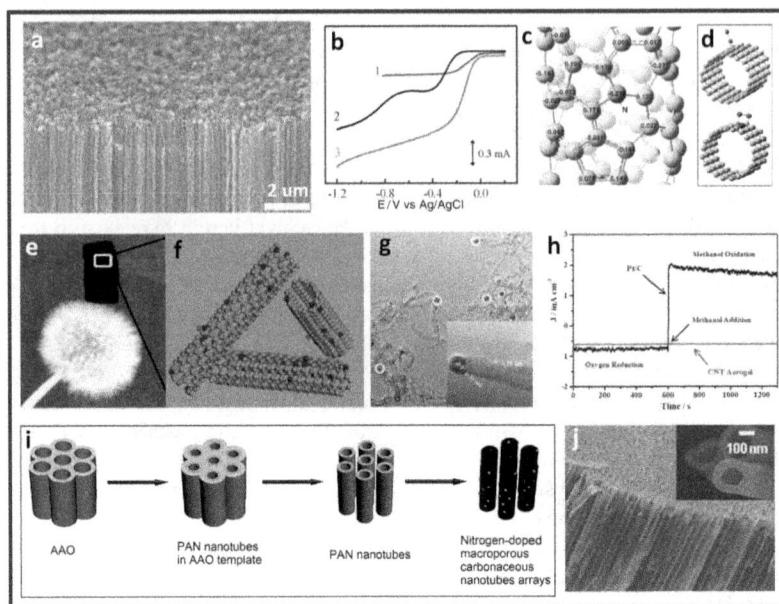

Figure 5. (**a**) Scanning electron microscopy (SEM) image of N-doped VA-CNTs; (**b**) ORR polarization curves of Pt/C (curve 1), VA-CNTs (curve 2), and N-doped VA-CNTs (curve 3); (**c**) Calculated charge density distribution for the N-doped CNTs; (**d**) Schematic representations of possible adsorption modes of an oxygen molecule at the CNTs (top) and N-doped CNTs (bottom). Reproduced from [27], Copyright 2009, American Association for the Advancement of Science. (**e**) Digital photograph; (**f**) model schematic diagram; and (**g**) Transmission electron microscopy (TEM) image of N-doped CNT aerogels; (**h**) Methanol crossover effect measurements of N-doped CNT and Pt/C catalyst at −0.4 V. Reproduced from [37], Copyright 2015, Wiley. (**i**) The fabrication process and (**j**) SEM image of N-doped macroporous carbonaceous nanotubes arrays. Reproduced from [65], Copyright 2014, The Royal Society of Chemistry.

Similar to N-doped CNTs, other heteroatom (e.g., phosphorus and boron)-doped CNTs also demonstrated improved electrocatalytic activity toward ORR, compared to its undoped counterpart [60,68]. Boron and phosphorus have a similar effect on the ORR activity as N, as they both disrupt the charge uniformity and change the charge density of the carbon network. However, the mechanisms of B-doped CNTs and N-doped CNTs are different. There are multiple active B moieties in B-doped CNTs, including BC_3, B_4C, BC_2O, and BCO_2. Owing to the lower electronegativity of B (2.04) than C (2.55), positively polarized B dopant in B-doped CNTs on one hand adsorbs O_2 on the other hand acts as a bridge to transport electrons from graphitic carbon p electrons to O_2, which can also improve the ORR activity. Although the ORR performance of B-doped CNT materials is not competitive to that of commercial Pt/C catalyst, the proportional relationship between the boron content and ORR performance suggests the great potential of B-CNTs for further improvement. P doped-CNTs are another interesting type of metal-free catalysts for improving ORR because P has the same number of valence electrons as N and often shows similar chemical properties. For example, p-doped MCNTs were reported to exhibit much higher ORR activity than commercial Pt/C in alkaline fuel cells [69]. Very recently, Zhang and co-authors also confirmed that porous P-doped CNTs exhibit better ORR catalytic activity than that of undoped-CNTs [68].

3.1.2. Multiple Heteroatom-Co-Doped 3D CNTs

Addition to the single heteroatom-doped CNTs, co-doped CNTs with different heteroatom were investigated to show much better electrocatalytic ORR performance duo to the synergistic effect between different heteroatoms. Vertically aligned MWCNT arrays co-doped with P atoms and N atoms were first synthesized by an injection-assisted CVD method [61]. Because of the synergetic effect arising from co-doping CNTs with both P and N, the obtained P, N co-doped MWCNT arrays significantly show outstanding electrocatalytic activity toward ORR comparable to the commercial Pt/C electrode and significantly better than that of CNTs doped by P or N only. Subsequently, another N, P-dual-doped CNT array was synthesized by a novel one-pot method with an aminophosphonic acid resin as the N, P, and C sources [70]. Compared with traditional bamboo-shaped N-CNTs and N, P-dual-doped CNTs, the as-obtained N, P-CNTs with unique architecture of which the large hollow channels and open ends provide abundant catalytic active sites in inner walls, being accessible to oxygen molecules exhibited comparable activity and much better CO and methanol tolerance towards ORR to Pt/C catalysts.

3.2. Heteroatom-Doped 3D Graphene for ORR

Graphene, which is composed of one monolayer of carbon atoms with a honeycomb structure, has been widely explored in different fields for its fascinating physical and chemical properties [71,72]. However, graphene is constructed of sp^2-bonded carbon atoms via hybridization of s, p_x, and p_y atomic orbitals, resulting in a zero-band gap semiconductor with the conduction and valence bands. The lack of intrinsic bandgap muchly limits the applications of pristine graphene in the areas of energy storage, electrocatalysis, and nanoelectronics [73]. Fortunately, chemical doping with foreign atoms has been demonstrated to be an effective method to tailor the electronic and electrochemical properties of graphene by changing the electronic density within the graphene sheet, thus opening the bandgap in graphene, and extending its applications [74,75]. For example, the increased active sites and enhanced catalytic activity of graphene towards ORR have been achieved by doping with foreign non-metallic atoms (e.g., N, B, P, or S) [36,76–78].

Other than heteroatom doping, morphology control, and structural design, which relate to the surface area, pore structure and electron donating/withdrawing capability, is perhaps the most effect way to enhance the ORR activity of graphene materials [29,33]. Especially, heteroatom doping, in company with 3D structure design, has been a popular and widely accepted strategy to develop graphene-based ORR electrocatalyst [79–82].

3.2.1. Single Heteroatom-Doped 3D Graphene

After their first discovery of metal-free VA-NCNTs as high-performance ORR electrocatalysts, Dai et al. used a modified CVD method to prepare N-doped graphene films on Ni-coating SiO_2/Si substrate [83]. They demonstrated that the N-graphene can act as a metal-free catalyst with a much better catalytic activity, tolerance to crossover effect, and long-term stability than Pt catalyst for ORR via a four-electron pathway in alkaline fuel cells. Subsequently, N-doped graphene was synthesized via catalyst-free thermal annealing graphite oxide and nitrogen source [76]. The synthesized N-doped graphene materials, which completely avoid the contamination of metal catalysts, have high nitrogen content and exhibit excellent catalytic activities toward ORR in alkaline electrolytes.

3D graphene structures can effectively restrain the restacking between graphene sheets, and therefore expose more active sites, heteroatom-doped 3D graphene materials are expected to show much better electrocatalytic performance for ORR than the 2D ones [84,85]. By using a hydrothermal self-assembly approach followed by high-temperature treatment, as shown in Figure 6a, Qiu et al. fabricated 3D porous N-doped graphene aerogels (NPGAs, Figure 6b), which exhibited good electrocatalytic activity and long-term stability in a Li–O_2 cell system [86]. The large void volume, interconnected porous channels, multidimensional electron transport pathways, less stacking of graphene sheets and the sufficient exposure of active sites originated from the doped N atoms within graphene sheets collectively contribute to the outstanding electrochemical performances of the as-made NPGA. Yi et al. prepared highly conductive and ultralight nitrogen-doped graphene nanoribbons aerogel (N-GNRs-A) by using a facile hydrothermal method (Figure 6c) [81]. Due to the synergistic effect of the nanoporous structure, high surface area, good conductivity and the N-doped structural integrity of the GNRs, the proposed aerogel as a novel ORR catalyst show comparable catalytic activity (Figure 6d), superb methanol tolerance (Figure 6e), and better stability (Figure 6f) than commercial Pt/C catalysts in both alkaline and acidic solutions. After that, various 3D N-doped graphene materials prepared through different methods have been developed as high-performance ORR catalysts [38,86–88]. S-doped 3D porous RGO hollow nanospheres framework (S-PGHS), prepared with GO and dibenzyl disulfide as precursors, exhibited superior electrocatalytic activity comparable with that of commercial Pt/C (40%), and much better durability and methanol tolerance [89]. 3D sulfur-doped graphene networks S-GFs were also prepared by using an ion-exchange/activation combination method, which showed outstanding ORR catalytic performance [90]. Recently, 3D P-doped graphene (3DPG) fabricated by CVD method with nickel foam as template and triphenylphosphine (TPP) as C and P sources was proposed as ORR catalyst, which exhibited better catalytic activity, long-term stability, and methanol tolerance than pristine 3D graphene and commercial Pt/C [91].

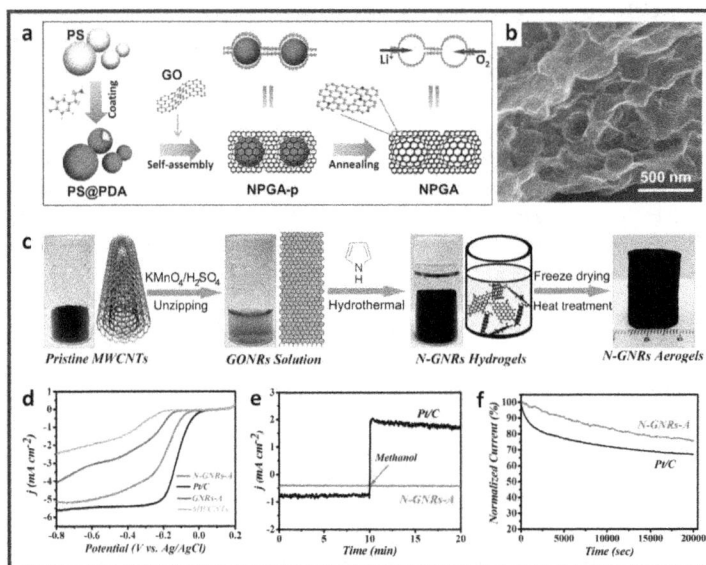

Figure 6. (**a**) Schematic process for synthesis of 3D NPGAs; (**b**) SEM image of 3D NPGAs. Reproduced from [86], Copyright 2015, Wiley. (**c**) Illustration of the synthetic route for N-GNRs-A; (**d**) Linear sweep voltammetry (LSV) curves of pristine MWCNTs, GNRs-A, N-GNRs-A, and Pt/C in an O_2-saturated 0.1 M KOH solution at a scan rate of 10 mV s^{-1} and a rotation speed of 1600 rpm; (**e**) Methanol crossover effect on N-GNRs-A and Pt/C upon addition of 3 M methanol after about 10 min in an O_2-saturated 0.1 M KOH solution at −0.4 V; (**f**) Current–time chronoamperometric response of N-GNRs-A and Pt/C catalysts at −0.4 V in O_2 saturated 0.1 M KOH aqueous solution at a rotation rate of 1600 rpm. Reproduced from [81], Copyright 2014, Wiley.

3.2.2. Multiple Heteroatom-Co-Doped 3D Graphene

Co-doped 3D graphene has been expected to possess better electrocatalytic performance compared to single doped 3D graphene due to the synergistic effect between different heteroatoms. Qiao and co-authors prepared N and S dual-doped mesoporous graphene (N-S-G) for the first time as a metal-free catalyst for ORR [92]. The obtained N-S-G showed outstanding ORR performance, which is comparable to commercial Pt/C and prominently better than that of graphene catalysts doped solely with S or with N. Furthermore, by using further DFT calculations, they elucidated that the synergistic performance improvement results from the redistribution of spin and charge densities caused by the co-doping of S and N, which leads to abundant carbon atom active sites. Soon after that, 3D N, S co-doped graphene frameworks (N/S-GFs) and 3D B, N co-doped graphene foams (BN-GFs) were prepared by one-pot hydrothermal approach and modified CVD method, respectively [79,80]. Both of which manifested superior ORR catalytic behavior with mainly four-electron transfer pathway in alkaline condition. 3D N, B-doped graphene aerogels (N, B-GAs) prepared via a two-step method involves a hydrothermal reaction and a pyrolysis procedure were also demonstrated to exhibit an outstanding catalytic activity for the ORR [93]. Among the dual-doped 3D graphene materials for ORR catalysts, N, S co-doped 3D graphene is the most popular one and has been widely prepared by various methods including biomass pyrolysis [94], hydrothermal method [95,96], hydrothermal reaction-pyrolysis two-step method [97], and soft template-assisted method [98].

In addition to dual doping, co-doping 3D graphene catalysts with more than two different heteroatoms is also an effective strategy to enhance the ORR performance, as exemplified by N-P-O co-doped 3D graphene [99]. The N-P-O co-doped free-standing 3D hierarchical porous graphene (3D-HPG) was fabricated through a one-pot gas-exfoliation assisted 'cutting-thin' technique from solid carbon sources (Figure 7a). The produced graphene exhibited continuously 3D hierarchical porous structure with heteroatoms of N, P, and O simultaneously doped into the carbon frameworks, which can effectively modulate the electronic characteristics and surface chemical feature (Figure 7b). The resultant N-P-O co-doped 3D-HPG catalysts exhibited excellent ORR activity. As shown in Figure 7c, in a 0.1 M KOH electrolyte, the ORR polarization curves reach well-defined diffusion limiting currents, and the Koutecky–Levich (K–L) plots suggest the good linearity at varied potentials with the electron transfer number calculated to be 3.83, which is comparable to the commercial Pt/C catalyst. Significantly, the durability of N-P-O co-doped 3D-HPG is much better than that of commercial Pt/C catalyst (Figure 7d).

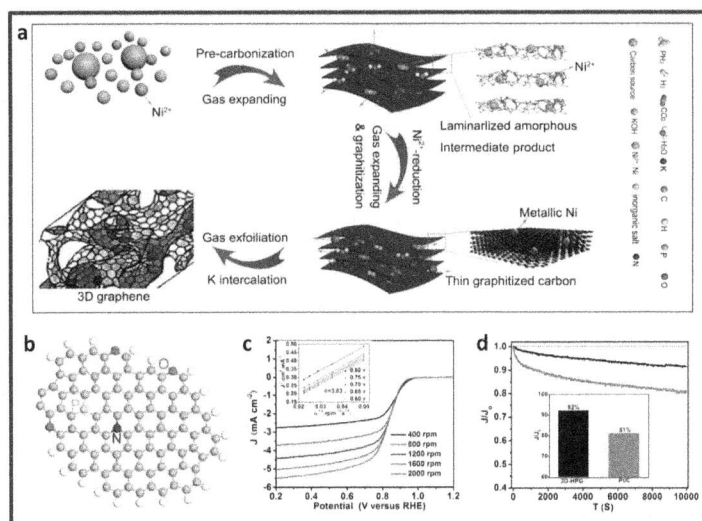

Figure 7. (a) Schematic process for synthesis of N-P-O co-doped 3D-HPG; (b) Schematic model of 3D-HPG; (c) ORR polarization curves for 3D-HPG at different rotating rates in O_2-Saturated 0.1 M KOH solution at scanning rates of 5 mV·s^{-1}, inset: K–L plots; (d) The current vs. time (i-t) chronoamperometric responses of 3D-HPG and Pt/C in O_2-saturated 0.1 M KOH at a constant potential at 0.65 V (versus RHE) and a rotation rate of 1600 rpm, inset: the ratio of the J/J0. Reproduced from [99], Copyright 2016, Elsevier.

3.3. Heteroatom-Doped 3D Porous Carbon for ORR

Engineering a 3D porous structure—which can provide good electrolyte permeability, mass transport, and an electron-transfer path—is considered the most promising approach to enhance the ORR performance of carbon-based non-precious metal ORR electrocatalysts. Up to now, various 3D porous carbon materials have been exploited as promising and efficient catalysts for their outstanding virtues such as low cost, high conductivity, high surface area with abundant porosity, designable carbon framework, as well as high chemical and mechanical stability [100–103]. Similar to 3D CNTs and 3D graphene, 3D porous carbon can also be doped with heteroatoms for ORR electrocatalysts [104].

3.3.1. Single Heteroatom-Doped 3D Porous Carbon

To develop N-doped carbon materials without any metal components, Feng and co-workers fabricated N-doped 3D ordered mesoporous carbons (N-OMCs) via a metal-free nanocasting technology by using SBA-15 as the template and N,N'-bis(2,6-diisopropyphenyl)-3,4,9,10-perylenetetracarboxylic diimide (PDI) as the precursors [100]. Owing to its high surface area and a graphitic framework with an appropriate nitrogen content, the obtained N-OMCs exhibited outstanding ORR performance with high catalytic activity, efficient resistance to crossover effects and excellent long-term stability. The ORR performance was superior to that observed for the commercial Pt/C catalyst, suggesting the superior ORR activity of N-OMCs. Soon after that, another nitrogen-doped carbon nanocages (NCNCs) with high nitrogen content and specific surface area were prepared by using in situ generated MgO as a template and pyridine as the source of both carbon an nitrogen [105]. The resulting NCNCs exhibited superior ORR performance with outstanding stability towards methanol crossover and CO poisoning in alkaline solution. Importantly, without the interference of metal impurities, this study clarifies that it is the N-doped carbon species rather than the metal-related active sites are responsible for the ORR activity of the NCNCs. By using a green biomass source method with fermented rice as starting materials, Qu and co-authors fabricated a porous N-doped carbon spheres (N-CSs) with high specific surface areas (2105.9 $m^2 \cdot g^{-1}$) and high porosity (1.14 $cm^3 \cdot g^{-1}$) [106]. When tested as ORR catalyst for fuel cells, the proposed N-CSs exhibit excellent catalytic activity with long-term stability and good resistance to crossover effects and CO poisoning superior to that of the commercially available catalyst Pt/C. Later, various N-doped 3D porous carbon derived from different biomass—such as malachium aquaticum [107], shrimp-shell [108], and porous cellulose [109]—have been demonstrated to exhibit excellent ORR performance.

In general, two crucial factors—including the doped element content/type and specific surface area/porous structure—govern the performance of the carbon-based ORR catalysts. Recently, to simultaneously optimize both surface functionalities and porous structures of the metal-free catalysts, Feng et al. developed N-doped carbon materials by using templating synthesis with nitrogen-enriched aromatic polymers as precursors and subsequent NH_3 activation (Figure 8a) [102]. The as-fabricated nitrogen-doped mesoporous carbon exhibit the outstanding ORR activity in alkaline media with half-wave potential of 0.85 V versus reversible hydrogen electrode with a loading of 0.1 $mg \cdot cm^{-2}$. More importantly, the H_2O_2 yield measured with meso/micro-P_oPD remained below 5% at all potentials, corresponding to a favorable high electron-transfer number of 3.97 (Figure 8b). Superior electrochemical durability was also observed for the meso/micro-P_oPD to the Pt/C catalyst under the same condition (Figure 8c). It should be noted that most N atoms are buried within the N-doped carbons, and these hidden N atoms are inaccessible to the reactants during ORR process. Recently, Wang and co-workers developed a 3D N-doped hierarchical porous carbon monolith (NHPCM) composed of branched mesoporous rods via an in situ source-template-interface reaction route by using furfuryl alcohol as the carbon source (Figure 8d) [103]. Owing to the increased exposure and achievability of the catalytic sites originates from the favorably activated O_2 at the edged groups, the resulting hybridized carbon nanowires possess an outstanding electrocatalytic ORR activity with a four-electron dominant reaction pathway. Interestingly, in spite of low N content, the NHPCM with 1.1 at% N shows not only superior ORR activity, but also improved MeOH crossover and high durability compared to commercial Pt/C (Figure 8e,f). This phenomenon is ascribed to the high ratio of graphitic to pyridinic N and the unique 3D macroporous scaffold with interconnected mesoporous rods, as well as the easily reachable catalytic sites. Uninterruptedly, various N-doped 3D porous carbon materials were exploited for ORR catalysts [110–115].

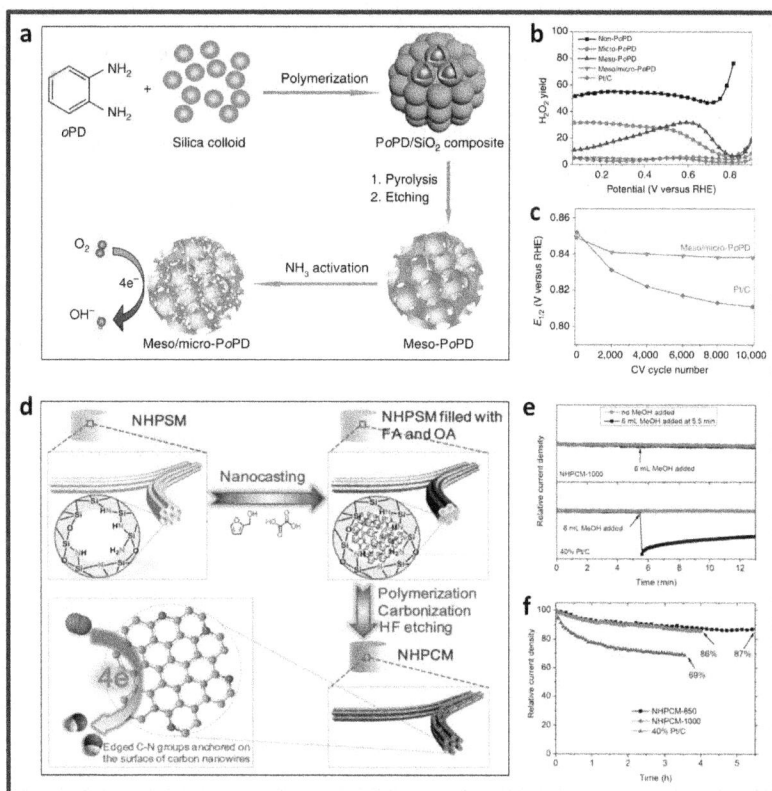

Figure 8. (**a**) Schematic illustration of the synthesis of meso/micro-N-doped carbon (P_oPD) electrocatalyst; (**b**) H_2O_2 yields plots of meso/micro-P_oPD, reference materials, and Pt/C catalyst; (**c**) Half-wave potential of meso/micro-P_oPD and Pt/C with the same loading of 0.1 mg cm^{-2} as a function of the number of potential cycles in O_2-saturated electrolyte. Reproduced from [102], Copyright 2014, Macmillan Publishers Limited; (**d**) Schematic illustration of the synthesis of 3D N-doped hierarchical porous carbon monolith (NHPCM) electrocatalyst; (**e**) Current–time chronoamperometric response of NHPCM-1000 and Pt/C with or without the addition of 6 mL MeOH into the electrochemical cell containing 100 mL electrolyte at 0.6 V (vs. RHE) with a rotating rate of 1600 rpm; (**f**) Current–time chronoamperometric response of NHPCM-850, NHPCM-1000, and Pt/C over 3.5 h at 0.6 V (vs. RHE) in O_2-saturated 0.1 M KOH solution at 1600 rpm. Reproduced from [103], Copyright 2015, Wiley.

3.3.2. Multiple Heteroatom-Co-Doped 3D Porous Carbon

As described above, the synergistic effect arising from the co-doping of heteroatoms significantly enhances the ORR activity of metal-free catalysts. For example, a 3D sulfur–nitrogen co-doped carbon foams (S–N–CF) with hierarchical pore structures were demonstrated to show better ORR performance with higher catalytic activity, higher methanol tolerance and longer-term stability than a commercial Pt/C catalyst [101]. The relationship between the catalyst properties and structures of metal-free carbon materials for ORR was also clarified: (1) the high heteroatom doping for S–N–CF can provide abundant active sites; (2) the hierarchical pore structures and 3D networks can ensure fast electron transfer and reactant transport within the electrodes. More recently, multiple heteroatom-co-doped 3D porous carbons—such as 3D S–N co-doped carbon foams [101], N and P dual-doped hierarchical porous carbon foams [39]; B, N co-doped 3D porous graphitic carbon [116]; N and P co-functionalized 3D porous carbon networks [117]; and N, S, and O co-doped hierarchically porous carbon [118]—have been developed as efficient metal-free electrocatalysts for ORR. As an example, the N, S, and O co-doped hierarchically porous carbon were fabricated via a one-pot pyrolysis reaction with silica as template, sucrose and trithiocyanuric acid (TA) as precursors [118], A hierarchically micro-, meso-, and macroporous carbon featured with abundant dopant species and high specific surface area were obtained (Figure 9a). The resulting product displays abundant low contrast holes with diverse sizes, suggesting the featured hierarchical porosity (Figure 9b). When tested in acidic electrolytes, the one-pot pyrolyzed metal-free electrocatalysts with optimized structure exhibits comparable or even better ORR activities than the commercial Pt/C catalyst (Figure 9c,d). The excellent electrocatalytic performance is ascribed to the abundant dopant species, good integrated conductivity, and hierarchically porous architecture.

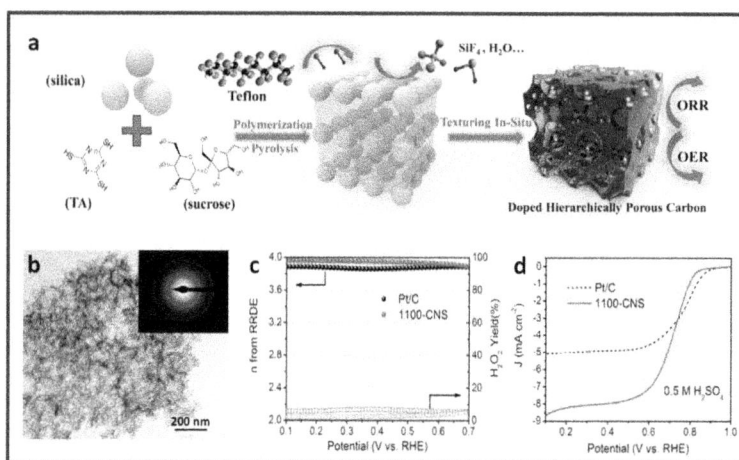

Figure 9. (**a**) Illustration of the one-pot fabrication process of N, S-doped porous carbon (CNS) materials; (**b**) TEM observation of the CNS sample, inset is the SAED patterns; (**c**) RRDE voltammograms of the 1100-CNS and Pt/C samples at 1600 rpm; (**d**) Electron transfer number (n) and HO_2^- yield derived from the RRDE test. Reproduced from [118], Copyright 2017, The Royal Society of Chemistry.

3.4. Nanocarbon Hybrid Materials for ORR

In addition to heteroatom-doped 3D CNT, graphene and porous carbon nanomaterials discussed above, nanocarbon hybrid materials with 3D structures also show superior ORR activity. To restrain the stacking interaction, which may bury active sites for ORR, between 2D heteroatom-doped graphene sheets, researchers tactfully incorporated 1D structured CNTs between graphene sheets [119–124]. The resulting CNTs/graphene hybrid exhibited good ORR activity comparable to and/or better than the commercial Pt/C catalysts under alkaline conditions. For example, Yu and coworkers, for the first time, proposed a nitrogen-doped graphene/carbon nanotube nanocomposite (NG-NCNT) as ORR catalyst. Herein, the NG-NCNT was synthesized via a hydrothermal process by using oxidized multiwalled carbon nanotube, graphene oxide, and ammonia as precursors (Figure 10a) [119]. The prepared electrode with NG-NCNT as catalyst displays much larger current and more positive onset potential than those of the NG, NCNT, G-CNT, and mixed product of GO and OCNT, respectively (Figure 10b). These indicate that the NG-NCNT possesses the best electrocatalytic ORR activity among the samples. Recently, a facile route by combining rapidly evaporating aerosol droplets with pyrolysis process was developed to fabricate N, P co-doped CNTs/graphene hybrid nanospheres (Figure 10c,d) [125]. The obtained hybrid material shows better ORR performance than a commercial Pt/C catalyst in alkaline condition (Figure 10e). When tested in acidic solution, a comparable ORR onset potential and much better stability than the commercial Pt/C catalyst were also achieved (Figure 10f). To date, carbon nanotube/graphene hybrid structures doped with heteroatom such as N [57,122,124,126,127] and N/S [121,123] have been fabricated by different methods as promising metal-free catalysts for ORR.

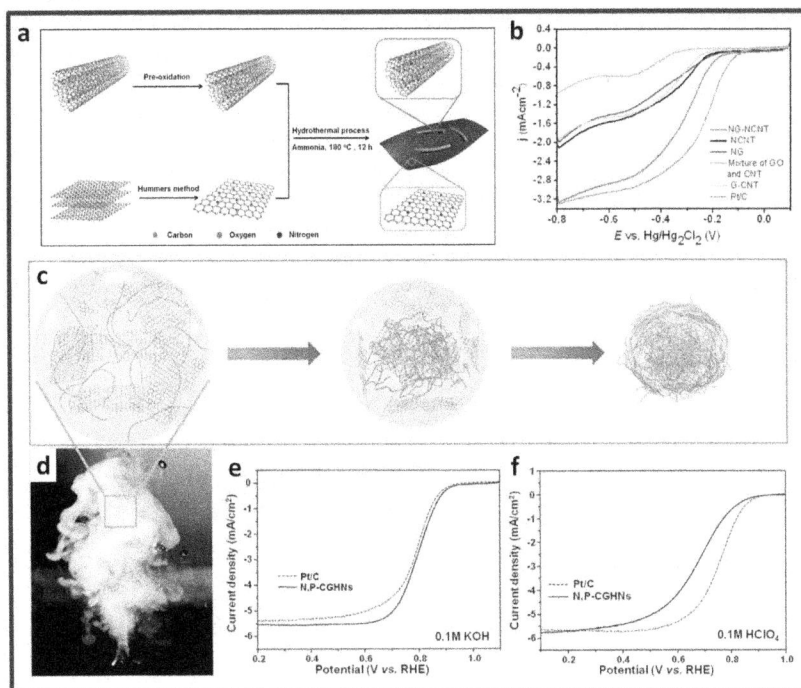

Figure 10. (**a**) Schematic illustration of the preparation of the NG-NCNT nanocomposites; (**b**) RDE voltammograms in O_2-saturated 0.1 M KOH solution at room temperature (rotation speed 1600 rpm, sweep rate 20 mV·s^{-1}) for the NG-NCNT, NCNT, NG, G-CNT, Pt/C and directly mixed product of GO and OCNT. Reproduced from [119], Copyright 2013, Wiley. (**c**) Schematic illustration of the process for co-assembling carbon nanotubes and graphene into hybrid nanospheres in rapidly evaporating aerosol droplets; (**d**) A photograph of the ultrasonic fountain and mist generated by a high-frequency ultrasound (1.7 MHz) from an aqueous dispersion containing oxidized carbon nanotubes and graphene oxides; (**e**) LSV curves of N, P-CGHNs and Pt/C in O_2-saturated 0.1 M KOH; (**f**) LSV curves of N, P-CGHNs and Pt/C in O_2-saturated 0.1 M HClO$_4$. Reproduced from [125], Copyright 2016, Wiley.

3.5. Other Kinds of Nanocarbon Materials for ORR

Aside from CNTs, graphene, porous carbon, and their hybrids mentioned above, other kinds of nanocarbon materials characterized with heteroatom doping and 3D structure also have been widely investigated as ORR electrocatalysts [128–131]. Nanocarbon networks especially N-doped nanocarbon networks can serve as excellent ORR catalysts. For instance, Hou et al. proposed a free-standing N-doped carbon nanotubes/carbon nanofibers hybrid (NCNT/CNFs) via simple pyrolysis of toluene or pyridine [132]. Due to the unique 3D hierarchical structure and pyridinic-N doping, the as-prepared NCNT/CNFs exhibited outstanding catalytic ORR performance with a favorable four-electron pathway, better selectivity and resistance to the methanol crossover, and long-term stability compared to the powder-form NCNTs and commercial Pt/C catalyst in an alkaline medium. Afterwards, Yan et al. proposed a N-doped carbon nanofiber aerogel (N-CNFA) as an efficient oxygen electrode catalyst for fuel cells [133]. The optimized N-CNFA follows a favorable four-electron ORR mechanism with more stability than commercial Pt/C catalyst. This excellent performance was attributed to the hierarchical porous structure, high specific surface area, and the abundance of catalytically active sites on N-CNFA. Recently, a metal-free N- and O-doped carbon nanowebs was also developed for use as an efficient ORR catalyst for hybrid Li-air batteries [131]. The 3D web structure shows good mass and electron transport properties, which render it a better framework support for the catalytically active sites, besides, the N and O groups together create highly ORR active pyridone groups on the nanoweb surface. It is well-known that 3D flexible electrodes are the fundamental requirement of flexible energy storage and conversion systems. By simply pyrolyzing the facial cotton under NH_3, Cheng et al. prepared a 3D flexible, porous N-doped carbon microtube (NCMT) sponge as a multifunctional ORR catalyst [134]. The flexible NCMT sponge consists of a mass of interconnected fiber-like structures with a micron-scale hollow core and porous well-graphitized walls (Figure 11a,b). Owing to the synergetic advantages of micron-scale hollow cores and the intimately-interconnected, porous tube walls, the sluggish three-phase (O_2, electrolyte, and electrode) reactions efficiently proceed as illustrated in Figure 11c. The exposed surface atoms, such as C and N, provide abundant active sites. The porous walls and hollow cores within the carbon fiber promote the fast and efficient transport of O_2 and electrolyte. The interconnected graphitic walls facilitate fast electron transfer. Therefore, the unique 3D structure of NCMT demonstrates excellent ORR activities with better durability than Pt/C (Figure 11d).

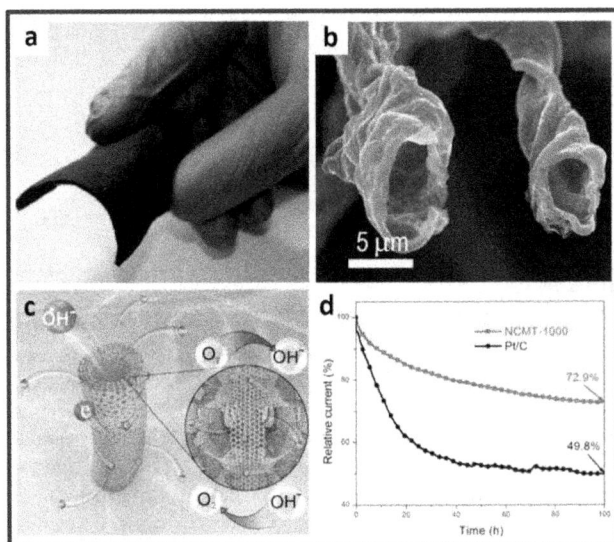

Figure 11. (**a**) Optical and (**b**) SEM images of NCMT-1000; (**c**) Schematic illustration showing the catalysis process on NCMT-1000; (**d**) Stability evaluation of NCMT-1000 and Pt/C tested by the chronamperometric responses. Reproduced from [134], Copyright 2016, The Royal Society of Chemistry.

More recently, N and S co-doped 3D hollow-structured carbon spheres (N,S-hcs) were synthesized via a facile and environmentally friendly route of soft template avenue as an efficient and stable metal free catalyst for the ORR [135]. Similar to the synthesis of N,S-hcs, cetyltrimethylammonium bromide (CTAB) was used as a typically pore-forming template to fabricate mesoporous 3D N-doped yolk-shelled carbon spheres (N-YS-CSs) via carbonization in the presence of carbon nitrogen precursors [136] . The mesoporous surface and particle size of N-YS-CSs can be well tuned by controlling the amount of ammonia a catalyst, and the optimized products exhibit outstanding cathode catalytic performance for direct methanol fuel cells. Another report is that Huang's group used urchin-like hierarchical silica spheres as templates for the synthesis of uniform 3D hierarchical N-doped carbon nanoflower (NCNF) and investigated its electrocatalytic activity towards ORR [137].

4. Conclusions and Perspectives

Nanocarbon-based metal-free catalysts are promising candidates originating from low cost and high-performance of ORR catalysts for fuel cells and metal–air batteries. In this review, we have summarized the recent development of advanced nanocarbon-based, metal-free ORR catalysts, including single and multiple heteroatom-doped carbon nanotubes, graphenes, and porous carbons, as well as their hybrids. The discussion of electrocatalysis has focused on the influence of 3D structure and heteroatoms on the electrochemical performance of nanocarbon catalysts. Compared with commercially available Pt/C, single nonmetal heteroatom (e.g., N, S, B, and P) or multiple heteroatom (e.g., NP, NS, NB NSO and NPO) doped nanocarbon materials show comparable or even higher electrocatalytic activity, better durability, and greater tolerance against fuel crossover and CO poisoning. The unique 3D structured nanocarbon materials can not only enhance the exposure and stability of ORR active sites, but also provide the mass transport and electron transfer pathways. Therefore, the synergetic effect between the 3D nanostructures and the doping-induced charge redistribution results in superior ORR activity.

Over the past decade, considerable progress has been made in the development of high-efficiency 3D structured nanocarbon-based ORR electrocatalysts. However, some important challenges may be addressed prior to practical applications: (1) The understanding of the activity mechanism of heteroatom-doped nanocarbon is challenging to rationally correlate the electron structure, adsorption properties, and apparent activities. For example, the nitrogen doping induces charge distribution, and parallel diatomic O_2 adsorption can effectively weaken the O–O bond and lower the ORR potential, facilitating oxygen reduction at the N-doped nanocarbon electrode. Some theoretical and experimental results indicate that different N doping configurations result in difference of the ORR activity and planar pyridinic N with a lone electron pair is claimed as the active type to improve the electron-donating capability and weaken the O–O bond. However, there is a debate that graphitic N rather than pyridinc N may be responsible for the ORR. Therefore, in-depth understanding of the type of active sites toward ORR and unambiguous identifying of different types of active configurations is imperative for developing advanced heteroatom-doped nanocarbon catalysts in terms of rationally selecting synthesis methods and precursors. Besides, more powerful and effective characterizations, including advanced electron microscopy and in situ or operando techniques, should be combined with theoretical calculations to identify the different active types and the actual active sites. (2) Except active sites, two other key factors of the mass transport and electrical conductivity together determine the ORR performance of a nanocarbon catalyst. Therefore, structure design and optimization of nanocarbon electrocatalysts, such as pore structure, surface area, and electrical conductivity, are significantly important to enhance their ORR performance. Generally, 3D porous structure can provide a high surface area with abundant exposed active sites and large pore volume with multidimensional electron transport pathways, and hence facilitate mass (e.g., ions, oxygen and discharge products) diffusion and electron transfer, further accelerating reaction kinetics. As discussed above, great progress has been made via designing 3D porous structures to achieve outstanding ORR performance. However, a detailed relationship between the pore structure and mass transport capability in different media is yet to be determined, and detailed models describing the transport of reactants and products within the active sites are still unclear. (3) Additionally, future efforts in the research and development of 3D nanocarbon catalysts toward ORR should focus on the tradeoffs between electrical conductivity and surface density of active sites. With continuous research in this promising field, we look forward to the bright future of 3D heteroatom-doped nanocarbon catalysts as well as the breakthroughs in the understanding of the nature of the ORR on these carbon-based metal-free ORR catalysts.

Acknowledgments: This work was supported by the National Natural Science Foundation of China (51572194 and 51672189), Academic Innovation Funding of Tianjin Normal University (52XC1404), Training Plan of Leader Talent of University in Tianjin, National Key R&D Program of China (2017YFB0310703), and the Fundamental Research Funds for the Central Universities (2652017369).

Conflicts of Interest: The authors declare no conflict of interest.

References

1. Larcher, D.; Tarascon, J.M. Towards greener and more sustainable batteries for electrical energy storage. *Nat. Chem.* **2015**, *7*, 19–29. [CrossRef] [PubMed]
2. Qin, P.; Tanaka, S.; Ito, S.; Tetreault, N.; Manabe, K.; Nishino, H.; Nazeeruddin, M.K.; Gratzel, M. Inorganic hole conductor-based lead halide perovskite solar cells with 12.4% conversion efficiency. *Nat. Commun.* **2014**, *5*, 3834. [CrossRef] [PubMed]
3. Xiong, D.; Li, X.; Bai, Z.; Lu, S. Recent Advances in Layered $Ti_3C_2T_x$ MXene for Electrochemical Energy Storage. *Small* **2018**, *14*, 1703419. [CrossRef] [PubMed]
4. Lu, J.; Chen, Z.; Pan, F.; Cui, Y.; Amine, K. High-Performance Anode Materials for Rechargeable Lithium-Ion Batteries. *Electrochem. Energy Rev.* **2018**, *1*, 35–53. [CrossRef]
5. Chou, S.L.; Dou, S.X. Next-Generation Batteries. *Adv. Mater.* **2017**, *29*, 1705871. [CrossRef] [PubMed]
6. Goodenough, J.B. Electrochemical energy storage in a sustainable modern society. *Energy Environ. Sci.* **2014**, *7*, 14–18. [CrossRef]
7. Cheng, F.; Chen, J. Metal-air batteries: From oxygen reduction electrochemistry to cathode catalysts. *Chem. Soc. Rev.* **2012**, *41*, 2172–2192. [CrossRef] [PubMed]
8. Li, J.C.; Hou, P.X.; Liu, C. Heteroatom-Doped Carbon Nanotube and Graphene-Based Electrocatalysts for Oxygen Reduction Reaction. *Small* **2017**, *13*, 1702002. [CrossRef] [PubMed]
9. Myles, T.; Bonville, L.; Maric, R. Catalyst, Membrane, Free Electrolyte Challenges, and Pathways to Resolutions in High Temperature Polymer Electrolyte Membrane Fuel Cells. *Catalysts* **2017**, *7*, 16. [CrossRef]
10. Suthirakun, S.; Ammal, S.C.; Munoz-Garcia, A.B.; Xiao, G.; Chen, F.; zur Loye, H.C.; Carter, E.A.; Heyden, A. Theoretical investigation of H_2 oxidation on the $Sr_2Fe_{1.5}Mo_{0.5}O_6$ (001) perovskite surface under anodic solid oxide fuel cell conditions. *J. Am. Chem. Soc.* **2014**, *136*, 8374–8386. [CrossRef] [PubMed]
11. Suntivich, J.; Gasteiger, H.A.; Yabuuchi, N.; Nakanishi, H.; Goodenough, J.B.; Shao-Horn, Y. Design principles for oxygen-reduction activity on perovskite oxide catalysts for fuel cells and metal-air batteries. *Nat. Chem.* **2011**, *3*, 546–550. [CrossRef] [PubMed]
12. Fu, S.; Zhu, C.; Song, J.; Du, D.; Lin, Y. Metal-Organic Framework-Derived Non-Precious Metal Nanocatalysts for Oxygen Reduction Reaction. *Adv. Energy Mater.* **2017**, *7*, 1700363. [CrossRef]
13. Marcel, R. Perovskite Electrocatalysts for the Oxygen Reduction Reaction in Alkaline Media. *Catalysts* **2017**, *7*, 154. [CrossRef]
14. Li, Q.; Cao, R.; Cho, J.; Wu, G. Nanocarbon Electrocatalysts for Oxygen Reduction in Alkaline Media for Advanced Energy Conversion and Storage. *Adv. Energy Mater.* **2014**, *4*, 1301415. [CrossRef]
15. Greeley, J.; Stephens, I.E.; Bondarenko, A.S.; Johansson, T.P.; Hansen, H.A.; Jaramillo, T.F.; Rossmeisl, J.; Chorkendorff, I.; Norskov, J.K. Alloys of platinum and early transition metals as oxygen reduction electrocatalysts. *Nat. Chem.* **2009**, *1*, 552–556. [CrossRef] [PubMed]
16. Cheng, N.; Banis, M.N.; Liu, J.; Riese, A.; Li, X.; Li, R.; Ye, S.; Knights, S.; Sun, X. Extremely stable platinum nanoparticles encapsulated in a zirconia nanocage by area-selective atomic layer deposition for the oxygen reduction reaction. *Adv. Mater.* **2015**, *27*, 277–281. [CrossRef] [PubMed]
17. Guo, S.; Li, D.; Zhu, H.; Zhang, S.; Markovic, N.M.; Stamenkovic, V.R.; Sun, S. FePt and CoPt nanowires as efficient catalysts for the oxygen reduction reaction. *Angew. Chem. Int. Ed. Engl.* **2013**, *125*, 3449–3552.
18. Tan, Y.; Xu, C.; Chen, G.; Zheng, N.; Xie, Q. A graphene–platinum nanoparticles–ionic liquid composite catalyst for methanol-tolerant oxygen reduction reaction. *Energy Environ. Sci.* **2012**, *5*, 6923–6927. [CrossRef]
19. Liang, Y.; Li, Y.; Wang, H.; Zhou, J.; Wang, J.; Regier, T.; Dai, H. Co_3O_4 nanocrystals on graphene as a synergistic catalyst for oxygen reduction reaction. *Nat. Mater.* **2011**, *10*, 780–786. [CrossRef] [PubMed]
20. Wang, D.; Chen, X.; Evans, D.G.; Yang, W. Well-dispersed Co_3O_4/Co_2MnO_4 nanocomposites as a synergistic bifunctional catalyst for oxygen reduction and oxygen evolution reactions. *Nanoscale* **2013**, *5*, 5312–5315. [CrossRef] [PubMed]

21. Wang, Y.C.; Lai, Y.J.; Song, L.; Zhou, Z.Y.; Liu, J.G.; Wang, Q.; Yang, X.D.; Chen, C.; Shi, W.; Zheng, Y.P.; et al. S-Doping of an Fe/N/C ORR Catalyst for Polymer Electrolyte Membrane Fuel Cells with High Power Density. *Angew. Chem. Int. Ed. Engl.* **2015**, *54*, 9907–9910. [CrossRef] [PubMed]

22. Zhou, M.; Wang, H.L.; Guo, S. Towards high-efficiency nanoelectrocatalysts for oxygen reduction through engineering advanced carbon nanomaterials. *Chem. Soc. Rev.* **2016**, *45*, 1273–1307. [CrossRef] [PubMed]

23. Wang, Y.-J.; Fang, B.; Zhang, D.; Li, A.; Wilkinson, D.P.; Ignaszak, A.; Zhang, L.; Zhang, J. A Review of Carbon-Composited Materials as Air-Electrode Bifunctional Electrocatalysts for Metal–Air Batteries. *Electrochem. Energy Rev.* **2018**, *1*, 1–34. [CrossRef]

24. Liu, X.; Dai, L. Carbon-based metal-free catalysts. *Nat. Rev. Mater.* **2016**, *1*, 16064. [CrossRef]

25. Li, Q.; Zhang, S.; Dai, L.; Li, L.S. Nitrogen-doped colloidal graphene quantum dots and their size-dependent electrocatalytic activity for the oxygen reduction reaction. *J. Am. Chem. Soc.* **2012**, *134*, 18932–18935. [CrossRef] [PubMed]

26. Lin, Z.; Waller, G.; Liu, Y.; Liu, M.; Wong, C.-P. Facile Synthesis of Nitrogen-Doped Graphene via Pyrolysis of Graphene Oxide and Urea, and its Electrocatalytic Activity toward the Oxygen-Reduction Reaction. *Adv. Energy Mater.* **2012**, *2*, 884–888. [CrossRef]

27. Gong, K.; Du, F.; Xia, Z.; Durstock, M.; Dai, L. Nitrogen-doped carbon nanotube arrays with high electrocatalytic activity for oxygen reduction. *Science* **2009**, *323*, 760–764. [CrossRef] [PubMed]

28. Dai, L. Carbon-based catalysts for metal-free electrocatalysis. *Curr. Opin. Electrochem.* **2017**, *4*, 18–25. [CrossRef]

29. Zheng, Y.; Jiao, Y.; Jaroniec, M.; Jin, Y.; Qiao, S.Z. Nanostructured metal-free electrochemical catalysts for highly efficient oxygen reduction. *Small* **2012**, *8*, 3550–3566. [CrossRef] [PubMed]

30. Zhu, Y.P.; Guo, C.; Zheng, Y.; Qiao, S.Z. Surface and Interface Engineering of Noble-Metal-Free Electrocatalysts for Efficient Energy Conversion Processes. *Acc. Chem. Res.* **2017**, *50*, 915–923. [CrossRef] [PubMed]

31. Zhang, J.; Dai, L. Heteroatom-Doped Graphitic Carbon Catalysts for Efficient Electrocatalysis of Oxygen Reduction Reaction. *ACS Catal.* **2015**, *5*, 7244–7253. [CrossRef]

32. Tang, C.; Zhang, Q. Nanocarbon for Oxygen Reduction Electrocatalysis: Dopants, Edges, and Defects. *Adv. Mater.* **2017**, *29*, 1604103. [CrossRef] [PubMed]

33. He, W.; Wang, Y.; Jiang, C.; Lu, L. Structural effects of a carbon matrix in non-precious metal O_2-reduction electrocatalysts. *Chem. Soc. Rev.* **2016**, *45*, 2396–2409. [CrossRef] [PubMed]

34. Tu, Y.; Deng, D.; Bao, X. Nanocarbons and their hybrids as catalysts for non-aqueous lithium–oxygen batteries. *J. Energy Chem.* **2016**, *25*, 957–966. [CrossRef]

35. Sawant, S.Y.; Han, T.H.; Cho, M.H. Metal-Free Carbon-Based Materials: Promising Electrocatalysts for Oxygen Reduction Reaction in Microbial Fuel Cells. *Int. J. Mol. Sci.* **2016**, *18*, 25. [CrossRef] [PubMed]

36. Higgins, D.; Zamani, P.; Yu, A.; Chen, Z. The application of graphene and its composites in oxygen reduction electrocatalysis: A perspective and review of recent progress. *Energy Environ. Sci.* **2016**, *9*, 357–390. [CrossRef]

37. Du, R.; Zhang, N.; Zhu, J.; Wang, Y.; Xu, C.; Hu, Y.; Mao, N.; Xu, H.; Duan, W.; Zhuang, L.; et al. Nitrogen-Doped Carbon Nanotube Aerogels for High-Performance ORR Catalysts. *Small* **2015**, *11*, 3903–3908. [CrossRef] [PubMed]

38. Tang, S.; Zhou, X.; Xu, N.; Bai, Z.; Qiao, J.; Zhang, J. Template-free synthesis of three-dimensional nanoporous N-doped graphene for high performance fuel cell oxygen reduction reaction in alkaline media. *Appl. Energy* **2016**, *175*, 405–413. [CrossRef]

39. Jiang, H.; Zhu, Y.; Feng, Q.; Su, Y.; Yang, X.; Li, C. Nitrogen and phosphorus dual-doped hierarchical porous carbon foams as efficient metal-free electrocatalysts for oxygen reduction reactions. *Chem. Eur. J.* **2014**, *20*, 3106–3112. [CrossRef] [PubMed]

40. Yang, Z.; Nie, H.; Chen, X.A.; Chen, X.; Huang, S. Recent progress in doped carbon nanomaterials as effective cathode catalysts for fuel cell oxygen reduction reaction. *J. Power Sources* **2013**, *236*, 238–249. [CrossRef]

41. Liu, J.; Song, P.; Ning, Z.; Xu, W. Recent Advances in Heteroatom-Doped Metal-Free Electrocatalysts for Highly Efficient Oxygen Reduction Reaction. *Electrocatalysis* **2015**, *6*, 132–147. [CrossRef]

42. Ma, R.; Ma, Y.; Dong, Y.; Lee, J.-M. Recent Advances in Heteroatom-Doped Graphene Materials as Efficient Electrocatalysts towards the Oxygen Reduction Reaction. *Nano Adv.* **2016**, *1*, 50–61. [CrossRef]

43. Zhang, L.; Xia, Z. Mechanisms of Oxygen Reduction Reaction on Nitrogen-Doped Graphene for Fuel Cells. *J. Phys. Chem. C* **2011**, *115*, 11170–11176. [CrossRef]

44. Carrette, L.; Friedrich, K.A.; Stimming, U. Fuel Cells–Fundamentals and Applications. *Fuel Cells* **2001**, *1*, 5–39. [CrossRef]

45. Liew, K.B.; Daud, W.R.W.; Ghasemi, M.; Leong, J.X.; Lim, S.S.; Ismail, M. Non-Pt catalyst as oxygen reduction reaction in microbial fuel cells: A review. *Int. J. Hydrog. Energy* **2014**, *39*, 4870–4883. [CrossRef]

46. Gu, W.; Hu, L.; Li, J.; Wang, E. Recent Advancements in Transition Metal-Nitrogen-Carbon Catalysts for Oxygen Reduction Reaction. *Electroanalysis* **2018**. [CrossRef]

47. Stacy, J.; Regmi, Y.N.; Leonard, B.; Fan, M. The recent progress and future of oxygen reduction reaction catalysis: A review. *Renew. Sustain. Energy Rev.* **2017**, *69*, 401–414. [CrossRef]

48. Shui, J.; Wang, M.; Du, F.; Dai, L. N-doped carbon nanomaterials are durable catalysts for oxygen reduction reaction in acidic fuel cells. *Sci. Adv.* **2015**, *1*, 1400129. [CrossRef] [PubMed]

49. Xiong, D.; Li, X.; Shan, H.; Yan, B.; Dong, L.; Cao, Y.; Li, D. Controllable oxygenic functional groups of metal-free cathodes for high performance lithium ion batteries. *J. Mater. Chem. A* **2015**, *3*, 11376–11386. [CrossRef]

50. Xiong, D.; Li, X.; Shan, H.; Yan, B.; Li, D.; Langford, C.; Sun, X. Scalable synthesis of functionalized graphene as cathodes in Li-ion electrochemical energy storage devices. *Appl. Energy* **2016**, *175*, 512–521. [CrossRef]

51. Li, C.; Zhang, X.; Wang, K.; Zhang, H.-T.; Sun, X.-Z.; Ma, Y.-W. Three dimensional graphene networks for supercapacitor electrode materials. *New Carbon Mater.* **2015**, *30*, 193–206. [CrossRef]

52. Lin, Z.; Zeng, Z.; Gui, X.; Tang, Z.; Zou, M.; Cao, A. Carbon Nanotube Sponges, Aerogels, and Hierarchical Composites: Synthesis, Properties, and Energy Applications. *Adv. Energy Mater.* **2016**, *6*, 1600554. [CrossRef]

53. Tang, J.; Liu, J.; Torad, N.L.; Kimura, T.; Yamauchi, Y. Tailored design of functional nanoporous carbon materials toward fuel cell applications. *Nano Today* **2014**, *9*, 305–323. [CrossRef]

54. Daems, N.; Sheng, X.; Vankelecom, F.J.; Pescarmona, P.P. Metal-free doped carbon materials as electrocatalysts for the oxygen reduction reaction. *J. Mater. Chem. A* **2014**, *2*, 4085–4110. [CrossRef]

55. Ito, Y.; Qiu, H.J.; Fujita, T.; Tanabe, Y.; Tanigaki, K.; Chen, M. Bicontinuous nanoporous N-doped graphene for the oxygen reduction reaction. *Adv. Mater.* **2014**, *26*, 4145–4150. [CrossRef] [PubMed]

56. Wang, Y.; Tao, L.; Xiao, Z.; Chen, R.; Jiang, Z.; Wang, S. 3D Carbon Electrocatalysts In Situ Constructed by Defect-Rich Nanosheets and Polyhedrons from NaCl-Sealed Zeolitic Imidazolate Frameworks. *Adv. Funct. Mater.* **2018**, *28*, 1705356. [CrossRef]

57. Tian, G.L.; Zhao, M.Q.; Yu, D.; Kong, X.Y.; Huang, J.Q.; Zhang, Q.; Wei, F. Nitrogen-doped graphene/carbon nanotube hybrids: In situ formation on bifunctional catalysts and their superior electrocatalytic activity for oxygen evolution/reduction reaction. *Small* **2014**, *10*, 2251–2259. [CrossRef] [PubMed]

58. Avouris, P.; Dimitrakopoulos, C. Graphene: Synthesis and applications. *Mater. Today* **2012**, *15*, 86–97. [CrossRef]

59. Xiao, X.; Peng, X.; Jin, H.; Li, T.; Zhang, C.; Gao, B.; Hu, B.; Huo, K.; Zhou, J. Freestanding mesoporous VN/CNT hybrid electrodes for flexible all-solid-state supercapacitors. *Adv. Mater.* **2013**, *25*, 5091–5097. [CrossRef] [PubMed]

60. Yang, L.; Jiang, S.; Zhao, Y.; Zhu, L.; Chen, S.; Wang, X.; Wu, Q.; Ma, J.; Ma, Y.; Hu, Z. Boron-doped carbon nanotubes as metal-free electrocatalysts for the oxygen reduction reaction. *Angew. Chem. Int. Ed. Engl.* **2011**, *50*, 7132–7135. [CrossRef] [PubMed]

61. Yu, D.; Xue, Y.; Dai, L. Vertically Aligned Carbon Nanotube Arrays Co-doped with Phosphorus and Nitrogen as Efficient Metal-Free Electrocatalysts for Oxygen Reduction. *J. Phys. Chem. Lett.* **2012**, *3*, 2863–2870. [CrossRef] [PubMed]

62. Yu, D.; Zhang, Q.; Dai, L. Highly efficient metal-free growth of nitrogen-doped single-walled carbon nanotubes on plasma-etched substrates for oxygen reduction. *J. Am. Chem. Soc.* **2010**, *132*, 15127–15129. [CrossRef] [PubMed]

63. Wang, Z.; Jia, R.; Zheng, J.; Zhao, J.; Li, L.; Song, J.; Zhu, Z. Nitrogen-promoted self-assembly of N-doped carbon nanotubes and their intrinsic catalysis for oxygen reduction in fuel cells. *ACS Nano* **2011**, *5*, 1677–1684. [CrossRef] [PubMed]

64. Pan, T.; Liu, H.; Ren, G.; Li, Y.; Lu, X.; Zhu, Y. Metal-free porous nitrogen-doped carbon nanotubes for enhanced oxygen reduction and evolution reactions. *Sci. Bull.* **2016**, *61*, 889–896. [CrossRef]

65. She, X.; Yang, D.; Jing, D.; Yuan, F.; Yang, W.; Guo, L.; Che, Y. Nitrogen-doped one-dimensional (1D) macroporous carbonaceous nanotube arrays and their application in electrocatalytic oxygen reduction reactions. *Nanoscale* **2014**, *6*, 11057–11061. [CrossRef] [PubMed]

66. Xiong, W.; Du, F.; Liu, Y.; Perez, A., Jr.; Supp, M.; Ramakrishnan, T.S.; Dai, L.; Jiang, L. 3-D carbon nanotube structures used as high performance catalyst for oxygen reduction reaction. *J. Am. Chem. Soc.* **2010**, *132*, 15839–15841. [CrossRef] [PubMed]

67. Qi, J.; Benipal, N.; Chadderdon, D.J.; Huo, J.; Jiang, Y.; Qiu, Y.; Han, X.; Hu, Y.H.; Shanks, B.H.; Li, W. Carbon nanotubes as catalysts for direct carbohydrazide fuel cells. *Carbon* **2015**, *89*, 142–147. [CrossRef]

68. Guo, M.-Q.; Huang, J.-Q.; Kong, X.-Y.; Peng, H.-J.; Shui, H.; Qian, F.-Y.; Zhu, L.; Zhu, W.-C.; Zhang, Q. Hydrothermal synthesis of porous phosphorus-doped carbon nanotubes and their use in the oxygen reduction reaction and lithium-sulfur batteries. *New Carbon Mater.* **2016**, *31*, 352–362. [CrossRef]

69. Liu, Z.; Peng, F.; Wang, H.; Yu, H.; Tan, J.; Zhu, L. Novel phosphorus-doped multiwalled nanotubes with high electrocatalytic activity for O_2 reduction in alkaline medium. *Catal. Commun.* **2011**, *16*, 35–38. [CrossRef]

70. Zhu, J.; Jiang, S.P.; Wang, R.; Shi, K.; Shen, P.K. One-pot synthesis of a nitrogen and phosphorus-dual-doped carbon nanotube array as a highly effective electrocatalyst for the oxygen reduction reaction. *J. Mater. Chem. A* **2014**, *2*, 15448–15453. [CrossRef]

71. Bonaccorso, F.; Colombo, L.; Yu, G.; Stoller, M.; Tozzini, V.; Ferrari, A.C.; Ruoff, R.S.; Pellegrini, V. 2D materials. Graphene, related two-dimensional crystals, and hybrid systems for energy conversion and storage. *Science* **2015**, *347*, 1246501. [CrossRef] [PubMed]

72. Cao, X.; Yin, Z.; Zhang, H. Three-dimensional graphene materials: Preparation, structures and application in supercapacitors. *Energy Environ. Sci.* **2014**, *7*, 1850–1865. [CrossRef]

73. Hu, C.; Liu, D.; Xiao, Y.; Dai, L. Functionalization of graphene materials by heteroatom-doping for energy conversion and storage. *Prog. Nat. Sci. Mater.* **2018**. [CrossRef]

74. Xiong, D.; Li, X.; Bai, Z.; Shan, H.; Fan, L.; Wu, C.; Li, D.; Lu, S. Superior Cathode Performance of Nitrogen-Doped Graphene Frameworks for Lithium Ion Batteries. *ACS Appl. Mater. Interfaces* **2017**, *9*, 10643–10651. [CrossRef] [PubMed]

75. Shan, H.; Li, X.; Cui, Y.; Xiong, D.; Yan, B.; Li, D.; Lushington, A.; Sun, X. Sulfur/Nitrogen Dual-doped Porous Graphene Aerogels Enhancing Anode Performance of Lithium Ion Batteries. *Electrochim. Acta* **2016**, *205*, 188–197. [CrossRef]

76. Sheng, Z.H.; Shao, L.; Chen, J.J.; Bao, W.J.; Wang, F.B.; Xia, X.H. Catalyst-free synthesis of nitrogen-doped graphene via thermal annealing graphite oxide with melamine and its excellent electrocatalysis. *ACS Nano* **2011**, *5*, 4350–4358. [CrossRef] [PubMed]

77. Yang, Z.; Yao, Z.; Li, G.; Fang, G.; Nie, H.; Liu, Z.; Zhou, X.; Chen, X.; Huang, S. Sulfur-doped graphene as an efficient metal-free cathode catalyst for oxygen reduction. *ACS Nano* **2011**, *6*, 205–211. [CrossRef] [PubMed]

78. Jo, G.; Sanetuntikul, J.; Shanmugam, S. Boron and phosphorous-doped graphene as a metal-free electrocatalyst for the oxygen reduction reaction in alkaline medium. *RSC Adv.* **2015**, *5*, 53637–53643. [CrossRef]

79. Su, Y.; Zhang, Y.; Zhuang, X.; Li, S.; Wu, D.; Zhang, F.; Feng, X. Low-temperature synthesis of nitrogen/sulfur co-doped three-dimensional graphene frameworks as efficient metal-free electrocatalyst for oxygen reduction reaction. *Carbon* **2013**, *62*, 296–301. [CrossRef]

80. Xue, Y.; Yu, D.; Dai, L.; Wang, R.; Li, D.; Roy, A.; Lu, F.; Chen, H.; Liu, Y.; Qu, J. Three-dimensional B,N-doped graphene foam as a metal-free catalyst for oxygen reduction reaction. *Phys. Chem. Chem. Phys.* **2013**, *15*, 12220–12226. [CrossRef] [PubMed]

81. Chen, L.; Du, R.; Zhu, J.; Mao, Y.; Xue, C.; Zhang, N.; Hou, Y.; Zhang, J.; Yi, T. Three-dimensional nitrogen-doped graphene nanoribbons aerogel as a highly efficient catalyst for the oxygen reduction reaction. *Small* **2015**, *11*, 1423–1429. [CrossRef] [PubMed]

82. Guan, Y.; Dou, Z.; Yang, Y.; Xue, J.; Zhu, Z.; Cui, L. Fabrication of functionalized 3D graphene with controllable micro/meso-pores as a superior electrocatalyst for enhanced oxygen reduction in both acidic and alkaline solutions. *RSC Adv.* **2016**, *6*, 79459–79469. [CrossRef]

83. Qu, L.; Liu, Y.; Baek, J.B.; Dai, L. Nitrogen-doped graphene as efficient metal-free electrocatalyst for oxygen reduction in fuel cells. *ACS Nano* **2010**, *4*, 1321–1326. [CrossRef] [PubMed]

84. Shan, H.; Xiong, D.; Li, X.; Sun, Y.; Yan, B.; Li, D.; Lawes, S.; Cui, Y.; Sun, X. Tailored lithium storage performance of graphene aerogel anodes with controlled surface defects for lithium-ion batteries. *Appl. Surf. Sci.* **2016**, *364*, 651–659. [CrossRef]

85. Kabir, S.; Artyushkova, K.; Serov, A.; Atanassov, P. Role of Nitrogen Moieties in N-Doped 3D-Graphene Nanosheets for Oxygen Electroreduction in Acidic and Alkaline Media. *ACS Appl. Mater. Interfaces* **2018**, *10*, 11623–11632. [CrossRef] [PubMed]

86. Zhao, C.; Yu, C.; Liu, S.; Yang, J.; Fan, X.; Huang, H.; Qiu, J. 3D Porous N-Doped Graphene Frameworks Made of Interconnected Nanocages for Ultrahigh-Rate and Long-Life Li–O₂ Batteries. *Adv. Funct. Mater.* **2015**, *25*, 6913–6920. [CrossRef]

87. Shi, J.-L.; Tang, C.; Huang, J.-Q.; Zhu, W.; Zhang, Q. Effective exposure of nitrogen heteroatoms in 3D porous graphene framework for oxygen reduction reaction and lithium–sulfur batteries. *J. Energy Chem.* **2018**, *27*, 167–175. [CrossRef]

88. Lu, X.; Li, Z.; Yin, X.; Wang, S.; Liu, Y.; Wang, Y. Controllable synthesis of three-dimensional nitrogen-doped graphene as a high performance electrocatalyst for oxygen reduction reaction. *Int. J. Hydrog. Energy* **2017**, *42*, 17504–17513. [CrossRef]

89. Chen, X.; Chen, X.; Xu, X.; Yang, Z.; Liu, Z.; Zhang, L.; Xu, X.; Chen, Y.; Huang, S. Sulfur-doped porous reduced graphene oxide hollow nanosphere frameworks as metal-free electrocatalysts for oxygen reduction reaction and as supercapacitor electrode materials. *Nanoscale* **2014**, *6*, 13740–13747. [CrossRef] [PubMed]

90. Zhang, Y.; Chu, M.; Yang, L.; Deng, W.; Tan, Y.; Ma, M.; Xie, Q. Synthesis and oxygen reduction properties of three-dimensional sulfur-doped graphene networks. *Chem. Commun.* **2014**, *50*, 6382–6385. [CrossRef] [PubMed]

91. Li, X.; Qiu, Y.; Hu, P.A. Three Dimensional P-doped Graphene Synthesized by Eco-Friendly Chemical Vapor Deposition for Oxygen Reduction Reactions. *J. Nanosci. Nanotechnol.* **2016**, *16*, 6216–6222. [CrossRef] [PubMed]

92. Liang, J.; Jiao, Y.; Jaroniec, M.; Qiao, S.Z. Sulfur and nitrogen dual-doped mesoporous graphene electrocatalyst for oxygen reduction with synergistically enhanced performance. *Angew. Chem. Int. Ed. Engl.* **2012**, *51*, 11496–11500. [CrossRef] [PubMed]

93. Xu, C.; Su, Y.; Liu, D.; He, X. Three-dimensional N,B-doped graphene aerogel as a synergistically enhanced metal-free catalyst for the oxygen reduction reaction. *Phys. Chem. Chem. Phys.* **2015**, *17*, 25440–25448. [CrossRef] [PubMed]

94. Amiinu, I.S.; Zhang, J.; Kou, Z.; Liu, X.; Asare, O.K.; Zhou, H.; Cheng, K.; Zhang, H.; Mai, L.; Pan, M.; et al. Self-Organized 3D Porous Graphene Dual-Doped with Biomass-Sponsored Nitrogen and Sulfur for Oxygen Reduction and Evolution. *ACS Appl. Mater. Interfaces* **2016**, *8*, 29408–29418. [CrossRef] [PubMed]

95. Rivera, L.M.; Fajardo, S.; Arévalo, M.D.C.; García, G.; Pastor, E. S- and N-Doped Graphene Nanomaterials for the Oxygen Reduction Reaction. *Catalysts* **2017**, *7*, 278. [CrossRef]

96. Chabu, J.M.; Wang, L.; Tang, F.-Y.; Zeng, K.; Sheng, J.; Walle, M.D.; Deng, L.; Liu, Y.-N. Synthesis of Three-Dimensional Nitrogen and Sulfur Dual-Doped Graphene Aerogels as an Efficient Metal-Free Electrocatalyst for the Oxygen Reduction Reaction. *ChemElectroChem* **2017**, *4*, 1885–1890. [CrossRef]

97. Wu, M.; Dou, Z.; Chang, J.; Cui, L. Nitrogen and sulfur co-doped graphene aerogels as an efficient metal-free catalyst for oxygen reduction reaction in an alkaline solution. *RSC Adv.* **2016**, *6*, 22781–22790. [CrossRef]

98. Li, Y.; Yang, J.; Huang, J.; Zhou, Y.; Xu, K.; Zhao, N.; Cheng, X. Soft template-assisted method for synthesis of nitrogen and sulfur co-doped three-dimensional reduced graphene oxide as an efficient metal free catalyst for oxygen reduction reaction. *Carbon* **2017**, *122*, 237–246. [CrossRef]

99. Zhao, Y.; Huang, S.; Xia, M.; Rehman, S.; Mu, S.; Kou, Z.; Zhang, Z.; Chen, Z.; Gao, F.; Hou, Y. N-P-O co-doped high performance 3D graphene prepared through red phosphorous-assisted "cutting-thin" technique: A universal synthesis and multifunctional applications. *Nano Energy* **2016**, *28*, 346–355. [CrossRef]

100. Liu, R.; Wu, D.; Feng, X.; Mullen, K. Nitrogen-doped ordered mesoporous graphitic arrays with high electrocatalytic activity for oxygen reduction. *Angew. Chem. Int. Ed. Engl.* **2010**, *49*, 2565–2569. [CrossRef] [PubMed]

101. Liu, Z.; Nie, H.; Yang, Z.; Zhang, J.; Jin, Z.; Lu, Y.; Xiao, Z.; Huang, S. Sulfur-nitrogen co-doped three-dimensional carbon foams with hierarchical pore structures as efficient metal-free electrocatalysts for oxygen reduction reactions. *Nanoscale* **2013**, *5*, 3283–3288. [CrossRef] [PubMed]

102. Liang, H.W.; Zhuang, X.; Bruller, S.; Feng, X.; Mullen, K. Hierarchically porous carbons with optimized nitrogen doping as highly active electrocatalysts for oxygen reduction. *Nat. Commun.* **2014**, *5*, 4973. [CrossRef] [PubMed]

103. Chen, Y.; Ma, R.; Zhou, Z.; Liu, G.; Zhou, Y.; Liu, Q.; Kaskel, S.; Wang, J. An In Situ Source-Template-Interface Reaction Route to 3D Nitrogen-Doped Hierarchical Porous Carbon as Oxygen Reduction Electrocatalyst. *Adv. Mater. Interfaces* **2015**, *2*, 1500199. [CrossRef]

104. Zhu, C.; Li, H.; Fu, S.; Du, D.; Lin, Y. Highly efficient nonprecious metal catalysts towards oxygen reduction reaction based on three-dimensional porous carbon nanostructures. *Chem. Soc. Rev.* **2016**, *45*, 517–531. [CrossRef] [PubMed]

105. Chen, S.; Bi, J.; Zhao, Y.; Yang, L.; Zhang, C.; Ma, Y.; Wu, Q.; Wang, X.; Hu, Z. Nitrogen-doped carbon nanocages as efficient metal-free electrocatalysts for oxygen reduction reaction. *Adv. Mater.* **2012**, *24*, 5593–5597. [CrossRef] [PubMed]

106. Gao, S.; Chen, Y.; Fan, H.; Wei, X.; Hu, C.; Luo, H.; Qu, L. Large scale production of biomass-derived N-doped porous carbon spheres for oxygen reduction and supercapacitors. *J. Mater. Chem. A* **2014**, *2*, 3317–3324. [CrossRef]

107. Huang, H.; Wei, X.; Gao, S. Nitrogen-Doped Porous Carbon Derived from Malachium Aquaticum Biomass as a Highly Efficient Electrocatalyst for Oxygen Reduction Reaction. *Electrochim. Acta* **2016**, *220*, 427–435. [CrossRef]

108. Liu, R.; Zhang, H.; Liu, S.; Zhang, X.; Wu, T.; Ge, X.; Zang, Y.; Zhao, H.; Wang, G. Shrimp-shell derived carbon nanodots as carbon and nitrogen sources to fabricate three-dimensional N-doped porous carbon electrocatalysts for the oxygen reduction reaction. *Phys. Chem. Chem. Phys.* **2016**, *18*, 4095–4101. [CrossRef] [PubMed]

109. Zhang, J.; Zhang, C.; Zhao, Y.; Amiinu, I.S.; Zhou, H.; Liu, X.; Tang, Y.; Mu, S. Three dimensional few-layer porous carbon nanosheets towards oxygen reduction. *Appl. Catal. B-Environ.* **2017**, *211*, 148–156. [CrossRef]

110. Zhong, H.X.; Wang, J.; Zhang, Y.W.; Xu, W.L.; Xing, W.; Xu, D.; Zhang, Y.F.; Zhang, X.B. ZIF-8 derived graphene-based nitrogen-doped porous carbon sheets as highly efficient and durable oxygen reduction electrocatalysts. *Angew. Chem. Int. Ed. Engl.* **2014**, *53*, 14235–14239. [CrossRef] [PubMed]

111. Yu, H.; Shang, L.; Bian, T.; Shi, R.; Waterhouse, G.I.; Zhao, Y.; Zhou, C.; Wu, L.Z.; Tung, C.H.; Zhang, T. Nitrogen-Doped Porous Carbon Nanosheets Templated from g-C$_3$N$_4$ as Metal-Free Electrocatalysts for Efficient Oxygen Reduction Reaction. *Adv. Mater.* **2016**, *28*, 5080–5086. [CrossRef] [PubMed]

112. Chen, C.; Sun, Z.; Li, Y.; Yi, L.; Hu, H. Self-assembly of N doped 3D porous carbon frameworks from carbon quantum dots and its application for oxygen reduction reaction. *J. Mater. Sci.-Mater. Electron.* **2017**, *28*, 12660–12669. [CrossRef]

113. Li, X.; Xue, X.; Fu, Y. Carbon Quantum Dots Derived N-Doped Porous Carbon Frameworks with High Electrocatalytic for Oxygen Reduction Reaction. *Nano* **2017**, *12*, 1750093. [CrossRef]

114. Luo, E.; Xiao, M.; Ge, J.; Liu, C.; Xing, W. Selectively doping pyridinic and pyrrolic nitrogen into a 3D porous carbon matrix through template-induced edge engineering: Enhanced catalytic activity towards the oxygen reduction reaction. *J. Mater. Chem. A* **2017**, *5*, 21709–21714. [CrossRef]

115. Wang, Y.; Liu, H.; Wang, K.; Song, S.; Tsiakaras, P. 3D interconnected hierarchically porous N-doped carbon with NH$_3$ activation for efficient oxygen reduction reaction. *Appl. Catal. B-Environ.* **2017**, *210*, 57–66. [CrossRef]

116. Huang, X.; Wang, Q.; Jiang, D.; Huang, Y. Facile synthesis of B, N co-doped three-dimensional porous graphitic carbon toward oxygen reduction reaction and oxygen evolution reaction. *Catal. Commun.* **2017**, *100*, 89–92. [CrossRef]

117. Jiang, H.; Wang, Y.; Hao, J.; Liu, Y.; Li, W.; Li, J. N and P co-functionalized three-dimensional porous carbon networks as efficient metal-free electrocatalysts for oxygen reduction reaction. *Carbon* **2017**, *122*, 64–73. [CrossRef]

118. Pei, Z.; Li, H.; Huang, Y.; Xue, Q.; Huang, Y.; Zhu, M.; Wang, Z.; Zhi, C. Texturing in situ: N,S-enriched hierarchically porous carbon as a highly active reversible oxygen electrocatalyst. *Energy Environ. Sci.* **2017**, *10*, 742–749. [CrossRef]

119. Chen, P.; Xiao, T.Y.; Qian, Y.H.; Li, S.S.; Yu, S.H. A nitrogen-doped graphene/carbon nanotube nanocomposite with synergistically enhanced electrochemical activity. *Adv. Mater.* **2013**, *25*, 3192–3196. [CrossRef] [PubMed]

120. Lee, J.-S.; Jo, K.; Lee, T.; Yun, T.; Cho, J.; Kim, B.-S. Facile synthesis of hybrid graphene and carbon nanotubes as a metal-free electrocatalyst with active dual interfaces for efficient oxygen reduction reaction. *J. Mater. Chem. A* **2013**, *1*, 9603–9607. [CrossRef]

121. Higgins, D.C.; Hoque, M.A.; Hassan, F.; Choi, J.-Y.; Kim, B.; Chen, Z. Oxygen Reduction on Graphene–Carbon Nanotube Composites Doped Sequentially with Nitrogen and Sulfur. *ACS Catal.* **2014**, *4*, 2734–2740. [CrossRef]

122. Liu, J.Y.; Wang, Z.; Chen, J.Y.; Wang, X. Nitrogen-Doped Carbon Nanotubes and Graphene Nanohybrid for Oxygen Reduction Reaction in Acidic, Alkaline and Neutral Solutions. *J. Nano Res.* **2015**, *30*, 50–58. [CrossRef]

123. Zhao, J.; Liu, Y.; Quan, X.; Chen, S.; Zhao, H.; Yu, H. Nitrogen and sulfur co-doped graphene/carbon nanotube as metal-free electrocatalyst for oxygen evolution reaction: The enhanced performance by sulfur doping. *Electrochim. Acta* **2016**, *204*, 169–175. [CrossRef]

124. Ma, Y.; Sun, L.; Huang, W.; Zhang, L.; Zhao, J.; Fan, Q.; Huang, W. Three-Dimensional Nitrogen-Doped Carbon Nanotubes/Graphene Structure Used as a Metal-Free Electrocatalyst for the Oxygen Reduction Reaction. *J. Phys. Chem. C* **2011**, *115*, 24592–24597. [CrossRef]

125. Yang, J.; Sun, H.; Liang, H.; Ji, H.; Song, L.; Gao, C.; Xu, H. A Highly Efficient Metal-Free Oxygen Reduction Electrocatalyst Assembled from Carbon Nanotubes and Graphene. *Adv. Mater.* **2016**, *28*, 4606–4613. [CrossRef] [PubMed]

126. Shi, Q.; Wang, Y.; Wang, Z.; Lei, Y.; Wang, B.; Wu, N.; Han, C.; Xie, S.; Gou, Y. Three-dimensional (3D) interconnected networks fabricated via in-situ growth of N-doped graphene/carbon nanotubes on Co-containing carbon nanofibers for enhanced oxygen reduction. *Nano Res.* **2015**, *9*, 317–328. [CrossRef]

127. Li, Y.; Zhou, W.; Wang, H.; Xie, L.; Liang, Y.; Wei, F.; Idrobo, J.C.; Pennycook, S.J.; Dai, H. An oxygen reduction electrocatalyst based on carbon nanotube-graphene complexes. *Nat. Nanotechnol.* **2012**, *7*, 394–400. [CrossRef] [PubMed]

128. Zhang, J.; Qu, L.; Shi, G.; Liu, J.; Chen, J.; Dai, L. N,P-Codoped Carbon Networks as Efficient Metal-free Bifunctional Catalysts for Oxygen Reduction and Hydrogen Evolution Reactions. *Angew. Chem. Int. Ed. Engl.* **2016**, *55*, 2230–2234. [CrossRef] [PubMed]

129. Chen, Y.; Wang, H.; Ji, S.; Lv, W.; Wang, R. Harvesting a 3D N-Doped Carbon Network from Waste Bean Dregs by Ionothermal Carbonization as an Electrocatalyst for an Oxygen Reduction Reaction. *Materials* **2017**, *10*, 1366. [CrossRef] [PubMed]

130. Mulyadi, A.; Zhang, Z.; Dutzer, M.; Liu, W.; Deng, Y. Facile approach for synthesis of doped carbon electrocatalyst from cellulose nanofibrils toward high-performance metal-free oxygen reduction and hydrogen evolution. *Nano Energy* **2017**, *32*, 336–346. [CrossRef]

131. Li, L.; Manthiram, A. O- and N-Doped Carbon Nanowebs as Metal-Free Catalysts for Hybrid Li-Air Batteries. *Adv. Energy Mater.* **2014**, *4*, 1301795. [CrossRef]

132. Guo, Q.; Zhao, D.; Liu, S.; Chen, S.; Hanif, M.; Hou, H. Free-standing nitrogen-doped carbon nanotubes at electrospun carbon nanofibers composite as an efficient electrocatalyst for oxygen reduction. *Electrochim. Acta* **2014**, *138*, 318–324. [CrossRef]

133. Meng, F.; Li, L.; Wu, Z.; Zhong, H.; Li, J.; Yan, J. Facile preparation of N-doped carbon nanofiber aerogels from bacterial cellulose as an efficient oxygen reduction reaction electrocatalyst. *Chin. J. Catal.* **2014**, *35*, 877–883. [CrossRef]

134. Li, J.-C.; Hou, P.-X.; Zhao, S.-Y.; Liu, C.; Tang, D.-M.; Cheng, M.; Zhang, F.; Cheng, H.-M. A 3D bi-functional porous N-doped carbon microtube sponge electrocatalyst for oxygen reduction and oxygen evolution reactions. *Energy Environ. Sci.* **2016**, *9*, 3079–3084. [CrossRef]

135. Wu, Z.; Liu, R.; Wang, J.; Zhu, J.; Xiao, W.; Xuan, C.; Lei, W.; Wang, D. Nitrogen and sulfur co-doping of 3D hollow-structured carbon spheres as an efficient and stable metal free catalyst for the oxygen reduction reaction. *Nanoscale* **2016**, *8*, 19086–19092. [CrossRef] [PubMed]

136. Shu, C.; Song, B.; Wei, X.; Liu, Y.; Tan, Q.; Chong, S.; Chen, Y.; Yang, X.-D.; Yang, W.-H.; Liu, Y. Mesoporous 3D nitrogen-doped yolk-shelled carbon spheres for direct methanol fuel cells with polymer fiber membranes. *Carbon* **2018**, *129*, 613–620. [CrossRef]

137. Guo, D.; Wei, H.; Chen, X.; Liu, M.; Ding, F.; Yang, Z.; Yang, Y.; Wang, S.; Yang, K.; Huang, S. 3D hierarchical nitrogen-doped carbon nanoflower derived from chitosan for efficient electrocatalytic oxygen reduction and high performance lithium–sulfur batteries. *J. Mater. Chem. A* **2017**, *5*, 18193–18206. [CrossRef]

catalysts

MDPI

Article

Preparation of $Ag_4Bi_2O_5$/MnO_2 Corn/Cob Like Nano Material as a Superior Catalyst for Oxygen Reduction Reaction in Alkaline Solution

Xun Zeng [1], Junqing Pan [1,*] and Yanzhi Sun [2,*]

[1] State Key Laboratory of Chemical Resource Engineering, Beijing Engineering Center for Hierarchical Catalysts, Beijing University of Chemical Technology, Beijing 100029, China; 2015201030@mail.buct.edu.cn
[2] National Fundamental Research Laboratory of New Hazardous Chemicals Assessment and Accident Analysis, Beijing University of Chemical Technology, Beijing 100029, China
* Correspondence: jqpan@mail.buct.edu.cn (J.P.); sunyz@mail.buct.edu.cn (Y.S.);
Tel./Fax: +86-10-6444-8461 (J.P. & Y.S.)

Received: 21 October 2017; Accepted: 30 November 2017; Published: 6 December 2017

Abstract: $Ag_4Bi_2O_5$/MnO_2 nano-sized material was synthesized by a co-precipitation method in concentrated KOH solution. The morphology characterization indicates that MnO_2 nanoparticles with a size of 20 nm are precipitated on the surface of nano $Ag_4Bi_2O_5$, forming a structure like corn on the cob. The obtained material with 60% Mn offers slightly higher initial potential (0.098 V vs. Hg/HgO) and limiting current density (-5.67 mA cm^{-2}) at a rotating speed of 1600 rpm compared to commercial Pt/C (-0.047 V and -5.35 mA cm^{-2}, respectively). Furthermore, the obtained material exhibits superior long-term durability and stronger methanol tolerance than commercial Pt/C. The remarkable features suggest that the $Ag_4Bi_2O_5$/MnO_2 nano-material is a very promising oxygen reduction reaction catalyst.

Keywords: fuel cells; manganese dioxide; silver bismuthate; alkaline; oxygen reduction reaction

1. Introduction

With the rapid consumption of fossil energy, the aggravated emission of carbon dioxide (CO_2) leads to severe environmental issues. In order to reduce the consumption of gasoline, fuel cells, especially Zn-O_2, Li-O_2, and the other metal-air batteries, have been considered as clean power sources, which can directly convert chemical energy into electrical energy with higher energy conversion efficiencies compared with traditional internal combustion engines [1]. However, the scarcity and high cost of platinum or platinum-based catalysts [2–4] commonly used in the cathodic oxygen reduction reaction (ORR) limit the commercialization of fuel cells. Additionally, the platinum (Pt) or platinum-based materials are sensitive to poisoning by methanol [5]. Recently, intensive research has been conducted to develop platinum free catalysts, such as silver (Ag) [6–9], manganese oxide [10–13], and silver/manganese oxide composite [14–16], which are promising catalysts with excellent performance at room temperature for alkaline fuel cells. Manganese oxides are promising catalysts for the oxygen reduction reaction owing to their outstanding activity toward oxygen reduction with the advantages of low cost and abundance [17]. Manganese dioxide (MnO_2) has good catalytic performance towards the oxygen reduction reaction through the integral four electrons mechanism, but its application is limited by the instability of the structure during the discharge process [18]. Silver and its complexes have a good ability to catalyze the oxygen reduction reaction. They were also reported to be more stable than Pt and Pt-based catalysts during long-term operation in alkaline media [7]. Ag and Ag-based materials are insensitive to methanol and can be applied in direct methanol fuel cells (DMFCs) [19]. In addition, Ag and its complexes have a lower price than Pt catalysts. The advantages of reasonably high electrochemical activity, long-term stability, methanol tolerance, and lower price compared with Pt make Ag and Ag-based composites attractive as catalysts for the oxygen reduction reaction in alkaline media, especially for DMFCs due to the high methanol tolerance of silver. However, the cost of silver and its complexes is still high compared with some other transition metals. It is urgent to develop a facile and green method to synthetize a new catalyst with low cost, relatively excellent electrocatalytic ability for the oxygen reduction reaction, and good stability and methanol tolerance in an alkaline electrolyte.

In this study, a facile method was proposed to prepare nano $Ag_4Bi_2O_5/MnO_2$ material. MnO_2 nano-particles were evenly precipitated on the surface of $Ag_4Bi_2O_5$ nano-rods to form a structure like corn on the cob. The synthesis method has the advantages of low temperature and short synthesis period, and more importantly, MnO_2 can be effectively anchored on the surface of $Ag_4Bi_2O_5$ nano-rods. This material has a comparable catalytic activity to the commercial Pt/C in terms of the oxygen reduction reaction. The catalyst displays superior stability and methanol tolerance compared with the commercial Pt/C. Therefore, nano $Ag_4Bi_2O_5/MnO_2$ material can be an effective catalyst for the oxygen reduction reaction in alkaline solution.

2. Results and Discussion

2.1. Illustration of the Synthesis Process

Figure 1 shows the schematic of the synthesis process of $Ag_4Bi_2O_5/MnO_2$ nano-material. This material was based on $Ag_4Bi_2O_5$ nano-rods with manganese oxide nanoparticles deposited on the surface to form a structure like corn on the cob. The synthesis method was so facile that manipulation at room temperature was enough to effectively precipitate MnO_2 on the surface of the $Ag_4Bi_2O_5$ rods. This structure is conducive to the synergistic effect of the three metals (Ag, Bi, and Mn) and the adsorption of O_2 on the surface of the catalyst as well as the disconnection of the O-O bonds. The nanostructure can effectively reduce the size of the catalyst and increase the specific surface area so that more catalytic activity sites are exposed and the oxygen reduction process can be catalyzed in a stable and efficient way [20]. Therefore, $Ag_4Bi_2O_5/MnO_2$ can be considered as a promising catalyst for the oxygen reduction reaction.

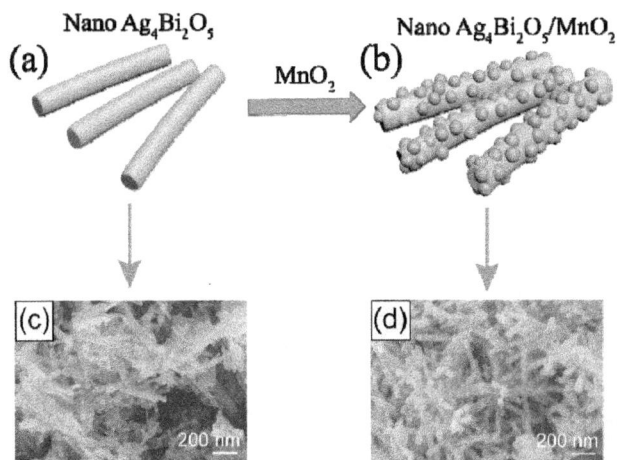

Figure 1. Schematic drawing of the synthesis process of nano $Ag_4Bi_2O_5/MnO_2$ material. (**a**) diagram of nano $Ag_4Bi_2O_5$; (**b**) diagram of nano $Ag_4Bi_2O_5/MnO_2$; (**c**) SEM image of nano $Ag_4Bi_2O_5$; (**d**) SEM image of nano $Ag_4Bi_2O_5$.

2.2. Structural and Morphological Characterizations

As is clearly illustrated in Figure 2, when $Ag_4Bi_2O_5$ is added to manganese dioxide, the X-ray powder diffraction (XRD) patterns of the samples (0–60% Mn) still have the typical characteristic peaks of $Ag_4Bi_2O_5$. From the XRD patterns, it was found that the samples were typical at $2\theta = 26.37°$, $31.25°$, $31.85°$, $37.76°$ and $56.19°$ corresponding to (112), (411), (312), (600), and (332) of the $Ag_4Bi_2O_5$, respectively, according to the standard JCPDS 87-0866 of $Ag_4Bi_2O_5$. The peaks of $Ag_4Bi_2O_5/MnO_2$ are broad, indicating that the incorporation of Mn affects the crystalline structure of $Ag_4Bi_2O_5$. There are no typical peaks of MnO_2 because the MnO_2 is amorphous.

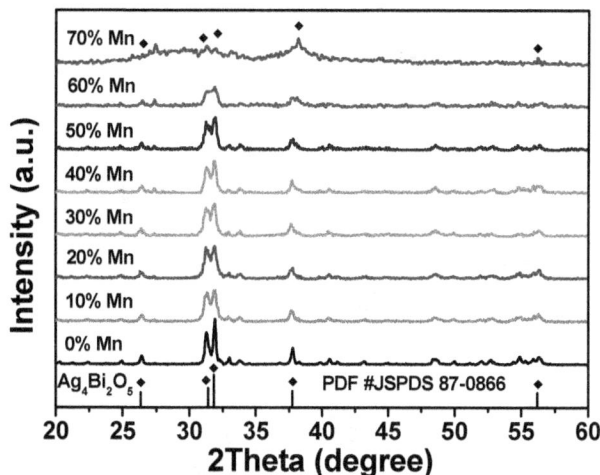

Figure 2. X-ray powder diffraction (XRD) patterns of $Ag_4Bi_2O_5/MnO_2$ with different ratios of manganese dioxide contents from 0% to 70%.

Scanning electron microscopy (SEM) images in Figure 3 of the $Ag_4Bi_2O_5/MnO_2$ samples show the morphological and structural information. It can be seen in Figure 3a that nano $Ag_4Bi_2O_5$ are smooth rods with length and width of 200 nm and 30 nm, respectively. When the amount of manganese dioxide is 10–60%, there are some nano particles on the rod-like $Ag_4Bi_2O_5$, and the length and width is still about 200 nm and 30 nm, respectively. With the increase of the ratio of manganese dioxide, there are more nano particles on the nano rods. The structure is beneficial to the synergistic effect of Ag, Bi, and Mn in catalyzing the oxygen reduction reaction. The nanostructure can reduce the size of the catalyst and more catalytic activity sites are exposed so that the oxygen reduction process can be catalyzed in a stable and efficient manner. When the amount of manganese oxide is 70%, the material particles become larger, and the length and width are about 400 nm and 100 nm, respectively.

The results of transmission electron microscope (TEM) and high-resolution transmission electron microscopy (HRTEM) are shown in Figure 4. It can be seen from Figure 4a that MnO_2 is distributed on the surface of rod-like $Ag_4Bi_2O_5$. The HRTEM image of $Ag_4Bi_2O_5$ with 60% Mn was used to further demonstrate the detailed structural features of the material. The distances between the lattice planes are 0.333 nm and 0.417 nm corresponding to the planes of $Ag_4Bi_2O_5$ (112) and $Ag_4Bi_2O_5$ (301), respectively (Figure 4b). There are no planes of MnO_2 because it is amorphous. This result is consistent with the result of XRD.

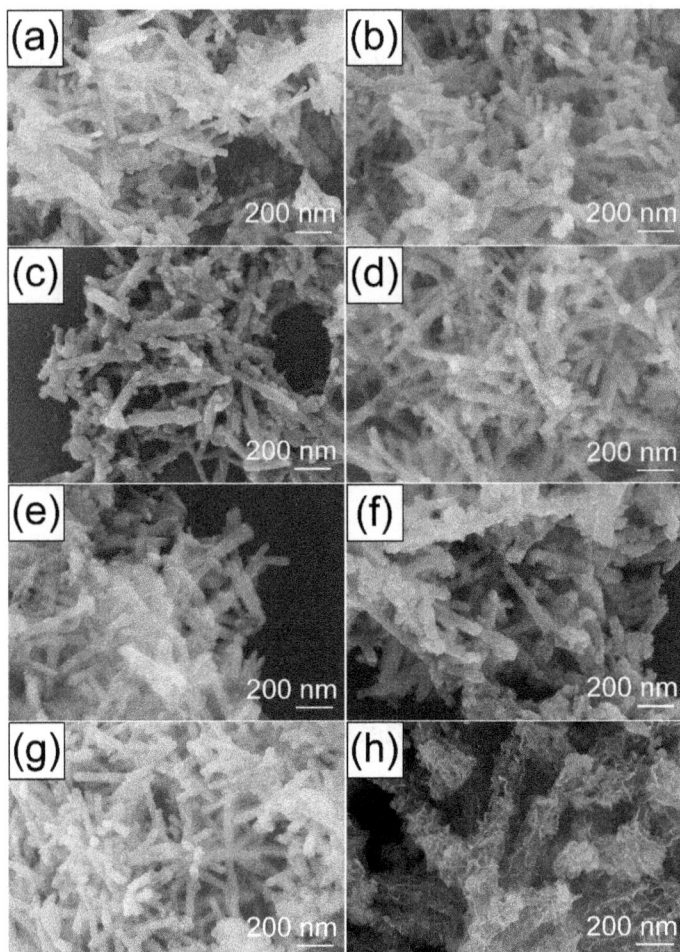

Figure 3. Scanning electron microscopy (SEM) images of $Ag_4Bi_2O_5/MnO_2$ with different ratios of manganese dioxides contents, (**a**) 0%; (**b**) 10%; (**c**) 20%; (**d**) 30%; (**e**) 40%; (**f**) 50%; (**g**) 60%; and (**h**) 70% Mn.

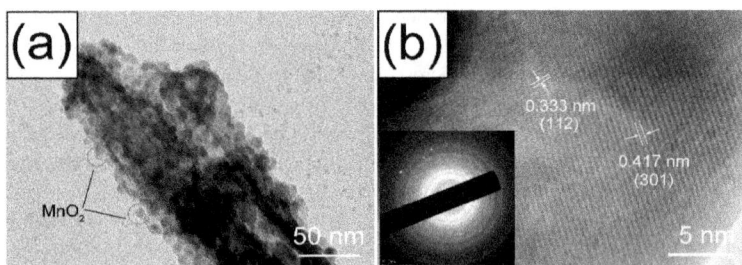

Figure 4. (**a**) Transmission electron microscopy (TEM) image of $Ag_4Bi_2O_5$ with 60% Mn, (**b**) high-resolution transmission electron microscopy (HRTEM) image of $Ag_4Bi_2O_5$ with 60% Mn.

In order to explore the elemental distribution of $Ag_4Bi_2O_5$ with 60% Mn, mapping analysis of $Ag_4Bi_2O_5$ with 60% Mn was performed as in Figure 5. It can be seen that the four elements of O, Ag, Bi, and Mn are evenly distributed. Energy dispersive spectrometer (EDS) analysis was also performed to obtain the chemical composition shown in Figure 6. It is proved that the presence of O, Ag, Bi, and Mn elements and the successful synthesis of $Ag_4Bi_2O_5$ with 60% Mn. The atomic ratios of Ag, Bi, and Mn are 15.56%, 7.57%, and 6.15%, respectively. The ratio of Bi and Mn is 1.23, which is close to the theoretical ratio of 1.33. The atomic ratio of Ag:Bi conforms to the atomic ratio 2:1 in the $Ag_4Bi_2O_5$.

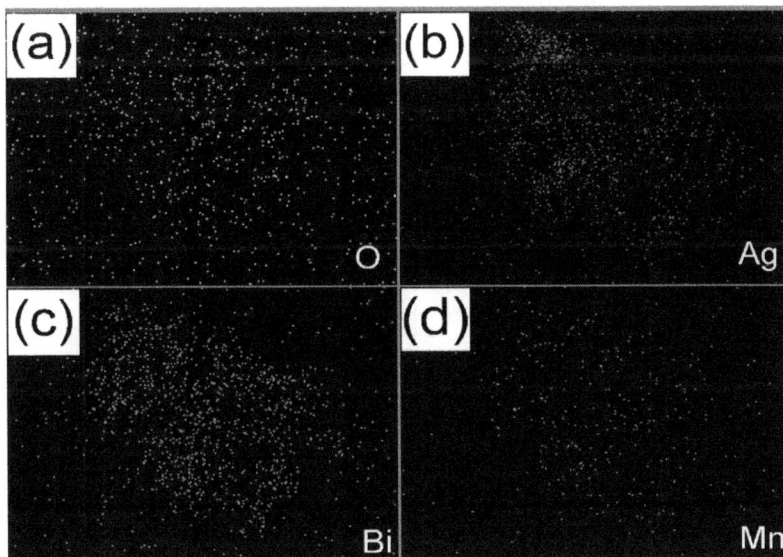

Figure 5. Elemental mapping images of $Ag_4Bi_2O_5$ with 60% Mn (**a**) O element, (**b**) Ag element, (**c**) Bi element, (**d**) Mn element.

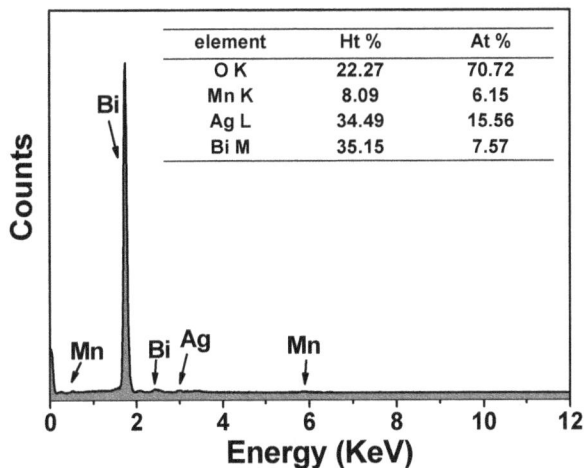

element	Ht %	At %
O K	22.27	70.72
Mn K	8.09	6.15
Ag L	34.49	15.56
Bi M	35.15	7.57

Figure 6. Energy dispersive spectrometry (EDS) analysis of $Ag_4Bi_2O_5$ with 60% Mn.

To investigate the elemental compositions and the valences of the elements in the as prepared $Ag_4Bi_2O_5$ with 60% Mn composite, X-ray photoelectron spectroscopy (XPS) was employed. The results are shown in Figure 7. Figure 7a shows the XPS survey spectra. Figure 7b–d shows the high resolution spectra of Ag 3d, Bi 4f, and Mn 2p of $Ag_4Bi_2O_5/MnO_2$, respectively. In Figure 7b, it can be seen that the peaks at 368.2 eV and 374.2 eV are Ag $3d_{5/2}$ and Ag $3d_{3/2}$, respectively. These two peaks show that the valence state of Ag is +1 [21]. The two peaks at 158.8 eV and 163.9 eV in Figure 7c correspond to Bi $3f_{7/2}$ and Bi $3f_{5/2}$, respectively, proving the presence of Bi of +3 [21]. Figure 7d shows the XPS spectra of the 2p orbital of the Mn element in $Ag_4Bi_2O_5/MnO_2$. The two peaks at 642 eV and 653.7 eV correspond to Mn $2p_{3/2}$ and Mn $2p_{1/2}$ of Mn^{4+}, respectively, which indicates that the valence state of the whole material is dominated by MnO_2 which is on the surface [22].

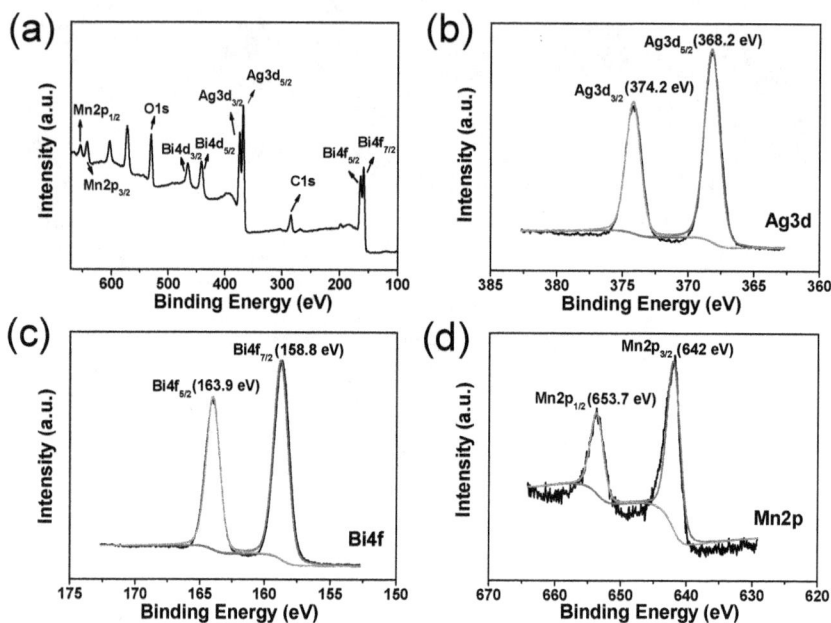

Figure 7. X-ray photoelectron spectroscopy (XPS) spectra of (a) $Ag_4Bi_2O_5$ with 60% Mn, (b) XPS spectra of Ag 3d, (c) XPS spectra of Bi 4f, (d) XPS spectra of Mn 2p.

2.3. Electrocatalytic Performance

To provide an insight into the activity of ORR on the $Ag_4Bi_2O_5/MnO_2$ material, linear sweep voltammetry (LSV) plots were tested by a rotating disk electrode (RDE) at a speed of 1600 rpm in O_2-saturated 0.1 mol L^{-1} KOH solution at a scanning rate of 5 mV s^{-1}. MnO_2, $Ag_4Bi_2O_5$, and commercial Pt/C was also investigated under the same conditions. The starting potential is the potential at the limiting current density of 0.1 mA cm^{-2} [23]. It is shown in Figure 8a and Table 1 that the initial potential (v_o), half-wave potential ($v_{1/2}$), and limiting current density (j) of $Ag_4Bi_2O_5$ is −0.09 V, −0.227 V, and −2.06 mA cm^{-2}, respectively. With the increase of content of MnO_2, the materials have a better ability to catalyze ORR due to the increase of the synergistic effects of Ag, Bi, and Mn. The initial potential, half-wave potential and limiting current density of $Ag_4Bi_2O_5$ with 60% Mn are 0.098 V, −0.047 V, and −5.67 mA cm^{-2}, respectively, which shows that this has the best ability to catalyze the oxygen reduction reaction compared to any other ratio of $Ag_4Bi_2O_5/MnO_2$. In addition, $Ag_4Bi_2O_5$ with 70% Mn is poor for catalyzing ORR although MnO_2 is increased. This result is due to the particles of the material becoming larger, leading to fewer catalytic activity sites thus reducing the activity for catalyzing the oxygen reduction reaction compared with 10–60% Mn. Figure 8b shows the RDE curves of $Ag_4Bi_2O_5$ with 60% Mn at different rotation speeds (400–2500 rpm). Figure 8c shows the Koutecky-Levich (K-L) plots which describe the relation between the inverse real current density (j^{-1}) and the inverse of the square root of the rotating rate ($\omega^{-1/2}$) [22]. The good linearity of the K-L plots reveals that the kinetics of the ORR is first-order with respect to the concentration of dissolved oxygen and similar electron transfer numbers at different potentials [24].

$$\frac{1}{i} = \frac{1}{i_k} + \frac{1}{i_d}$$

$$i_k = nfkC_0$$

$$i_d = 0.62nFD_0^{2/3}v^{-1/6}C_0\omega^{1/2}$$

Here, n is number of electrons involved in the reaction, F is the Faraday constant (96,500 C mol^{-1}), D_0 is the diffusion coefficient of oxygen in the electrolyte (0.1 mol L^{-1} KOH solution) (1.93×10^{-5} cm^2 s^{-1}), v is the viscosity coefficient of the solution (0.1 cm^2 s^{-1}), C_0 is the concentration of O_2 in 0.1 mol L^{-1} KOH solution (1.26×10^{-3} mol L^{-1}), ω is the speed of the disc (rad s^{-1}) [25,26]. As calculated from Figure 8d in the potential range from −0.25 V to −0.35 V, the average number of transferring electrons of ORR on the $Ag_4Bi_2O_5$ with 60% Mn is about 3.9, indicating that the ORR catalyzed by the $Ag_4Bi_2O_5$ with 60% Mn follows the most efficient four-electron mechanism.

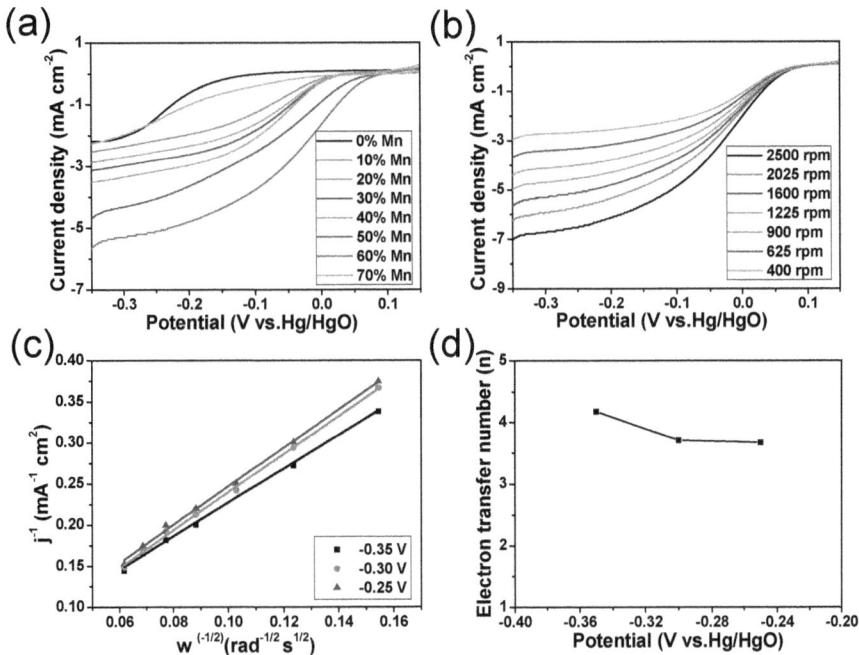

Figure 8. In 0.1 mol L^{-1} KOH solution, (a) linear sweep voltammetry (LSV) curves of $Ag_4Bi_2O_5$ with different ratios of manganese dioxides (0% Mn to 70% Mn) with O_2 saturation at a potential scanning rate of 5 mV s^{-1} and a rotation speed of 1600 rpm, (b) LSV curves of $Ag_4Bi_2O_5$ with 60% Mn at different rotation speeds (400–2500 rpm) at scanning rate of 5 mV s^{-1}, (c) Koutecky-Levich curves of $Ag_4Bi_2O_5$ with 60% Mn at different potentials, (d) Number of transferring electrons at the corresponding potentials.

Table 1. Initial potential (v_o), half-wave potential ($v_{1/2}$) and limiting current density (j) of commercial Pt/C and $Ag_4Bi_2O_5$ with different contents of MnO_2.

Parameter	v_o (V vs. Hg/HgO)	$v_{1/2}$ (V vs. Hg/HgO)	J (mA cm^{-2})
0% Mn	−0.09	−0.227	−2.06
10% Mn	0.042	−0.123	−3.00
20% Mn	0.046	−0.099	−3.086
30% Mn	0.052	−0.090	−3.23
40% Mn	0.052	−0.078	−3.52
50% Mn	0.111	−0.084	−4.70
60% Mn	0.098	−0.047	−5.67
70% Mn	0.107	−0.219	−2.41
100% Mn	−0.014	−0.203	−2.19
Pt/C	0.179	−0.014	−5.09

To further elucidate the electrochemical performance of the $Ag_4Bi_2O_5/MnO_2$ materials, the electrochemically active surface area (ECSA) can be evaluated through a simple cyclic voltammetry (CV) measurement as in Figure S1 [27]. The calculated electrochemically effective double-layer capacitances (C_{dl}) shown in Figure S2 are 0.92, 1.39, 1.51, 2.07, 2.47, 3.78, 5.02, and 1.34 mF cm^{-2} for 0%, 10%, 20%, 30%, 40%, 50%, 60%, and 70% Mn, respectively. This result indicates that $Ag_4Bi_2O_5$ with 60% Mn can expose most electrochemically active sites during the ORR process.

It is inferred that the obtained $Ag_4Bi_2O_5$ with 60% Mn catalyst offers a larger enhanced electrocatalytic activity than $Ag_4Bi_2O_5$ and MnO_2 in Figure 9a,b. The initial potential (v_o), half-wave potential ($v_{1/2}$) and limiting current density (j) of $Ag_4Bi_2O_5$ with 60% Mn are 0.098 V, −0.047 V, and −5.67 mA cm^{-2}, respectively. Compared with commercial Pt/C, the initial potential and half-wave potential of $Ag_4Bi_2O_5$ with 60% Mn are 0.005 V more positive (Pt/C, 0.093 V) and 0.017 V more negative (Pt/C, −0.03 V), respectively, and the limiting current density is slightly higher than Pt/C, showing the performance of $Ag_4Bi_2O_5$ with 60% Mn toward oxygen reduction reaction is close to Pt/C in potential and slightly superior in current density. Previous papers showed that the doping of Bi_2O_3 and $NaBiO_3$ into MnO_2 would enhance the durability during the discharge process [21,28,29]. The new material of $Ag_4Bi_2O_5/MnO_2$ displays two beneficial functions for both couples of Mn-Bi and Ag-Mn. The former plays an important role of strong support of the structure of MnO_2, which will greatly enhance the durability of the catalyst, and the latter offers remarkable catalytic performance compared to the single Ag or Mn compound. The electrocatalytic activity of $Ag_4Bi_2O_5$ with 60% Mn towards ORR was examined by CV in 0.1 mol L^{-1} KOH solution saturated with O_2 and Ar, respectively, at a sweep rate of 50 mV s^{-1}. The oxygen reduction peak of $Ag_4Bi_2O_5$ with 60% Mn is at −0.09 V and the peak current density is 1.53 mA cm^{-2} in electrolyte solution saturated with O_2, as shown in Figure 9c,d shows the Tafel of $Ag_4Bi_2O_5$ with 60% Mn and commercial Pt/C. The Tafel slope of the $Ag_4Bi_2O_5$ with 60% Mn and commercial Pt/C are 73 and 68 mV dec^{-1}, respectively. The low ORR Tafel slope of $Ag_4Bi_2O_5$ with 60% Mn implies a transition from Langmuirian adsorption to Temkin adsorption of adsorbed O/OH groups [30]. Therefore, the electrochemical catalyst activation of $Ag_4Bi_2O_5$ with 60% Mn is close to that of Pt/C [15]. The long-term stability was assessed in O_2-saturated 0.1 mol L^{-1} KOH solution by chrono-amperometry. $Ag_4Bi_2O_5/MnO_2$ has a small decay of −13% in ORR activity after 10,800 s, while Pt/C has an attenuation of 18% shown in Figure 9e. In addition, the current density of the Pt/C shows a much sharper decrease when methanol is introduced in the electrolyte (Figure 9f), while the current density of $Ag_4Bi_2O_5/MnO_2$ only exhibits a small decrease under the same conditions. This means that $Ag_4Bi_2O_5/MnO_2$ displays superior stability and methanol tolerance compared to the commercial Pt/C.

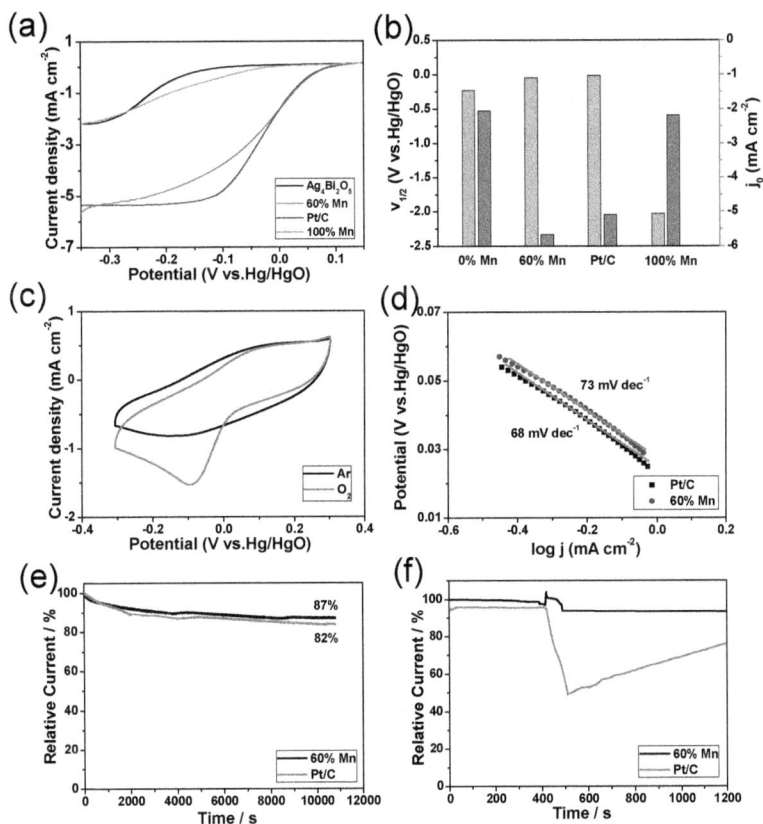

Figure 9. (**a**) LSV curves of $Ag_4Bi_2O_5$ with 60% Mn, Pt/C and MnO_2 by rotating disk electrode (RDE) in KOH solution with O_2 saturation at a speed of 5 mV s^{-1} and a rotation speed of 1600 rpm, (**b**) comparison between commercial Pt/C and $Ag_4Bi_2O_5$ with 0% Mn, 60% Mn, 100% Mn, at half-wave potential ($v_{1/2}$) and current density (j), (**c**) cyclic voltammetry (CV) curves of $Ag_4Bi_2O_5$ with 60% Mn at a scan rate of 50 mV s^{-1} with Ar and O_2 saturation, respectively, (**d**) Tafel curves of $Ag_4Bi_2O_5$ with 60% Mn and commercial Pt/C, (**e**) i-t chrono-amperometric curves of $Ag_4Bi_2O_5$ with 60% Mn and Pt/C at -0.26 V with O_2 saturation at a rotation speed of 1600 rpm, (**f**) methanol tolerance of $Ag_4Bi_2O_5$ with 60% Mn and commercial Pt/C by chronoamperometric response with O_2 saturation for 1200 s and adding 3.0 mol L^{-1} CH$_3$OH at about the 420th second.

Table 2. Comparison of oxygen reduction reaction (ORR) catalytic activities between $Ag_4Bi_2O_5$ with 60% Mn and the other relevant catalysts.

Catalyst	E_0 (V vs. Hg/HgO)	$E_{1/2}$ (V vs. Hg/HgO)	Currents (mA cm^{-2})	Stability	Reference
$Ag_4Bi_2O_5$ with 60% Mn	0.098	-0.047	5.67	88% (10,800 s)	This work
Mn_3O_4/NrGO	-0.1	-0.2	4.4	63% (21,600 s)	[10]
Ag@MnFe$_2$O$_4$/C	-0.08	-0.171	4.8	88.7% (15,000 s)	[31]
50% Ag-MnO$_2$/C	-0.036	-0.216	5.5	91% (50,000 s)	[32]
SC-PMO	-0.037	-0.237	5.0	82% (16,000 s)	[33]
C$_{PANI}$/Mn$_2$O$_3$	0.108	-0.082	5.61	91.1% (80,000 s)	[34]
rGO/MnO$_2$/Ag	0.034	-0.126	3.4	94% (10,000 s)	[35]
Ag-PBMO$_5$	0.054	-0.056	5.0	91% (50,000 s)	[36]
Ag/Cu$_{37}$Pd$_{63}$	0.05	-0.04	2.96	77.6% (48,000 s)	[37]

Recently, several papers all from prestigious journals have reported several catalysts with good ORR performance. The catalytic properties of these materials and the as-prepared $Ag_4Bi_2O_5$ with 60% Mn in this study are compared in Table 2. Among all the listed catalysts, the $Ag_4Bi_2O_5$ with 60% Mn shows the most efficient ORR performance with positive initial potential and half-wave potential, high limiting current density, and excellent long-term stability [33–37]. The comparison shows that $Ag_4Bi_2O_5$ with 60% Mn is a highly efficient catalyst for ORR.

3. Experimental

3.1. Reagents

All the chemicals were analytical grade and used as received without further purification. Ag_2O, Bi_2O_3, $Mn(NO_3)_2$, HNO_3, and isopropanol were commercially available from Tianjin Fuchen Industry Co. Ltd. (Tianjin, China). KOH was purchased from Chengdu Huarong Chemical Reagent Co. Ltd. (Chengdu, China). Commercial Pt/C (20 wt%, Hispec 3000) was obtained from Johnson Matthey Company (London, UK). Conductive graphite (Timcal-ks6) was available Timical Company (Changzhou, China). Nafion solution (5 wt%, D520) was from DuPont (Wilmington, NC, USA).

3.2. Synthesis and Physical Characterizations

An amount of 100 mL 6.5 mol L^{-1} KOH was employed as basic solution, marked as solution A. Then 1.16 g Ag_2O and 1.17 g Bi_2O_3 were dissolved in 50 mL 1 mol L^{-1} HNO_3, and the mixture solution was marked as solution B. The solution B was added into the solution A at a rate of 3 mL min^{-1}. After complete reaction of the solutions A and B, the resulting solution was aged for 1 h and marked as solution C.

Amounts of 0 g, 0.1 g, 0.22 g, 0.38 g, 0.60 g, 0.89 g, 1.34 g, and 2.09 g of 50% $Mn(NO_3)_2$ solutions were dissolved in deionized water to prepare portions of 20 mL solutions, respectively. These eight $Mn(NO_3)_2$ solutions were dripped into the above solution C at a rate of 3 mL min^{-1}, respectively. After crystallization for 2 h, the samples were washed with deionized water until pH = 7, and then dried in vacuum at 303.15 K for 8 h. The materials were named 0% Mn, 10% Mn, 20% Mn, 30% Mn, 40% Mn, 50% Mn, 60% Mn, and 70% Mn by ratio of moles.

All the preparation was conducted under conditions of room temperature and a strong agitation at 2000 rpm.

3.3. Physicochemical Characterization

The phase structures of samples were analyzed by a Rigaku D/max2500VB2+/PC X-ray diffractometer (XRD) with a Cu Kα anticathode (40 kV, 200 mA) (Rigaku Corporation, Tokyo, Japan). The morphology and surface structure of $Ag_4Bi_2O_5/MnO_2$ were investigated by using scanning electron microscopy (SEM, ZEISS, SUPRA 55, Carl Zeiss AG, Oberkochen, Germany), transmission electron microscope (TEM, Hitachi, H-7700, Hitachi Company, Tokyo, Japan) and high resolution transmission electron microscope (HR-TEM, JEOL JEM-2100F, JEOL Company, Tokyo, Japan). The elemental mapping and energy dispersive spectrometer (EDS) was carried out on an X-ray energy instrument (LINK-ISIS300, Oxford, UK). The valences of elements were analyzed by using an X-ray photoelectron spectrometer (XPS, Thermo VG Scientic ESCALAB 250, Thermo Fisher Scientific, Waltham, MA, USA).

3.4. Electrochemical Measurements

The electrochemical performance of the $Ag_4Bi_2O_5/MnO_2$ samples was tested on a rotating disk electrode with a diameter of 5 mm (RDE, AFMRSCE, Pine Instrument, Grove, PA, USA) at rotation speeds of 400–2500 rpm and a potential sweeping rate of 5 mV s^{-1} by linear sweep voltammetry (LSV). A Hg/HgO electrode was used as the reference electrode, and a platinum wire as the auxiliary electrode. The working electrode was prepared as follows: 8 mg catalyst and 4 mg conductive graphite were dispersed ultrasonically for 30 min in a mixture of 55 μL Nafion solution, 430 μL isopropanol and 650 μL deionized water to form catalyst ink. Then 5.6 μL of the ink was spread on the glassy carbon electrode (GCE) and dried at room temperature. The working electrode was loaded with a catalyst amount of 0.2 mg cm^{-2}. Cyclic voltammetry (CV) of $Ag_4Bi_2O_5/MnO_2$ was conducted using an electrochemical workstation (CHI 760d, Chenhua Instruments, Shanghai, China). The potential range of the CV was between −0.3 V and 0.3 V (V vs. Hg/HgO). The electrochemical tests of LSV, CV, and Tafel were performed in 0.1 mol L^{-1} KOH solution saturated with oxygen (O_2) or argon (Ar) gas. The i-t chronoamperometric was tested at −0.26 V with O_2 saturation for 10,800 s. The methanol tolerance test was performed in 0.1 mol L^{-1} KOH solution saturated with O_2 for 1200 s and 3.0 mol L^{-1} CH_3OH was added at about the 420th second. The results were shown by chronoamperometry curves. For comparison, $Ag_4Bi_2O_5$, and MnO_2 were also tested by the same procedure.

4. Conclusions

$Ag_4Bi_2O_5/MnO_2$ nano-material was successfully synthesized by a co-precipitation method. The MnO_2 nano particles were distributed on the surface of rod-like $Ag_4Bi_2O_5$ crystals to form a structure like corn on the cob. Compared with a commercial Pt/C catalyst, the $Ag_4Bi_2O_5$ with 60% Mn nano-material has an equivalent ORR activity in alkaline media, but a better stability and methanol tolerance. The excellent electrochemical activity of $Ag_4Bi_2O_5/MnO_2$ might benefit from the synergistic effects of Ag, Bi, and Mn as well as more active sites. The synergistic effects of the Ag-Mn elements remarkably enhanced the catalytic activity. The Bi-Mn elements and the corn-cob like structure of $Ag_4Bi_2O_5/MnO_2$ are propitious to strengthen the stability of the catalyst. More active sites contribute to the adsorption of O_2 on the catalyst surface and the breakage of O-O bonds. In alkaline conditions, the catalyst can catalyze the oxygen reduction reaction through the four-electron mechanism. Therefore, nano $Ag_4Bi_2O_5/MnO_2$ has proved to be a very promising oxygen reduction catalyst under alkaline conditions.

Supplementary Materials: The following are available online at www.mdpi.com/2073-4344/7/12/379/s1, Figure S1: CV curves of $Ag_4Bi_2O_5$ with different content of MnO_2, (a) 0%; (b) 10%; (c) 20%; (d) 30%; (e) 40%; (f) 50%; (g) 60%; (h) 70%, Figure S2: Extraction of the C_{dl} for $Ag_4Bi_2O_5/MnO_2$ with different ratios of manganese dioxides.

Acknowledgments: This work was supported by the National Natural Science Foundation of China (21676022 & 21706004), the State Key Program of National Natural Science of China (21236003), and the Fundamental Research Funds for the Central Universities (BHYC1701A & JD1701).

Author Contributions: Junqing Pan proposed the concept, and supervised the research work at BUCT. Xun Zeng, Junqing Pan, and Yanzhi Sun designed the experiments; Xun Zeng performed the experiments. All authors analyzed the data and wrote the paper.

Conflicts of Interest: The authors declare no conflict of interest.

References

1. Steele, B.C.; Heinzel, A. Materials for fuel-cell technologies. *Nature* **2001**, *414*, 345–352. [CrossRef] [PubMed]
2. Chauhan, S.; Mori, T.; Masuda, T.; Ueda, S.; Richards, G.J.; Hill, J.P.; Ariga, K.; Isaka, N.; Auchterlonie, G.; Drennan, J. Design of Low Pt Concentration Electrocatalyst Surfaces with High Oxygen Reduction Reaction Activity Promoted by Formation of a Heterogeneous Interface between Pt and CeO$_x$ Nanowire. *ACS Appl. Mater. Interfaces* **2016**, *8*, 9059–9070. [CrossRef] [PubMed]
3. Escuderoescribano, M.; Malacrida, P.; Hansen, M.H.; Vejhansen, U.G.; Velazquezpalenzuela, A.; Tripkovic, V.; Schiotz, J.; Rossmeisl, J.; Stephens, I.E.; Chorkendorff, I. Tuning the activity of Pt alloy electrocatalysts by means of the lanthanide contraction. *Science* **2016**, *352*, 73–76. [CrossRef] [PubMed]
4. Gan, L.; Heggen, M.; Cui, C.; Strasser, P. Thermal Facet Healing of Concave Octahedral Pt-Ni Nanoparticles Imaged in Situ at the Atomic Scale: Implications for the Rational Synthesis of Durable High-Performance ORR Electrocatalysts. *ACS Catal.* **2016**, *6*, 692–695. [CrossRef]
5. Gautam, R.K.; Bhattacharjee, H.; Venkata Mohan, S.; Verma, A. Nitrogen doped graphene supported α-MnO$_2$ nanorods for efficient ORR in a microbial fuel cell. *RSC Adv.* **2016**, *6*, 110091–110101. [CrossRef]

6. Lu, Y.; Chen, W. Size effect of silver nanoclusters on their catalytic activity for oxygen electro-reduction. *J. Power Sources* **2012**, *197*, 107–110. [CrossRef]
7. Liu, M.; Chen, W. Green synthesis of silver nanoclusters supported on carbon nanodots: Enhanced photoluminescence and high catalytic activity for oxygen reduction reaction. *Nanoscale* **2013**, *5*, 12558–12564. [CrossRef] [PubMed]
8. Sekol, R.C.; Li, X.; Cohen, P.; Doubek, G.; Carmo, M.; Taylor, A.D. Silver palladium core-shell electrocatalyst supported on MWNTs for ORR in alkaline media. *Appl. Catal. B Environ.* **2013**, *138*, 285–293. [CrossRef]
9. Wang, Q.; Cui, X.; Guan, W.; Zhang, L.; Fan, X.; Shi, Z.; Zheng, W. Shape-dependent catalytic activity of oxygen reduction reaction (ORR) on silver nanodecahedra and nanocubes. *J. Power Sources* **2014**, *269*, 152–157. [CrossRef]
10. Bikkarolla, S.K.; Yu, F.; Zhou, W.; Joseph, P.; Cumpson, P.; Papakonstantinou, P. A three-dimensional Mn_3O_4 network supported on a nitrogenated graphene electrocatalyst for efficient oxygen reduction reaction in alkaline media. *J. Mater. Chem. A* **2014**, *2*, 14493. [CrossRef]
11. Meng, Y.; Song, W.; Huang, H.; Ren, Z.; Chen, S.Y.; Suib, S.L. Structure-property relationship of bifunctional MnO_2 nanostructures: Highly efficient, ultra-stable electrochemical water oxidation and oxygen reduction reaction catalysts identified in alkaline media. *J. Am. Chem. Soc.* **2014**, *136*, 11452–11464. [CrossRef] [PubMed]
12. Wei, C.; Yu, L.; Cui, C.; Lin, J.; Wei, C.; Mathews, N.; Huo, F.; Sritharan, T.; Xu, Z. Ultrathin MnO_2 nanoflakes as efficient catalysts for oxygen reduction reaction. *Chem. Commun.* **2014**, *50*, 7885–7888. [CrossRef] [PubMed]
13. Du, J.; Chen, C.; Cheng, F.; Chen, J. Rapid Synthesis and Efficient Electrocatalytic Oxygen Reduction/Evolution Reaction of $CoMn_2O_4$ Nanodots Supported on Graphene. *Inorg. Chem.* **2015**, *54*, 5467–5474. [CrossRef] [PubMed]
14. Liu, J.; Liu, J.; Song, W.; Wang, F.; Song, Y. The role of electronic interaction in the use of Ag and Mn_3O_4 hybrid nanocrystals covalently coupled with carbon as advanced oxygen reduction electrocatalysts. *J. Mater. Chem. A* **2014**, *2*, 17477–17488. [CrossRef]
15. Liu, S.; Qin, X. Preparation of a $Ag-MnO_2$/graphene composite for the oxygen reduction reaction in alkaline solution. *RSC Adv.* **2015**, *5*, 15627–15633. [CrossRef]
16. Park, S.A.; Lim, H.; Kim, Y.T. Enhanced Oxygen Reduction Reaction Activity Due to Electronic Effects between Ag and Mn_3O_4 in Alkaline Media. *ACS Catal.* **2015**, *5*, 3995–4002. [CrossRef]
17. Ryabova, A.S.; Napolskiy, F.S.; Poux, T.; Istomin, S.Y.; Bonnefont, A.; Antipin, D.M.; Baranchikov, A.Y.; Levin, E.E.; Abakumov, A.M.; Kéranguéven, G.; et al. Rationalizing the Influence of the $Mn(IV)/Mn(III)$ Red-Ox Transition on the Electrocatalytic Activity of Manganese Oxides in the Oxygen Reduction Reaction. *Electrochim. Acta* **2016**, *187*, 161–172. [CrossRef]
18. Li, L.; Hu, Z.A.; An, N.; Yang, Y.Y.; Li, Z.M.; Wu, H.Y. Facile Synthesis of MnO_2/CNTs Composite for Supercapacitor Electrodes with Long Cycle Stability. *J. Phys. Chem. C* **2014**, *118*, 22865–22872. [CrossRef]
19. Tang, Q.; Jiang, L.; Qi, J.; Jiang, Q.; Wang, S.; Sun, G. One step synthesis of carbon-supported Ag/Mn_yO_x composites for oxygen reduction reaction in alkaline media. *Appl. Catal. B Environ.* **2011**, *104*, 337–345. [CrossRef]
20. Cheng, F.; Su, Y.; Liang, J.; Tao, Z.; Chen, J. MnO_2-Based Nanostructures as Catalysts for Electrochemical Oxygen Reduction in Alkaline Media. *Chem. Mater.* **2014**, *22*, 898–905. [CrossRef]
21. Sun, Y.; Yang, M.; Pan, J.; Wang, P.; Li, W.; Wan, P. Manganese dioxide-supported silver bismuthate as an efficient electrocatalyst for oxygen reduction reaction in zinc-oxygen batteries. *Electrochim. Acta* **2016**, *197*, 68–76. [CrossRef]
22. Guo, D.; Dou, S.; Li, X.; Xu, J.; Wang, S.; Lai, L.; Liu, H.K.; Ma, J.; Dou, S.X. Hierarchical MnO_2/rGO hybrid nanosheets as an efficient electrocatalyst for the oxygen reduction reaction. *Int. J. Hydrogen Energy* **2016**, *41*, 5260–5268. [CrossRef]
23. Xia, W.; Mahmood, A.; Liang, Z.; Zou, R.; Guo, S. Earth-Abundant Nanomaterials for Oxygen Reduction. *Angew. Chem. Int. Ed.* **2016**, *55*, 2650–2676. [CrossRef] [PubMed]
24. Liang, Y.; Li, Y.; Wang, H.; Zhou, J.; Wang, J.; Regier, T.; Dai, H. Co_3O_4 nanocrystals on graphene as a synergistic catalyst for oxygen reduction reaction. *Nat. Mater.* **2011**, *10*, 780–786. [CrossRef] [PubMed]
25. Hu, F.P.; Zhang, X.G.; Xiao, F.; Zhang, J.L. Oxygen reduction on $Ag-MnO_2$/SWNT and $Ag-MnO_2$/AB electrodes. *Carbon* **2005**, *43*, 2931–2936. [CrossRef]
26. Sheng, Z.H.; Gao, H.L.; Bao, W.J.; Wang, F.B.; Xia, X.H. Synthesis of boron doped graphene for oxygen reduction reaction in fuel cells. *J. Mater. Chem.* **2011**, *22*, 390–395. [CrossRef]

27. He, J.; He, Y.; Fan, Y.; Zhang, B.; Du, Y.; Wang, J.; Xu, P. Conjugated polymer-mediated synthesis of nitrogen-doped carbon nanoribbons for oxygen reduction reaction. *Carbon* **2017**, *124*, 630–636. [CrossRef]

28. Pan, J.; Wang, Q.; Sun, Y.; Wang, Z. Analysis of electrochemical mechanism of coprecipitated nano-$Ag_4Bi_2O_5$ as super high charge–discharge rate cathode materials for aqueous rechargeable battery. *Electrochim. Acta* **2012**, *59*, 515–521. [CrossRef]

29. Wang, Q.; Pan, J.; Sun, Y.; Wang, Z. A high capacity cathode material-MnO_2 doped with nano $Ag_4Bi_2O_5$ for alkaline secondary batteries. *J. Power Sources* **2012**, *199*, 355–359. [CrossRef]

30. Goh, F.W.T.; Liu, Z.; Ge, X.; Zong, Y.; Du, G.; Hor, T.S.A. Ag nanoparticle-modified MnO_2 nanorods catalyst for use as an air electrode in zinc-air battery. *Electrochim. Acta* **2013**, *114*, 598–604. [CrossRef]

31. Chen, Y.; Liu, S.; Yu, L.; Liu, Q.; Wang, Y.; Dong, L. Efficient carbon-supported $Ag-MFe_2O_4$ (M = Co, Mn) core–shell catalysts for oxygen reduction reactions in alkaline media. *Int. J. Hydrogen Energy* **2017**, *42*, 11304–11311. [CrossRef]

32. Sun, S.; Miao, H.; Xue, Y.; Wang, Q.; Li, S.; Liu, Z. Oxygen reduction reaction catalysts of manganese oxide decorated by silver nanoparticles for aluminum-air batteries. *Electrochim. Acta* **2016**, *214*, 49–55. [CrossRef]

33. Zuo, L.X.; Jiang, L.P.; Abdel-Halim, E.S.; Zhu, J.J. Sonochemical preparation of stable porous MnO_2 and its application as an efficient electrocatalyst for oxygen reduction reaction. *Ultrason. Sonochem.* **2017**, *35*, 219–225. [CrossRef] [PubMed]

34. Cao, S.; Han, N.; Han, J.; Hu, Y.; Fan, L.; Zhou, C.; Guo, R. Mesoporous Hybrid Shells of Carbonized Polyaniline/Mn_2O_3 as Non-Precious Efficient Oxygen Reduction Reaction Catalyst. *ACS Appl. Mater. Interfaces* **2016**, *8*, 6040–6050. [CrossRef] [PubMed]

35. Lee, K.; Ahmed, M.S.; Jeon, S. Electrochemical deposition of silver on manganese dioxide coated reduced graphene oxide for enhanced oxygen reduction reaction. *J. Power Sources* **2015**, *288*, 261–269. [CrossRef]

36. Zhang, Y.Q.; Tao, H.B.; Liu, J.; Sun, Y.F.; Chen, J.; Hua, B.; Thundat, T.; Luo, J.-L. A rational design for enhanced oxygen reduction: Strongly coupled silver nanoparticles and engineered perovskite nanofibers. *Nano Energy* **2017**, *38*, 392–400. [CrossRef]

37. Guo, S.; Zhang, X.; Zhu, W.; He, K.; Su, D.; Mendoza-Garcia, A.; Ho, S.F.; Lu, G.; Sun, S. Nanocatalyst superior to Pt for oxygen reduction reactions: The case of core/shell Ag(Au)/CuPd nanoparticles. *J. Am. Chem. Soc.* **2014**, *136*, 15026–15033. [CrossRef] [PubMed]

catalysts

MDPI

Review

Recent Advances of Cobalt-Based Electrocatalysts for Oxygen Electrode Reactions and Hydrogen Evolution Reaction

Haihong Zhong [1], Carlos A. Campos-Roldán [2], Yuan Zhao [2,3], Shuwei Zhang [1], Yongjun Feng [1,*] and Nicolas Alonso-Vante [2,*]

1   State Key Laboratory of Chemical Resource Engineering, Beijing University of Chemical Technology, No. 15 Beisanhuan East Road, Beijing 100029, China; 2014200922@grad.buct.edu.cn (H.Z.); zhangshuwei024@gmail.com (S.Z.)
2   IC2MP, UMR-CNRS 7285, University of Poitiers, F-86022 Poitiers cedex, France; carlos.campos.roldan@univ-poitiers.fr (C.A.C.-R.); yuan.zhao@univ-poitiers.fr (Y.Z.)
3   College of Material Science and Science Technology, Nanjing University of Aeronautics and Astronautics, Nanjing 211106, China
*   Correspondence: yjfeng@mail.buct.edu.cn (Y.F.); nicolas.alonso.vante@univ-poitiers.fr (N.A.-V.); Tel.: +86-106-443-6992 (Y.F.); +33-549-453-625 (N.A.-V.)

Received: 1 September 2018; Accepted: 14 November 2018; Published: 19 November 2018

Abstract: This review summarizes recent progress in the development of cobalt-based catalytic centers as the most potentially useful alternatives to noble metal-based electrocatalysts (Pt-, Ir-, and Ru-based) towards the oxygen reduction reaction (ORR), oxygen evolution reaction (OER), and hydrogen evolution reaction (HER) in acid and alkaline media. A series of cobalt-based high-performance electrocatalysts have been designed and synthesized including cobalt oxides/chalcogenides, Co–N$_x$/C, Co-layered double hydroxides (LDH), and Co–metal-organic frameworks (MOFs). The strategies of controllable synthesis, the structural properties, ligand effect, defects, oxygen vacancies, and support materials are thoroughly discussed as a function of the electrocatalytic performance of cobalt-based electrocatalysts. Finally, prospects for the design of novel, efficient cobalt-based materials, for large-scale application and opportunities, are encouraged.

Keywords: electrocatalysis; cobalt-based electrocatalysts; oxygen reduction reaction; oxygen evolution reaction; hydrogen evolution reaction; non-precious metal

1. Introduction

Considering the continuous decrease of fossil fuels and deteriorated environments, it is of great importance and urgency to explore abundant, eco-friendly and renewable energy sources. Many energy conversion and storage technologies, e.g., proton exchange membrane fuel cells (PEMFCs, in which the free energy of a chemical reaction is directly converted into electrical energy), water electrolyzers (WEs, where oxygen and hydrogen from water are produced), unitized regenerative cells (URCs, a system comprising an electrolyzer and a fuel cell), due to their high efficiency and friendly environments, have been extensively developed [1,2]. To some extent, electrochemical processes play an essential role in these systems, for example, the oxygen reduction reaction (ORR) and hydrogen oxidation reaction (HOR) occurring on the cathode and anode of a H$_2$–O$_2$ fuel cell, whereas the oxygen evolution reaction (OER) and hydrogen evolution reaction (HER) are, respectively, the anodic and cathodic reactions in an electrolyzer [3–5]. Platinum-based materials perform efficiently the HOR with much less Pt mass loading (0.05 mg cm^{-2}) at the anode [6]. The complex oxygen electrode reactions (oxygen reduction reaction and oxygen evolution reaction), which involve various elementary steps, are kinetically intrinsic sluggish reactions, and primarily catalyzed by precious metal centers, e.g., Pt,

Ir, Ru. The limited resources and high cost of these precious metals are an unavoidably obstacle for widespread commercial applications [7]. Therefore, it is of increasing interest to reduce the usage of precious metals, or completely replace precious metals with abundant, cheap and highly active ones. In the past few years, various kinds of novel non-precious metal nanomaterials have been explored as alternatives to precious metal-based electrocatalysts, including strongly coupled transition metals (oxides, phosphides, chalcogenides, hydroxides, double perovskites, and so on) [8–14], nanocarbon hybrids [15], and free-metal carbon-based materials [7,16].

Cobalt (Co), the 32nd most abundant element in the Earth's crust, has emerged as an attractive non-precious metal for electrochemical reactions due to its catalytic performance. The price of cobalt per mass fluctuates over the years; currently its average-price is estimated at 72.82–48.07 €/kg (January–October 2018). From 2005 to 2018, the price has been subjected to small or important changes, meaning that the market is rather sensitive to the use purposes and to the localized reserve in the world. Regarding selenium and sulfur, for the period January to October 2018, the average-price ranges are 23.208–36.109 €/kg, and 0.063–0.067 €/kg, respectively. A series of cobalt-based catalytic center materials such as chalcogenides [17–22], oxides [23–25], metal–organic frameworks (MOFs) [26–30] and layered double hydroxides (LDHs) [31–33], have been recognized as potential candidates because of their parallel or even better activities, and superior electrochemical stability, compared with precious metals (Pt, Ir and Ru) [34]. Additionally, recent research results have established that cobalt-containing compounds supported onto conducting carbonaceous materials, e.g., Vulcan-XC-72 [35], nitrogen-doped carbon nanotubes (CNTs) [36], carbon nanowebs (CNWs) [37], graphene [38], reduced graphene oxide (RGO) [39,40] and so on, represent a valid way to endow the catalysts with rich exposed catalytic sites, high surface area, high electrical conductivity and fast mass transport, thus enhancing the catalytic activities.

In this review, we stress this novel cobalt-based material (CoCat) associated with some electrochemical processes, e.g., ORR, OER and HER in acid and alkaline electrolytes from the surface electrochemistry perspective. All these carbons supported cobalt-centered catalysts are organized into several categories, namely, cobalt oxides, cobalt chalcogenides (selenides, sulfides), Co–LDH, Co–MOFs, and Co–N_x/C, see Figure 1. In what follows, we firstly illustrate the basic reaction mechanism of ORR, OER and HER, then summarize recent progress in the development of cobalt-based electrocatalysts towards ORR, OER, HER. For the Co-centered electrocatalysts, particular attention is paid to the design, synthesis strategies, and electrocatalytic performance. Finally, we further discuss the challenges ahead in designing novel, highly efficient cobalt-based electrocatalysts for large-scale applications and opportunities.

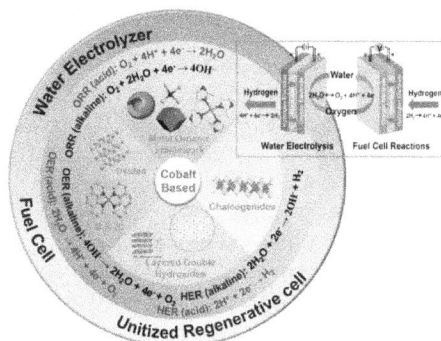

Figure 1. Schematic organization of the cobalt-based catalysts (CoCat) including layered double hydroxides, chalcogenides, oxides, M–N_x/C and metal organic framework with some major applications.

2. Electrocatalysts for Oxygen Reduction Reaction (ORR)

2.1. Mechanism of ORR

The oxygen reduction reaction is the cathodic electrode reaction. The ORR electrochemical properties can be evaluated from rotating disc electrode (RDE) measurements (namely, the onset potential (E_{onset}), half-wave potential ($E_{1/2}$), overpotential (η), and diffusion-limiting current density (j_L)). The ORR electrochemical reactions, in acid and alkaline medium, are shown below. The adsorbed molecular oxygen is reduced by a "direct" four-electron charge transfer process or reduced to water (acid medium) or OH^- (alkaline medium) via the formation of HO_2^- and H_2O_2 intermediates with the consumption of two electrons [41]. In acid condition, oxygen can be reduced to water with a standard thermodynamic potential at 1.229 V vs. SHE (Standard Hydrogen Electrode) for the four-electron pathway (Equation (1)); while in the alkaline medium, hydroxide is produced with a standard thermodynamic potential at 0.401 V vs. SHE in the four-electron reaction (Equation (4)). The charge-transfer reaction depends on the electrolyte nature and the surface properties of the catalytic centers. Even for Pt, considered as the best ORR catalyst, a substantial cathodic overpotential of 300 mV [41] is observed in acid electrolyte. Clearly, a selectivity for the four-e$^-$ reduction pathway is highly desirable to improve the electrocatalytic ORR efficiency.

Acid medium:

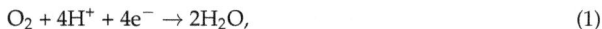
$$O_2 + 4H^+ + 4e^- \rightarrow 2H_2O, \tag{1}$$

or

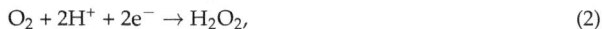
$$O_2 + 2H^+ + 2e^- \rightarrow H_2O_2, \tag{2}$$

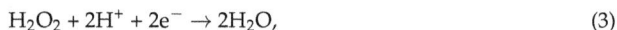
$$H_2O_2 + 2H^+ + 2e^- \rightarrow 2H_2O, \tag{3}$$

Alkaline medium:

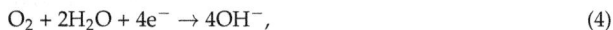
$$O_2 + 2H_2O + 4e^- \rightarrow 4OH^-, \tag{4}$$

or

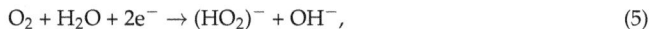
$$O_2 + H_2O + 2e^- \rightarrow (HO_2)^- + OH^-, \tag{5}$$

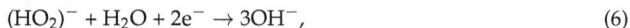
$$(HO_2)^- + H_2O + 2e^- \rightarrow 3OH^-, \tag{6}$$

2.2. Oxygen Reduction on Cobalt Chalcogenides Catalysts

2.2.1. Bond Ionicity or Covalency of S, Se, Te

In transition metals (TM) chalcogenides, the increase of covalency is obtained when d-state of TM interact with p-state of chalcogenides take place in MX_2 compounds (e.g., M: Co; X: S, Se). The hybridization degree of chalcogenides p-state with d-state of TM favors the metal–metal interaction. Moreover, metal cluster chalcogenides catalysts, $Ru_2Mo_4X_8$ (Chevrel phase), where X: Se- and S-based, have motivated a strong interest since Alonso-Vante and Tributsch [42] in 1986 reported the comparable ORR activity of $Ru_2Mo_4Se_8$ to that of Pt in H_2SO_4 [43–46]. In binary metal clusters (e.g., Ru) coordinated with X, they proposed that the activity of Ru_xX_y catalysts depends on the chalcogen, and increases according to: $Ru_xS_y < Ru_xTe_y < Ru_xSe_y \sim Mo_xRu_ySe_z$ in acid solution [47]. The structure analysis from the EXAFS (Extended X-ray Absorption Fine Structure) data suggested that that the variation of the chalcogen nature led to a change in its amount in the first co-ordination sphere of ruthenium. The latter apparently affected the strength of ruthenium interaction with oxygen, which is evidenced by significant differences observed in the Ru/O distances. For Ru_xS_y clusters, the high co-ordination number of Ru to sulfur resulted in blocking the ruthenium active sites towards adsorption of molecular oxygen, which showed lowest ORR catalytic activity. In this respect, the coordinating strength of S, Se, Te on Ru chalcogenides was demonstrated (ligand effect). This is in contrast to cobalt dichalcogenides compound (CoX_2) demonstrated by Behret et al. [48] that reported a trend of ORR activity basically followed the sequence: $M_xS_y > M_xSe_y > M_xTe_y$. In addition, the activity was also related to the catalytic centers with Co. Both theoretical and experimental results confirmed that Co–selenides are less active than its sulfides by ~0.2 V [49]. The considerable decrease in activity was observed when S was partly or totally substituted by O, Se, and Te, probably attributed to the geometric and electrostatic conditions in spinel structures. It was claimed that the selenium and tellurium with higher atomic radii give a weaker electrostatic repulsion to the reduction product (O^{2-}) and also to the intermediate reduction products. Therefore, the reaction products on the surface of seleno- and tellurospinels were not as easily desorbed as those on thiospinels [48].

Furthermore, a series of transition metal chalcogenides were reported by Behret et al. [48] consisting of the anionic substitution (S, Se and Te), and the cationic substitution (Fe, Co, Ni). These primary results disclosed a trend for ORR electrochemical performance with the metal cations as follows: Co > Ni > Fe, and for the chalcogenide anions: S > Se > Te. They found that Co–S and Co–NiS systems possessed very high catalytic activity in acidic medium. They declaimed that better catalytic activity may be caused by the minimal energy difference between the oxygen 2p orbital and the highest occupied d orbital of sulfides [50]. CoS was, apparently, the most promising ORR non-precious metal catalyst in alkaline media. Besides, Co_9S_8 was predicted to have similar ORR activity with that of Pt via a four-electron ORR pathway in acidic solution [51]. Zhu et al. [52] developed a simple and scalable route to synthesize 3D hybrid nanocatalyst—the Co_9S_8 incorporated in N-, S-doped porous carbon exhibited an excellent catalytic activity, superior long-term stability, and good tolerance against methanol. Recently, Dai et al. [19], for the first time, designed and prepared the etched and doped Co_9S_8/graphene hybrid as an advanced bifunctional oxygen electrode reactions catalyst.

2.2.2. Crystal Structure and Particle Size Effect

The different crystal structures with the same chemical composition led to the different ORR catalytic activity. For instance, $CoSe_2$ has two common crystal structures: cubic (pyrite-type), and orthorhombic marcasite-type. Alonso-Vante and Feng [53,54] discovered that the orthorhombic $CoSe_2$ was obtained after a heat treatment of 250–300 °C. The cubic $CoSe_2$ phase was obtained at high temperature (400–430 °C). The latter showed a higher ORR activity by 30 mV in 0.5 M H_2SO_4. Wu et al. [55] prepared cubic Co_9S_8 particles surrounded by nitrogen-doped graphene sheets for the ORR in alkaline medium with an improved ORR activity and stability comparable to Pt/C. Similarly, cobalt sulfides with different chemical compositions and crystal structures, namely, $Co_{1-x}S$, CoS (hexagonal phase) [56], CoS_2, Co_9S_8 and Co_3S_4 (cubic phase) [56], are the most promising type of chalcogenides for the ORR. As an illustration, Dai et al. [18] obtained the Co_2S/RGO with an average particle size of ca. 50 nm and $Co_{1-x}S$/RGO with an average particle size of 10–20 nm. The latter displayed a better ORR activity in terms of the onset potential of ca. 0.8 V vs. RHE (Reversible Hydrogen Electrode) in 0.5 M H_2SO_4. The authors noticed that the ORR performance depends on the particle size and on the crystal structure.

Generally, the mass activity (MA, activity mass^{-1}) of a catalyst is defined as the current normalized by the noble metal loading or catalyst loading as measured at a specific electrode potential. In order to increase MA, some strategies can be adopted, i.e., increasing the specific surface area of the catalysts by decreasing the particle size [57]. The particle size effect on the specific activity of catalysts has been attributed to different factors, such as, the structural sensitivity, i.e., the dependence of surface geometry, the electronic state, and the metal–support interaction [58]. Moreover, the particle size of the catalysts on the oxygen reduction reaction can be influenced by the adsorption of oxygen during the reaction, which is mainly associated with the fraction of active sites on the surface of the catalysts. Wang et al. [59] demonstrated that the Pd_3Co alloy had a slightly smaller lattice spacing than that of Pd, and thus a further shifting to larger size was expected due to the stronger lattice-induced compression. The 3-fold enhancement in the specific activity of Pd_3Co could be attributed to the nanosized-, and lattice mismatch-induced contraction in (111) facets, based on the DFT (density Functional Theory) calculation using a nanoparticle model. Feng et al. [46] developed cubic CoS_2 nanocatalysts with different particle size from 30 nm to 80 nm by adjusting the initial Co^{2+} concentration in the presence of hexa-decylcetyl-trimethylammonium (CTAB), and further demonstrated that the ORR activity depends on the average particle size, see Figure 2. The ORR activity remained similar to the particle size from 30 nm to 50 nm, and then significantly decreased with size from 50 nm to 80 nm. The CoS_2 catalyst with an average particle size of 30.7 nm demonstrated an excellent electrocatalytic performance with an onset potential (E_{onset}) of 0.94 V vs. RHE, a half-wave potential of 0.71 V vs. RHE, and high tolerance in methanol-containing 0.1 M KOH. Such a trend was also found for 20 wt% Co_3S_4/C nanoparticles and 20 wt% $CoSe_2$/C [60]. The particle size can be approximately adjusted within a certain range by controlling the initial feeding concentrations, which has strong effect on the crystal nucleation and crystal growth.

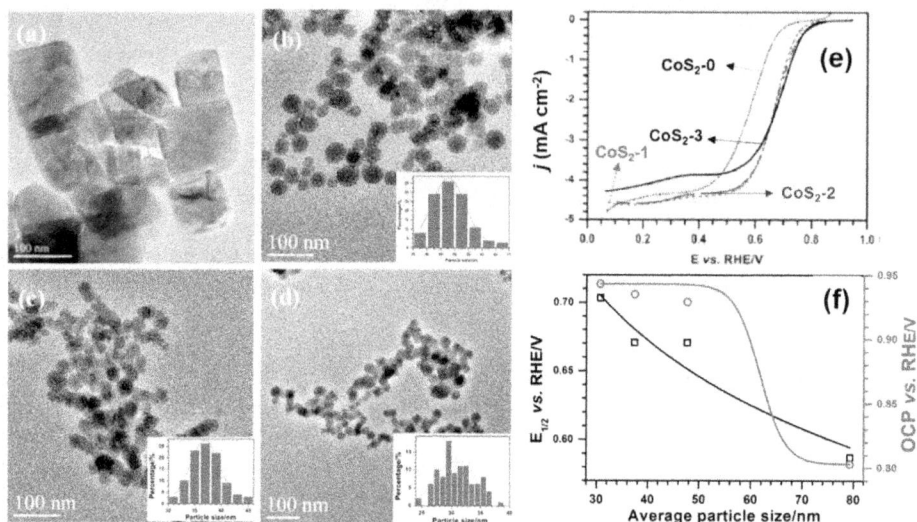

Figure 2. (**a–d**) High-resolution transmission electron microscope (HRTEM) images of CoS_2-0 (without addition of hexa-decylcetyl-trimethylammonium (CTAB)), CoS_2-1, -2, and -3 nanoparticles with the initial Co^{2+} concentrations from 0.025 M to 0.075 M in the presence of 0.5 mM CTAB; (**e**) rotating disc electrode (RDE) curves of the four CoS_2 samples in O_2-saturated 0.1 KOH at 1600 rpm at a scan rate of 5 mV s^{-1} (cathodic sweep) at room temperature. The catalyst loading was 0.1 mg cm^{-2}; (**f**) half-wave potential ($E_{1/2}$) and OCP (Open Circuit Potential) extracted from Tafel plots as a function of the average CoS_2 particle size. Reproduced from [46], Copyright © Royal Society of Chemistry, 2013.

2.2.3. Synthesis and Support Effect

Cobalt selenides have received extensive attention for their ORR activity [61,62]. This electrocatalyst deposited on highly conductive supporting material is very important to enhance the electrocatalytic performance. The ideal supporting materials for catalytic centers have high surface area, high electrical conductivity, and high chemical stability [63]. Taking advantage of the high surface area of amorphous carbon, Feng et al. [62] synthetized via a surfactant-free way the $CoSe_2$ orthorhombic-phase nanoparticle supported onto XC-72 Vulcan with a promising ORR performance in acid media. Zhou et al. [64] developed supported $CoSe_2$ nanostructures by a hydrothermal approach using the increased disordered domains of carbon nanotubes (CNTs) derived from a MOF with excellent ORR performance in 0.1 M KOH with a Tafel slope of 45 mV dec^{-1}, onset potential of ca. 0.8 V vs. RHE, and a long-term stability. Although a relative success was obtained, the major drawback of the hybrid catalysts is the weak interaction between the catalytic centers and the support leading to low activity and stability. The formation of interfacial bonds between the catalytic center and the support favors the adsorption properties, therefore, enhancing the kinetics toward the electrocatalytic reaction [65].

The catalytic–support interaction may weaken the adsorption events, and then decrease the energy barriers for the reaction. This phenomenon, known as strong metal–support interaction (SMSI) is an important ingredient to tailor, and tune highly active and stable catalyst centers. Unni et al. [66] fabricated high-surface area N-doped carbon nanohorns (NCNH) supported cubic-phase $CoSe_2$ via a simple $NaBH_4$-chemical reduction process of $CoCl_2$ and SeO_2, Figure 3a,b. The $CoSe_2$/NCNH demonstrated considerable ORR activity in alkaline medium, cf. Figure 3c,d, as a result of the electronic structure modification of chalcogenide ($CoSe_2$) centers throughout its interaction with NCNH via the nitrogen moieties, Figure 3e.

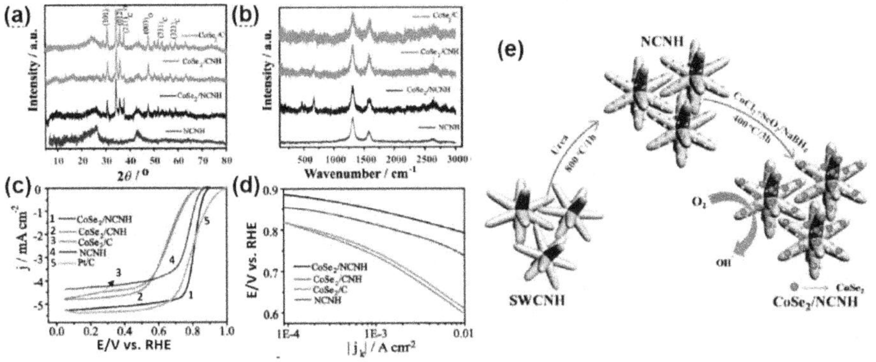

Figure 3. (**a**) Raman spectra; (**b**) X-ray diffraction (XRD) patterns; (**c**) oxygen reduction reaction (ORR) polarization curves of $CoSe_2$/N-doped carbon nanohorns (NCNH), $CoSe_2$/carbon nanohorns (CNH), $CoSe_2$/C, NCNH and Pt/C catalysts in O_2-saturated 0.1 KOH at 1600 rpm at a scan rate of 5 mV s^{-1} (cathodic sweep) at 25 °C. The catalyst loading was 0.214 mg cm^{-2}; (**d**) ORR Tafel plots of supported $CoSe_2$ onto carbon Vulcan, CNH and NCNH; (**e**) schematics of the synthesis procedure. Reproduced from [66], Copyright © Wiley, 2015.

Similarly, García-Rosado et al. [67] prepared a series of N-doped reduced graphene oxide (N-RGO) as supports for hexagonal cobalt selenide (CoSe) by a $NaBH_4$-assisted chemical reduction, Figure 4. The carbon support after the surface reduction supplied more available active and anchor sites due to the increase in the pore size and surface area. The ORR performance is concomitant with the properties of the supports. The graphitic and pyridinic nitrogen moieties of N-RGO acted as electrochemical active sites for the ORR in alkaline media.

Additionally, Pan et al. [68] used porous g–C_3N_4 as a template and a N-source to successfully synthesize mesoporous S-, and N-co-doped carbon matrix coupled with Co@Co_9S_8 nanoparticles, cf. Figure 5. Herein, the g–C_3N_4 carbon matrixes not only enhance the conductivity, but also suppress the aggregation phenomenon during the electrochemical reactions. Besides, the strong coupling between Co@Co_9S_8 and N–, S–carbon promoted significantly the ORR electrocatalytic performance with a more positive half-wave potential and lower Tafel slope value, as compared with the commercial Pt/C catalyst.

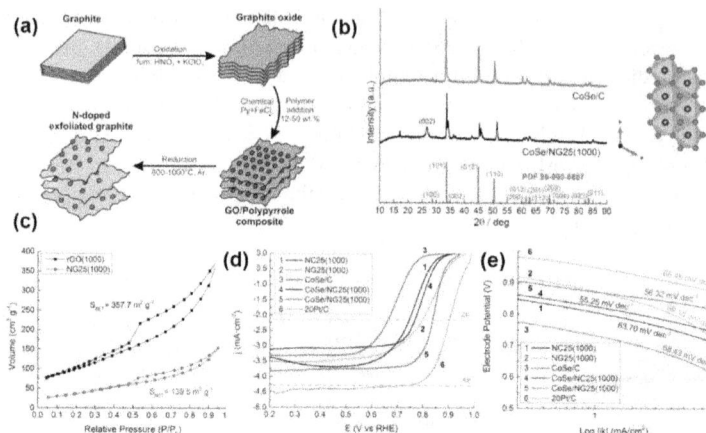

Figure 4. (**a**) scheme of the synthesis of N-doped reduced graphene oxide (N-RGO); (**b**) XRD patterns of supported hexagonal CoSe onto GO and N-RGO; (**c**) BET–nitrogen adsorption–desorption isotherms of rGO(1000), and NG25(1000); (**d**) ORR polarization curves of NC25(1000), NG25(1000), CoSe/C, CoSe/NC25(1000), CoSe/NG25(1000) and Pt/C electrocatalysts at 900 rpm (cathodic sweep) in O_2-saturated 0.1 M KOH at 25 °C with the catalyst loading of 0.286 mg cm^{-2}; and (**e**) ORR Tafel plots of supported CoSe catalysts in alkaline medium. Reproduced from [67], Copyright © The Electrochemical Society, 2017.

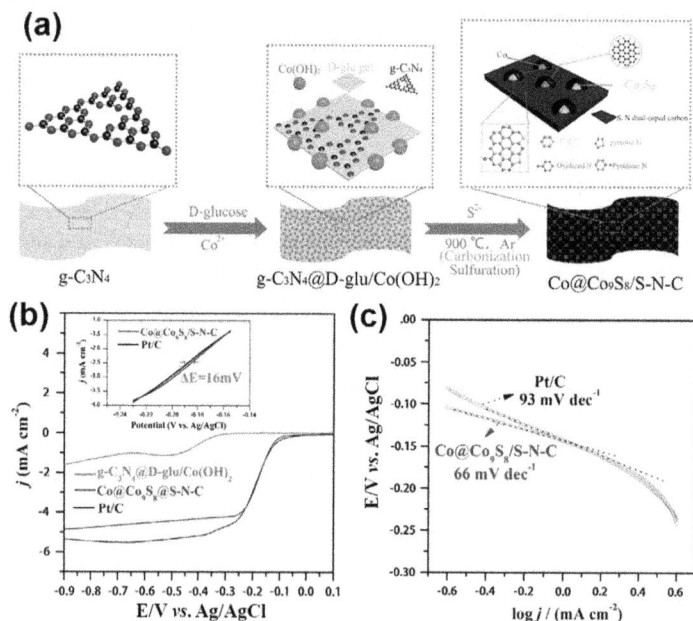

Figure 5. (**a**) The schematic chemical synthesis route of Co@Co$_9$S$_8$/S–N–C; (**b**) linear sweep voltammetry (LSV) curves for the g–C$_3$N$_4$@D–glu/Co(OH)$_2$, Co@Co$_9$S$_8$/S–N–C, and Pt/C electrodes recorded in O_2-saturated 0.1 KOH solution at 1600 rpm with scan rate of 10 mV s^{-1} (cathodic sweep) at 25 °C (catalyst mass loading: 0.2 mg cm^{-2}); (**c**) Tafel plots of Co@Co$_9$S$_8$/S–N–C and Pt/C catalysts. Reproduced from [68], Copyright © Elsevier, 2018.

2.3. Oxygen Reduction on Metal–Organic Frameworks (MOFs) Catalysts

MOFs and their derivatives have also been used as efficient precursors and self-sacrificing templates because of their well-tunable physical and chemical properties. Benefiting from the unique properties of MOF, such as, large surface area, tailoring porosity, and easy functionalization with other heteroatoms or metal/metal oxides, various carbon-based nanomaterials (as support) have been prepared. Based on organic ligands and cobalt metal centers, MOFs compounds possess special superiority to prepare various cobalt-based functionalized carbon nanomaterials, including heteroatom-doped porous carbons, and metal/metal oxide decorated porous carbons, via thermal decomposition under controlled atmospheres. For instance, Lou et al. [69] used ZIF-67 as a template and thioacetamide as a sulfur source to prepare a double-shelled Co–C@Co$_9$S$_8$ nanocages electrocatalyst for the ORR. By adjusting the reaction time, the amorphous CoS nanocages (a–CoS NCs) and ZIF-67@a–CoS yolk-shelled hollow nanostructures were fabricated. The Co$_9$S$_8$ shells, served as a nanoreactor, effectively prevented the Co–C active centers from aggregation, Figure 6a. Interestingly, the unique Co–C hollow cages significantly shortened the diffusion pathway of the electrolyte, and thus promoted the electrocatalytic activity, durability, and tolerant to methanol toward the ORR of Co–C@Co$_9$S$_8$ catalyst. In Figure 6b–e, the onset potential remarkably reached 0.96 V vs. RHE with a limiting current of 4.5 mA cm^{-2} and a four-electron transfer route with OH$^-$ production in alkaline media.

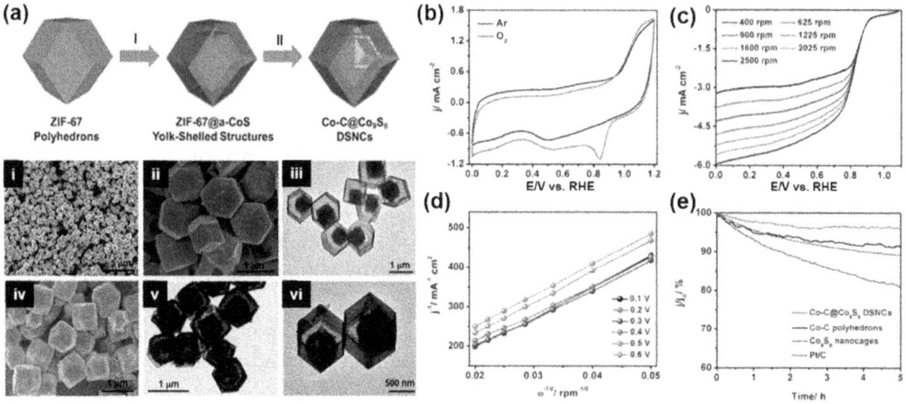

Figure 6. (a) The synthesis route of Co–C@Co$_9$S$_8$ DSNCs (double-shelled nanocages); (i–vi) field-emission scanning electron microscopy (FESEM) and TEM micrographs of the catalyst; (b) CV curves of Co–C@Co$_9$S$_8$ DSNCs in Ar- or O$_2$-saturated 0.1 M KOH electrolyte with a scan rate of 10 mV cm s^{-1}; (c) LSV (cathodic sweep) curves of Co–C@Co$_9$S$_8$ DSNCs catalyst recorded at different rotation speeds from 400 to 2500 rpm in O$_2$-saturated 0.1 M KOH; (d) the K–L plots of Co–C@Co$_9$S$_8$ DSNCs at various potentials (0.1–0.6 V vs. RHE (Reversible Hydrogen Electrode); (e) the chronoamperometric responses (j/j$_0$ (%) vs. t) of Co–C@Co$_9$S$_8$ DSNCs, Co–C polyhedrons, Co$_9$S$_8$ nanocages and Pt/C electrodes at 0.5 V in O$_2$-saturated 0.1 M KOH solution at 1600 rpm. Reproduced from [69], Copyright © Royal Society of Chemistry, 2015.

Additionally, a simple and efficient method to produce homogeneously dispersed cobalt sulfide/N-, S-co-doped porous carbon electrocatalysts, using ZIF-67 as precursor and template, was reported by Xia et al. [70]. Due to a unique core–shell structure, high porosity, homogeneous dispersion of active components together with N-, and S-doping effects, the electrocatalyst not only showed an excellent electrocatalytic activity towards ORR with a high onset potential (ca. −0.04 V compared with −0.02 V for the benchmark Pt/C catalyst), but also revealed a superior stability (92%) compared with the commercial Pt/C catalyst (74%) in ORR.

Table 1 summarizes some transition metal chalcogenides as ORR catalysts investigated so far. One can conclude that: (1) Co-based chalcogenides are promising catalysts for the ORR in alkaline media; (2) the carbon support having a pore morphological structure, and hetero-atom doping plays a key role to promote the ORR performance of Co-based chalcogenides.

2.4. Oxygen Reduction on Cobalt Oxide Catalysts

2.4.1. Nanostructure

Intensive efforts have been devoted to fabricate various surface-tuned cobalt oxides materials with different nanostructures and morphologies, including Co_3O_4 nanorods [71], nanowires [72], core–shell [73], and hollow structures [74], which are responsible for the electrochemical performance. For example, Kurungot et al. [75] prepared a series of surface-tuned Co_3O_4 nanoparticles with different morphologies (cubic (Co_3O_4–NC/NGr–9h), blunt-edged cubic (Co_3O_4–BC/NGr–12h), and rough edged spherical (Co_3O_4–SP/NGr–24h)) dispersed on nitrogen-doped graphene (NGr) electrocatalysts for ORR by varying the reaction time. The transformations of the morphologies of Co_3O_4 could be assigned to the transformation of the higher to the lower surface energy crystal plane structure, which was expected to obtain the mixed facets of the exposed Co-oxide crystals. Among them, the NGr supported spherical Co_3O_4 (Co_3O_4–SP) nanoparticles (Co_3O_4–SP/NGr–24h) with the highest roughness factor and increased surface area exhibited the best activity in alkaline medium, and attributed to: (1) homogenous dispersion of Co_3O_4 nanoparticles on NGr support; (2) NGr served as nucleation sites, efficiently controlling the growth kinetics of Co_3O_4 nanoparticles; and (3) a strong synergistic interaction between the active sites and the support.

2.4.2. Particle Size Effect and Chemical Composition

Different particle size also influences the electrocatalytic active surface. For example, Feng's group [76] synthesized a series of Co_3O_4/N–RGO composites with different particle size by controlling Co^{2+} initial feeding concentrations, and carefully investigated the particle size effect of Co_3O_4 nanoparticles on the bifunctional oxygen electrocatalytic performance. The Co_3O_4/N–RGO with the smallest particle size of 12.2 nm revealed the best bifunctional oxygen activity (ORR and OER) with lower $\Delta E = 0.75$ V ($\Delta E = E_{OER, j@10\,mA\,cm^{-2}} - E_{ORR, j@-3\,mA\,cm^{-2}}$) ($E_{OER, j@10\,mA\,cm^{-2}}$, the potential at a current density of 10 mA cm^{-2} for OER; $E_{ORR, j@-3\,mA\,cm^{-2}}$, the potential at a current density of -3 mA cm^{-2} for ORR) than other larger particle sized samples. Sun and co-workers [77] investigated, on the other hand, the impact of the CoO particle size on the ORR activity via a combined controllable hydrolysis and thermal treatment process. The promoted ORR current and the increased content of HO_2^- produced over the smallest CoO particles benefited from the increase of the interface between carbon support and CoO nanoparticles (NPs). Thus, the interface between carbon and CoO NPs could be identified as the most active site.

Besides the control of particles' size, fine-tuning of the chemical composition (e.g., Co^{3+}/Co^{2+} ratio) was an effective strategy to boost the electrocatalytic activity of cobalt oxide catalysts. Liu et al. [78] proposed an interesting approach to enhance the electrocatalytic activity of Co_3O_4 nanosheets through the modulation of the inner oxygen vacancy concentration, and the Co^{3+}/Co^{2+} ratio. Based on the synergistic effect of the fashioned 2D nanosheets, the presence of oxygen vacancies, and the Co^{3+}/Co^{2+} ratio, the catalyst showed a much lower overpotential towards ORR and OER for Li–O_2 batteries. Furthermore, Zhao's group [79] reported that the ORR catalytic activity of the prepared catalysts is sensitive to the number and activity of surface-exposed Co^{3+} ions that could be tailored by the morphology of cobalt oxides.

2.4.3. Support Effect

Cobalt oxide (Co_3O_4 and CoO) nanoparticles are recognized as a class of non-precious catalysts with high activity, stability and durability in alkaline solution. For example, Co_3O_4/graphene hybrids had a remarkable ORR performance comparable to that of commercial Pt/C in alkaline, which was attributed to the synergistic catalyst-support coupling [25]; as for graphene supported 3D "sheet-on-sheet" interleaved Co_3O_4 nanosheets (Co–S/G), the strong electron transport and charge transfer from graphene to Co_3O_4 nanosheets significantly enhanced the ORR electrocatalytic properties of Co–S/G [80].

Bao et al. [81] proposed that the excessive ascorbic acid (AA) could form abundant negative functional groups on the surface of 3D graphene aerogel (3DG, Figure 7a), with Co^{2+} ions uniformly anchored on 3DG by electrostatic or coordination interaction. The hollow Co_3O_4/3DG was synthesized by a direct oxidation of C@Co/3DG precursor through the Kirkendall effect, Figure 7b. The prepared Co_3O_4/3DG (22 wt% Co_3O_4) electrocatalyst showed a positive onset potential of 0.82 V vs. RHE, and a limiting current density of 5.12 mA cm^{-2} (1600 rpm), Figure 7c. The rotating ring disk electrode (RRDE) method further confirmed a charge transfer pathway close to 4-electron with a production of hydrogen peroxide (HO$_2^-$) of 4.27% to 14.2% in Figure 7d.

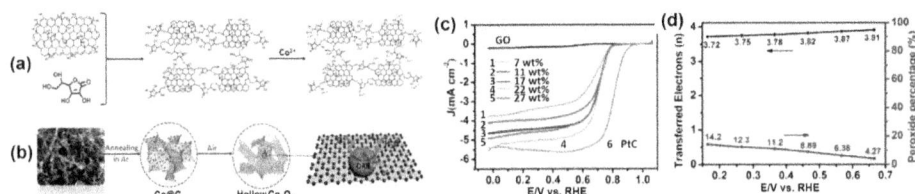

Figure 7. (a) Schematic representation of how an excess of ascorbic acid reduced graphene oxide (GO) and formed 3DG (three-dimensional graphene). The added Co^{2+} are anchored on 3DG; (b) Schematic showing the formation of HNPs (hollow nanoparticles) Co_3O_4/3DG material; (c) LSVs (cathodic sweep) curves of GO, Co_3O_4/3DG (with various weight loading) and Pt/C electrodes in O_2-saturated 0.1 M KOH electrolyte at 5 mV s^{-1} at 1600 rpm; (d) the transferred electrons and peroxide percentage of Co_3O_4/3DG (22 wt%). Reproduced from [81], Copyright © Wiley, 2017.

Wang et al. [82] developed a supramolecular gel-assisted method to manufacture N-doped carbon shell coated Co@CoO nanoparticles on carbon Vulcan XC-72 (Co@CoO@N–C/C) for ORR. Herein, melamine not only acted as a chelating agent interacting with Co^{2+}, but also as a nitrogen source to dope at high-temperature pyrolysis treatment. Due to the synergistic effect, the double-shelled Co@CoO@N–C/C nanoparticles displayed the excellent ORR performance with E_{onset} = 0.92 V vs. RHE and $E_{1/2}$ = 0.81 V vs. RHE, comparable with the commercial Pt/C in 0.1 M KOH.

2.5. Oxygen Reduction on Cobalt-Based Layered Double-Hydroxides Catalysts

Synthesis Strategy

Layered double hydroxides (LDHs), a class of two-dimensional (2D) layered material, consist of host sheets with divalent (M^{II}) and trivalent (M^{III}) metal cations coordinated to hydroxide anions, and the guest anions in the interlayer regions. This kind of structure has attracted interest in various energy conversion systems [31]. Generally, hydroxides have a good affinity to aqueous electrolytes and their layered structures offer an enlarged surface area and improved dispersion degree based on the confinement effect. This latter property favors the accessibility of catalytic sites. Nevertheless, the positively charged host sheets in LDHs favor the oxygen adsorption and its oxygen reduction. Currently, Co-containing LDHs, as ORR catalysts, have been the object of intense research [83–85]. For instance, Li et al. [85] investigated the interaction of LDHs with RGO toward the ORR via a four-electron transfer pathway and found that the Co^{2+} in the LDHs, as active sites, catalyzed the disproportionation of the peroxide species to form H_2O and O_2. Additionally, the RGO support increased the electrical conductivity.

Apart from the direct preparation of the LDHs precursor, the materials derived from LDHs, after thermal treatment, showed similar promising electrocatalytic activity in alkaline solution. For example, $Co_3Mn–CO_3–LDH/RGO$ precursor was calcined to produce reduced graphene oxide supported Co–Mn oxides under nitrogen atmosphere with enhanced catalytic performance (E_{onset} = 0.95 V vs. RHE, $E_{1/2}$ = 0.76 V vs. RHE, and j_{limit} = 4.2 mA cm^{-2} at 0.2 V vs. RHE, 1600 rpm) [84]. Similarly, Co_3O_4/Co_2MnO_4 composites, resulting from CoMn–LDHs precursor, displayed excellent bifunctional activities for ORR and OER with ΔE = 1.09 V.

Herein, the activity of the material was attributed to the large specific surface area and well-dispersed heterogeneous structure [83]. Xu et al. [86], on the other hand, fabricated a $Co/CoO/CoFe_2O_4$ as ORR electrocatalyst based on the utilization of CoFe–LDHs as precursor by separate nucleation and aging steps (SNAS). The same group later reported Co@N–CNTs obtained by calcination of melamine/CoAl–LDH mixture, cf. Figure 8 [32]. Here, the CoAl–LDH precursor operated as: (i) catalyst for carbonization and formation of N–CNTs; (ii) detacher of active Co nanoparticles from the Co component in the host sheet; (iii) enabler for the growth of long N–CNTs with the aid of the confinement effect of non-active Al_2O_3 matrix formed.

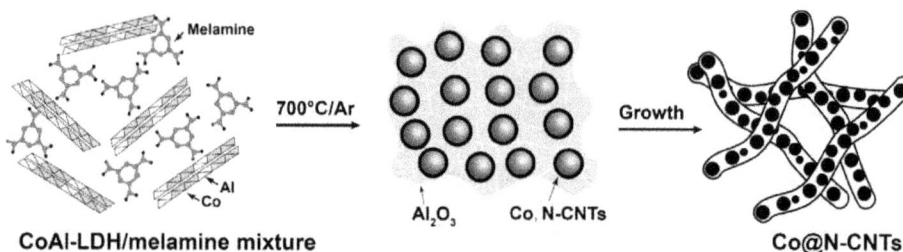

Figure 8. Representation of Co@N–CNTs–m resulting from CoAl–layered double hydroxides (LDH)/melamine mixture. Reproduced from [32], Copyright © Elsevier, 2016.

2.6. Oxygen Reduction on Co–N_x/C Catalysts

2.6.1. Co–N_x Active center

Besides the aforementioned Co-based electrocatalysts, Co–N_x/C has also been considered as another potential candidate to substitute Pt for ORR. In 1964, Jasinski [87] for the first time found that cobalt–phthalocyanine (CoPc) activated the ORR process in alkaline electrolyte. However, the catalytic sites in macrocyclic cathode catalysts for ORR are still controversial. Apropos Co–N_x/C catalysts, various type of active sites were proposed, including Co–N_4/C and Co–N_2/C [88–92], N_xC_y [93], pyridinic N [91,94], and graphitic N [95,96]. For example, Kiefer et al. [97] explained the origin and mechanism of ORR on Co–N_x (x = 2, 4) based on self-assembled carbon supported electrocatalysts in alkaline and acidic media via first-principle DFT calculations. The graphitic Co–N_4 defect was energetically more favorable than the graphitic Co–N_2 defects, and the former was predicted to be the dominant in-plane graphitic defect in Co–N_x/C electrocatalysts.

Nevertheless, Li et al. [98] believed that the dominant reactive sites for cobalt single atoms (Co–SAs) in Co–SAs/N–C could be postulated as Co–N_4 and Co–N_2 moieties, and Co–N_2 species had a stronger interaction with peroxide than Co–N_4, promoting the ORR four-electron reduction process. Tang et al. [99] considered that both of Co–N_x and pyri–N species were very important and their coupling effect (Co–pyri–N) was of paramount importance in the high electrocatalytic activity due to the high lying HOMO energy levels. Besides, Liao et al. [100] proposed that doped transition metals could act, on one hand, as a catalyst for the formation of stable active sites in the pyrolysis process, and on the other hand, the residual metal compounds could serve as active sites. In fact, the ORR catalytic performance improvement of transition metals might be the result of joint effects: (i) the overall N content/active N content; (ii) metal residue; and (iii) the surface area and pore structures. Metal-free carbon structure was originally suggested as active sites by Wiesener [101]. Yet, Masa et al. [102] argued for the concept of a metal-free catalyst. Indeed, truly metal-free nitrogen-doped carbon demonstrated much lower ORR activity and reduced O_2 through a two-electron pathway in acidic solution, so that undetectable metal residues may play a crucial role for the ORR.

2.6.2. Co–N_xSynthesis Strategy

Generally, the electrocatalytic activity and stability of Co–N_x/C, or nitrogen-modified carbon supports (NCSs) or templates materials, basically rely on the morphology, pore structure, dispersion of the active sites, and nitrogen species, which are up to the crucial synthetic route. Traditional methods involving the pyrolysis of carbon-supported cobalt N_4–macrocycles, e.g., phthalocyanine (Pc), tetra-azaannulene (TAA), tetra-phenylpophyrin (TPP) and tetra-methoxyphenlyporphyrin (TMPP), at temperatures between 550 °C and 800 °C preserved or created Co–N_x active sites, although the macrocyclic structure of the complex was partially or completely destroyed [103–106]. Above 800 °C, the metal–nitrogen species are known to decompose with a concomitant decrease in the ORR performance. For example, Popov et al. [107] found, via EXAFS, that the content of Co–N species decreased when the heat treatment exceeded 800 °C. Yet, Dodelet's group [108] proposed that Co metal and/or CoPc fragments containing Co could be responsible for the catalytic activity of CoPc/XC-72 pyrolyzed at 600 or 700 °C, a treatment where the maximum of the electrocatalytic activity was obtained [109]. Jahnke et al. [110] improved the stability as well as the electrochemical activity of transition metal porphyrins deposited on carbon support by a pyrolytic treatment step in the range of 450 to 900 °C, in an inert atmosphere. Qiao et al. [35] synthesized carbon-supported cobalt catalysts, namely, Py–CoPc/C, heat-treated at 600–900 °C, and investigated its heat-treatment effect onto the ORR activity in alkaline electrolyte. The catalysts that annealed at 700 °C exhibited the best activity towards ORR in 0.1 M KOH. Such a phenomenon was related with Co–N_x/C, pyridinic–N and graphitic–N active sites. Shan et al. [111] fabricated novel mesoporous Co–N_x–C, by a nanocasting-pyrolysis method, using Co–phenanthroline as the only precursor, permitting that abundant Co–N_x moieties remained embedded in the graphitic framework during pyrolysis.

Recently, most Co–N_x/C catalysts were fabricated by pyrolysis treatment of Co salts mixed with carbon support and nitrogen-rich precursor [100]. Zelenay et al. [88] used polyaniline (PANI) as a nitrogen source for high-temperature synthesis of catalysts incorporating iron or cobalt centers. The most active catalysts: FeCo/N/C, showed a very positive onset potential of ~0.93 V with a long-term stability test of 700 h at a fuel cell voltage of 0.4 V in 0.5 M H_2SO_4. Zhang et al. [112] demonstrated the direct utilization of the intrinsic structural defects in nanocarbon to generate atomically dispersed Co–N_x–C active sites via defect engineering. The as-obtained Co/N/O tri-doped graphene mesh (denoted as NGM–Co) were prepared via the carbonization of a mechanical mixture of gelatinized amylopectin, melamine, cobalt nitrate, and in situ generated $Mg(OH)_2$ nanoflakes. The remarkable bifunctional electroactivity for ORR and OER in alkaline medium of NGM–Co catalyst was ascribed to Co–N_x–C moieties, nitrogen-doping, oxygen functional groups, and topological defects. Miao et al. [113] reported a noble-free Co–N–C catalyst, derived from cobalt–phenanthroline complexes on CMK-3 under heat-treatment. By careful examination of the Co–N–C catalyst with sub-Ångström-resolution aberration-corrected scanning transmission electron microscopy (HAADF-STEM), the authors suggested that the Co single atoms bonded to N within the graphitic sheets served as the active sites. Wang's group [114] developed a new type of Co–N/C catalysts that contained two forms of active components, namely, coordinating cobalt moieties (CoC_xN_y or CoN_x) and Co nanoparticles encapsulated in mesoporous N-doped carbon hollow spheres (Co–N–mC) by pyrolyzing the polystyrene@polydopamine–Co (PS/PDA–Co) precursor. The Co–N–mC catalysts revealed striking ORR performance, comparable with commercial Pt/C, with onset potential of 0.94 V vs. RHE, half-wave potential of 0.851 V vs. RHE, and a Tafel slope of 45 mV dec^{-1} in 0.1 M KOH. Similarly, Tsiakaras et al. [115] designed and fabricated a 3D hollow carbon spheres (HCS) with co-doping of cobalt and nitrogen, using dopamine and poly-methyl-methacrylate

(PMMA) as a template and a vacuum-assisted impregnation method, Figure 9a. The Co–N@HCS catalyst with a large specific surface area (347.3 m^2 g^{-1}) displayed excellent catalytic activity for both ORR and OER with a ΔE (ΔE = $E_{OER, j@10\,mA\,cm^{-2}} - E_{1/2,\,ORR}$) of 0.856 V, much lower than those of the N@HCS (1.233 V) and the benchmark Pt/C catalysts (1.044 V).

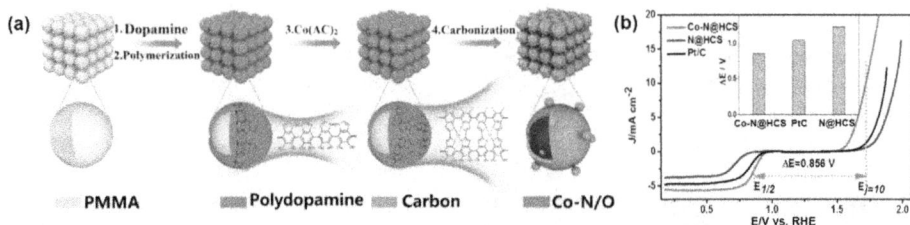

Figure 9. (a) The schematic chemical synthesis route of Co–N@HCS catalyst; (b) the LSV curves within ORR (cathodic sweep) and OER (anodic sweep) potential window of Co–N@HCS, N@HCS and Pt/C catalysts in O$_2$-saturated 0.1 KOH at a rotation speed of 1600 rpm with a mass loading of 0.3 mg cm^{-2}. Inset: the comparison of the ΔE values. Reproduced from [115], Copyright © Elsevier, 2017.

Besides, more recently, some new synthesis strategies have been proposed to precisely control and fabricate the Co–N$_x$ catalytic materials, under the confinement effect, with well-engineered nanostructure derived from metal-organic framework precursors (MOFs). The Co–N$_x$/C catalysts synthesized by a simple pyrolysis of MOF precursor, inherited the large surface area and satisfactory porosity of MOFs, which led to dense active sites on the surface of materials. Feng's group [28] fabricated the Co/CoN$_x$/N–CNT/C electrocatalysts by heat treatment of MOF (Co–mela–BDC) at different temperatures under N$_2$ atmosphere. The Co/CoN$_x$/N–CNT/C catalyst obtained at 800 °C contained N-doped carbon nanotubes, which were generated by the catalysis of cobalt species. Owing to the synergistic effect between Co–N$_x$–C and NCNTs, the Co/CoN$_x$/N–CNT/C composite boosted the much higher ORR performance. Aijaz et al. [116] reported highly active bifunctional electrocatalysts for oxygen electrodes containing core–shell Co@Co$_3$O$_4$ nanoparticles embedded in CNT-grafted N-doped carbon-polyhedra from the pyrolysis of ZIF-67, involved in the sequential reduction and oxidation steps. The use of ZIF-67, as sacrificial precursor, was advantageous to build a core–shell nanostructured polyhedral with large surface area, and high dispersion of Co–N$_x$ and Co$_3$O$_4$–N$_x$ active sites. Furthermore, Zheng et al. [117] developed a bimetal (Cu and Co) embedded N-doped carbon framework, using the in situ growth of ZIF-67 polyhedrons on Cu(OH)$_2$ nanowires, followed by pyrolysis treatment, Figure 10. The pyrolytic decomposition proceeded successively for Cu(OH)$_2$ and ZIF-67, and thus the hierarchical porosity of ZIF-67 enabled the confinement of Cu nanocrystals and particles' size. The authors proposed that the existence of Cu ions not only provided extra active sites, but also led to an increased nitrogen content in the carbon frameworks via Cu–N coordination, optimized porous structure and large specific surface area, favorable to enhance the ORR electrocatalytic activity, Figure 10b. The CuCo@NC material showed an outstanding ORR activity, with much more positive E_{onset} = 0.96 V and $E_{1/2}$ = 0.88 V, comparable to commercial 30% Pt/C (1.04 V and 0.84 V, for E_{onset}, and $E_{1/2}$, respectively).

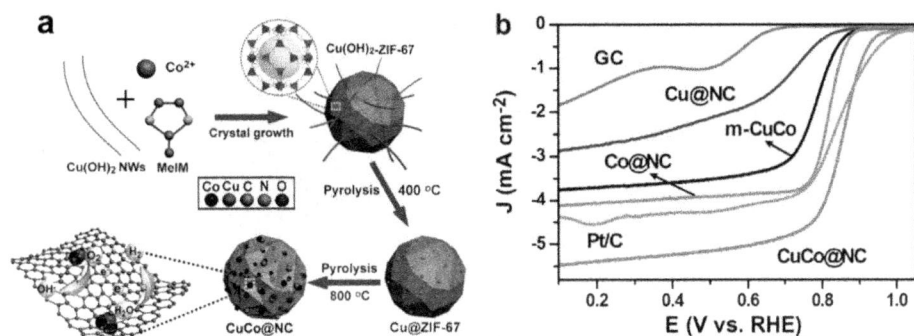

Figure 10. (**a**) The chemical synthesis route of Co–N@HCS catalyst; (**b**) ORR current-potential characteristics on (Cu and Co) embedded N-doped carbon materials in O_2-saturated 0.1 M KOH solution at a scan rate of 10 mV s^{-1} and rotation speed of 1600 rpm (cathodic sweep) at room temperature (25 ± 1 °C). The catalyst loadings of all samples were 0.182 mg cm^{-2}. Reproduced from [117], Copyright © Wiley, 2017.

Because of some interesting properties, such as, tunable chemical composition, permanent porosity, and a high thermal/chemical stability, the porous covalent networks (PCNs), also known as covalent organic frameworks (COFs), are chemical precursors of interest for electrocatalysis. In this context, Bu et al. [118] proved that the porphyrinic conjugated network PCN–FeCo/C, a carbonized product of heterometalloporphyrinic PCN–FeCo at 800 °C, showed a spectacular ORR activity and electrochemical stability in alkaline and acid electrolytes, Figure 11. The PCN–FeCo/C electrode displayed a very positive onset potential (1.00 V vs. RHE), and a half-wave potential of 0.85 V vs. RHE comparable to that of Pt/C (0.84 V vs. RHE) in 0.1 M KOH. Again, the binary PCN–FeCo/C catalyst was found to give the highest activity, i.e., the most positive E_{onset} = 0.90 V vs. RHE, and $E_{1/2}$ = 0.76 V vs. RHE in 0.1 M HClO$_4$ solution. The prominent performance of PCN–FeCo/C was attributed to the high homogeneity of active components derived from ordered distribution of Fe and Co covalent network, i.e., a generated hierarchical porosity.

Table 1. Selected Co-based catalyst for the oxygen reduction reaction.

Catalysts	Mass Loading (mg cm^{-2})	Electrolyte	RPM (rpm)	j_k (mA cm^{-2})	$E_{1/2}$ (V/RHE)	Tafel Slope (mV dec^{-1})	Refs.
Cubic CoSe$_2$/NCNH [a]	0.214	0.1 M KOH	1600	8.1 @ 0.8 V	0.81	52	[66]
Hexagonal CoSe/N–RGO [b]	0.286	0.1 M KOH	1600	2.9 @ 0.85 V	0.86	56	[67]
Co$_3$C–GNRs [c]	-	0.1 M KOH	1600	4.6 @ 0.5 V	0.77	41	[119]
Cubic CoSe$_2$/XC-72 Vulcan	0.1	0.5 M H$_2$SO$_4$	2000	0.1 @ 0.8 V	-	113	[120]
Cubic CoSe$_2$/XC-72 Vulcan	0.22	0.1 M KOH	1600	-	0.71	-	[61]
Co$_{1-x}$S/RGO	0.285	0.1 M KOH / 0.5 M H$_2$SO$_4$	1600	3.8 @ 0.7 V / 1.1 @ 0.7 V	0.75 / 0.59	- / -	[18]
Co$_{1-x}$S/N–S–G [d]	0.5	0.1 M KOH	1600	15% higher than Pt/C @ 0.6 V	0.86	58	[121]
Co$_{1-x}$S/SNG/CF [e]	0.153	0.1 M KOH	1600	4.3 @ 0.2 V	0.83	85	[122]
Co–S/NS–RGO [f]	0.38	0.1 M KOH	900	-	0.81	-	[123]
CoS$_2$/NS–GO	0.25	0.1 M KOH	1600	7.7	0.79	30	[124]
CoS$_2$/XC-72	0.1	0.1 M KOH	1600	4.2 @ 0.4 V	0.71	73	[46]
Co$_3$S$_4$/G [g]	0.051	0.1 M KOH	1600	4.5 @ −1.1 V (vs. Ag/AgCl)	-	-	[125]

Table 1. *Cont.*

Catalysts	Mass Loading (mg cm^{-2})	Electrolyte	RPM (rpm)	j_k (mA cm^{-2})	$E_{1/2}$ (V/RHE)	Tafel Slope (mV dec^{-1})	Refs.
Co$_3$S$_4$ nanosheets	-	0.1 M KOH	1600	-	−0.19 (vs. Hg/HgO)	-	[126]
Co$_3$S$_4$/C	0.011	0.5 M H$_2$SO$_4$	1600	-	0.26	-	[60]
Co$_9$S$_8$/G	0.6	0.5 M H$_2$SO$_4$	1600	3.7 @ −0.1 V (vs. Ag/AgCl)	−0.11 (vs. Ag/AgCl)	52	[127]
Co$_9$S$_8$/N–S–C [h]	0.1	0.1 M KOH	1600	-	0.90	74	[128]
Hollow Co$_9$S$_8$ microspheres	0.61	0.5 M H$_2$SO$_4$	1600	-	~0.18	-	[129]
Co$_9$S$_8$/N–S–G$_{gC3N4}$	0.612	0.1 M KOH	1600	-	−0.10 (vs. Ag/AgCl)	-	[130]
Co$_3$O$_4$/N–rmGO [i]	0.24	0.1 M KOH	1600	5.0 @ 0.4 V	0.83	42	[25]
Co@N–CNTs–m [j]	0.6	0.1 M KOH	1600	6.0	0.85	-	[32]
Co–S/G–3	0.08	0.1 M KOH	1600	7.0 @ 0.85 V	0.83	38	[80]
Co$_3$O$_4$–SP/NGr–24h [k]	-	0.1 M KOH	1600	-	0.82	76	[75]
Co@CoO@N–C/C	-	0.1 M KOH	1600	-	0.81	69	[82]
Co$_3$O$_4$/N–RGO–3	0.1	0.1 M KOH	1600	5.24	0.82	55	[76]
L$_1$G$_5$ [l] CoAl–LDHs/RGO	0.255	0.1 M KOH	1600	5.1 @ 0.2 V	0.71	-	[85]
CoPc/C	0.071	0.1 M KOH	1500	-	0.03 (vs. SHE)	62	[35]
NGM–Co	0.25	0.1 M KOH	1600	-	-	58	[112]
Co–N–mC	0.285	0.1 M KOH	1600	4.5 @ 0.8 V	0.85	45	[114]
Co–N@HCS [m]	0.3	0.1 M KOH	1600	4.8 @ 0.8 V	-	56	[115]
MOFs–800	0.335	0.1 M KOH	1600	3.7 @ 0.7 V	0.80	42	[28]
Co@Co$_3$O$_4$/NC–1	0.21	0.1 M KOH	1600	-	0.80	92	[116]
CuCo@NC	0.182	0.1 M KOH	1600	4.4 @ 0.8 V	0.88	80	[117]

j_k: kinetically current density; $E_{1/2}$: half-wave potential. [a] NCNH, N-doped carbon nanohorns; [b] N–RGO, N-doped reduced graphene oxides; [c] GNRs, graphene nanoribbons; [d] N–S–G, N, S-doped graphene; [e] CF, carbon fiber; [f] NS–RGO, N, S-doped reduced graphene oxides; [g] G, graphene; [h] N–S–C, N, S-doped carbon; [i] N–rmGO, N-doped reduced mildly graphene oxides; [j] N–CNTs–m, N-doped carbon nanotubes mixture; [k] NGr–24h, N-doped graphene; [l] L$_1$G$_5$, the mass ratio of LDHs/GO (1:5); [m] HCS, hollow carbon spheres.

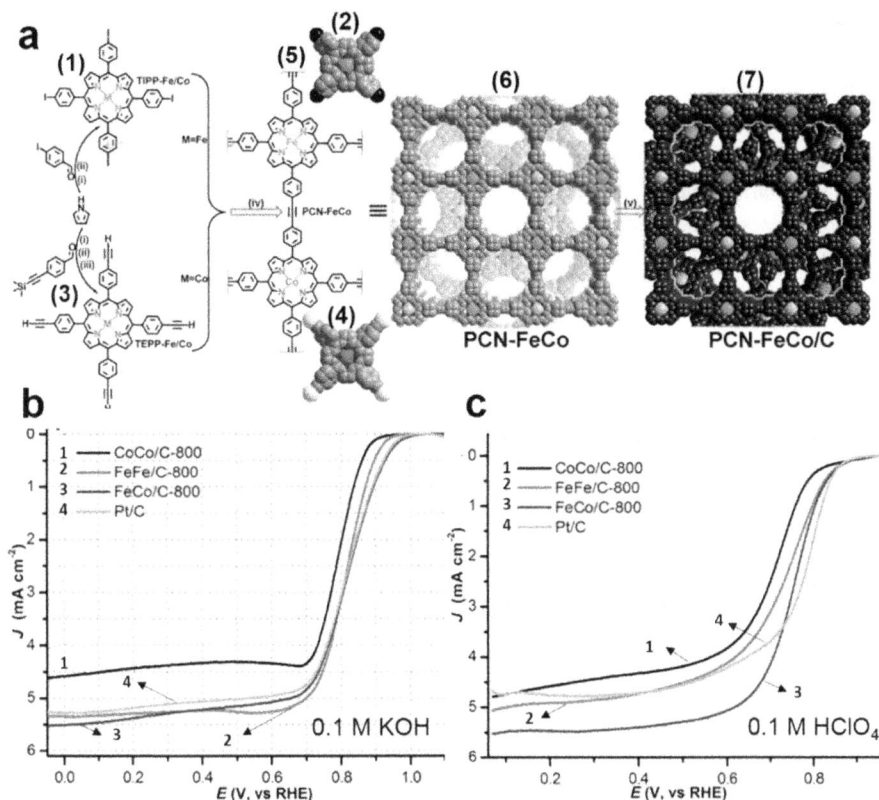

Figure 11. (**a**) Synthesis representation of porous covalent network (PCN)–FeCo/C by carbonization. Monomers: (1,2) TIPP–M, and (3,4) TEPP–M (M = Fe, Co). (5,6) porphyrinic conjugated network PCN–FeCo, and (7) the PCN–FeCo/C product. The reagents and conditions were: (i) propionic acid, reflux, 3 h; (ii) Co(OAc)$_2$·4H$_2$O or FeCl$_2$·4H$_2$O, CHCl$_3$·CH$_3$OH, reflux, 12 h; (iii) tetrabutylammonium fluoride (TBAF), THF–CH$_2$Cl$_2$, at R.T. 1 h; (iv) Pd$_2$(dba)$_3$, AsPh$_3$, THF/Et$_3$N, 50 °C, 72 h. TIPP stands for: 5, 10, 15, 20-tetrakis (4-iodophenyl) porphyrin, and TEPP for: 5, 10, 15, 20-tetrakis (4-ethynylphenyl) porphyrin; (**b**) ORR current-potential curves on PCN/C and 20% Pt/C (Alfa) at 1600 rpm (cathodic sweep) in O$_2$-saturated 0.1 M KOH solution, with the catalyst loading on PCN/C electrode of 0.2 mg cm^{-2} and on Pt/C electrode of 0.1 mg cm^{-2}; (**c**) in O$_2$-saturated 0.1 M HClO$_4$ with the catalyst loading on PCN/C electrode of 0.6 mg cm^{-2} and on Pt/C electrode of 0.1 mg cm^{-2}. Reproduced from [118], Copyright © Wiley, 2015.

3. Electrocatalysts for Oxygen Evolution Reaction (OER)

3.1. Mechanistic Approach of OER

OER kinetics are also a multi-electron charge transfer process in acid and alkaline media. The kinetic parameters such as overpotential (η), exchange current density (i), Tafel slope (b), turnover frequency (TOF), and so on, are employed to evaluate the OER performance of electrocatalytic materials. Particularly, the overpotential at a current density of 10 mA cm^{-2} is a crucial criterion to examine the OER performance (indicated as $\eta_{@10}$). These parameters play a key role for obtaining an insight into the mechanism of this electrocatalytic process. In general, the electrochemical reaction that occurs at the anode (OER) in acid, and alkaline electrolytes are:

Acid medium:

$$2H_2O_{(aq)} \rightarrow 4H^+{}_{(aq)} + 4e^- + O_{2(g)}, \tag{7}$$

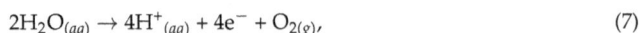

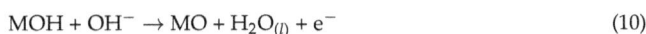

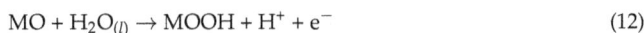

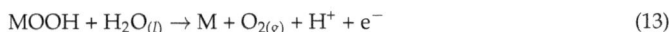

Alkaline medium:

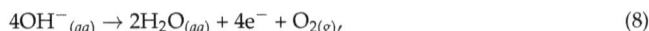

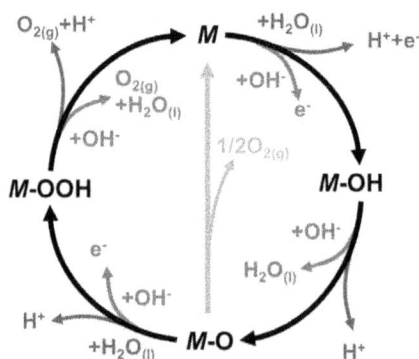

$$4OH^-{}_{(aq)} \rightarrow 2H_2O_{(aq)} + 4e^- + O_{2(g)}, \tag{8}$$

Various research groups have proposed possible OER mechanisms in acid (Equations (9)–(13)) and alkaline medium (Equations (14)–(18)). Most of the proposed mechanisms involve MOH and MO intermediates. The diagram in Figure 12 displays two different routes to form oxygen from a MO intermediate. One, the green route, via the direct combination of 2MO to produce $O_{2(g)}$ (Equation (11)), and that involving the generation of the MOOH intermediate (Equations (12) and (17)) which subsequently decomposes, black route, to $O_{2(g)}$ (Equations (13) and (18)). During the heterogeneous OER process, all the bonding interactions (M–O) within the intermediates (MOH, MO and MOOH) are crucial to determining the overall electrocatalytic performance.

Acid medium:

$$M + H_2O_{(l)} \rightarrow MOH + H^+ + e^- \tag{9}$$

$$MOH + OH^- \rightarrow MO + H_2O_{(l)} + e^- \tag{10}$$

$$MO \rightarrow 2M + O_{2(g)} \tag{11}$$

$$MO + H_2O_{(l)} \rightarrow MOOH + H^+ + e^- \tag{12}$$

$$MOOH + H_2O_{(l)} \rightarrow M + O_{2(g)} + H^+ + e^- \tag{13}$$

Alkaline medium:

$$M + OH^- \rightarrow MOH \tag{14}$$

$$MOH + OH^- \rightarrow MO + H_2O_{(l)} \tag{15}$$

$$2MO \rightarrow 2M + O_{2(g)} \tag{16}$$

$$MO + OH^- \rightarrow MOOH + e^- \tag{17}$$

$$MOOH + OH^- \rightarrow M + O_{2(g)} + H_2O_{(l)} \tag{18}$$

Figure 12. The oxygen evolution reaction (OER) mechanism in acid (blue line) and alkaline (red line) medium. Two reaction routes of oxygen evolution take place: (1) black line indicates that the process involves the formation of a peroxide (M–OOH) intermediate; (2) green line indicates that the direct reaction of two adjacent oxo (M–O) intermediates to produce molecular oxygen. Reproduced from [4], Copyright © Royal Society of Chemistry, 2017.

Until now, the Ni- and Co-based materials (both free support and supported on, e.g., carbon) have been intensively investigated as promising non-precious OER electrocatalysts. The catalysts derived from cobalt metal centers can activate the OER in alkaline medium rather than in acid medium. Particularly for CoO_x and $CoOOH$, the nature of their OER activities, stabilities in alkaline solution and OER mechanism have been thoroughly investigated [131–133]. For example, Mattioli group [131] provided insightful information into the pathways towards oxygen evolution of a cobalt-based catalyst (CoCat) by performing ab initio DFT + U molecular dynamics calculations of cluster models in water solution. The reaction pathways of CoCat were proposed as follows: (1) the fast H^+ mobility at the CoCat/water interface is responsible for an optimal distribution of terminal Co(III)–OH groups, as sites of injected holes. These sites are preferred in the case of complete cubane units; (2) the oxygen evolution process starts with the release of a proton from one of such terminal Co–OH sites, a process favored by the proton-acceptor species in solution, leading to the formation of a Co(IV)=O• oxyl radical; (3) the coupling of Co=O radicals with germinal (i.e., bonded to the same Co atom) Co–OH or Co–μO–Co groups to form hydroperoxo and peroxo intermediates.

3.2. Oxygen Evolution on Cobalt Chalcogenides Catalysts

3.2.1. Synthesis Strategy

One of the attractive electrocatalysts surface engineering strategies, e.g., etching and edging, could greatly improve the catalytic performance. Dai et al. [19], for the first time, developed an novel bifunctional oxygen electrode catalysts by using NH_3–plasma to simultaneously etch and dope the cobalt sulfides-graphene hybrid (N–Co_9S_8/G). NH_3–plasma treatment, not only could induce the N doping into both Co_9S_8 and graphene to enhance the activity, but also realized etching on the surface to expose more active sites for electrocatalysis. The N–Co_9S_8/G catalyst exhibited a low onset potential at ca. 1.51 V vs. RHE and a small Tafel slope of 82.7 mV dec^{-1}. Notably, the required overpotential of N–Co_9S_8/G catalyst to reach the current density of 10 mA cm^{-2} was only 0.409 V in 0.1 M KOH. Additionally, Xie's group [134] proposed that reducing the thickness of bulk $CoSe_2$ into the atomic scale, rather than doping or hybridizing, was inclined to form a great deal of exposed active sites e.g., V_{Co}" vacancies, which could effectively catalyze the OER process evidenced by the positron annihilation spectrometry, XAFS (X-ray Absorption Fine Structure) spectra and DFT calculations. The ultrathin $CoSe_2$ nanosheets with rich V_{Co}" vacancies manifested an OER overpotential of 0.32 V at 10 mA cm^{-2} in alkaline electrolyte (pH = 13) much less than that of the bulk $CoSe_2$.

3.2.2. Support Effect

Clear evidence of the strong metal–support interaction (SMSI) was given by Gao et al. [135] with synthesized $CoSe_2$ nanobelts on N-doped reduced graphene oxide (RGO) sheets. This system showed an exceptional OER activity and stability in alkaline environment, with a Tafel slope of 40 mV dec^{-1} and $\eta_{@10}$ (the overpotential required to achieve the current density of 10 mA cm^{-2}) of 0.366 V. The authors concluded that the high performance is dependent on the interaction between N-doped carbon domains and $CoSe_2$ nanobelts. Similarly, Liao et al. [121] embedded $Co_{1-x}S$ hollow nanospheres in N and S co-doped graphene holes to create an efficient bifunctional catalysts ($Co_{1-x}S$/N–S–G) with hierarchical meso-macroporous structures for ORR and OER. The $Co_{1-x}S$/N–S–G material with a large specific surface area (390.6 m^2 g^{-1}), showed a small overpotential of 371 mV for 10 mA cm^{-2} in 0.1 KOH, and the ΔE value is 0.706 V, which was much smaller than those of many reported non-precious metal catalysts. The investigated results demonstrated that the excellent bifunctional performance was mainly attributed to a synergistic effect of the multiple active sites consisting of $Co_{1-x}S$, N and S dopants, and possible Co–N–C sites. Interestingly, Guo et al. [136] used $CoSe_2$ nanosheets as support to grow Co_2B at room temperature. The resulting Co_2B/$CoSe_2$ hybrid catalysts showed $\eta_{@10}$ values for OER and HER in alkaline of 0.32 V, and 0.30 V, respectively. Apparently, $CoSe_2$

nanosheets supplied nucleation sites for Co_2B, as well as high electrical conductivity that promoted a high stability through water splitting reactions in alkaline medium.

3.3. Oxygen Evolution on Cobalt Oxides Catalysts

3.3.1. Mechanism of Cobalt Oxides

Cobalt (II) commonly undergoes the oxidation ($Co^{II} \rightarrow Co^{III}$) under an anodic potential prior to the OER in an aqueous solution, forming a layered oxidic cobalt species [137,138]. Some researchers pointed out that Co^{IV} species were generated on the outermost surface of the electrode [137,139]. So that the catalytic site formed on the Co_3O_4 electrode was the quasi-reversible redox couple Co^{III}/Co^{IV} that accelerated the one-step or two-step OER [131,133,140,141]. In order to investigate the role of the peroxo process in the oxygen evolution reaction, Fu et al. [142] prepared ultrathin Co_3O_4 nanosheets (NSs) with abundant active centers with a large electroactive surface to gain an insight into the OER performance of Co_3O_4 NSs. The possible mechanism of Co_3O_4 nanosheets towards OER included essentially double two-electron steps: (1) oxidation of OH^- to OOH_{ad} (thermodynamic rate-limiting step); and (2) fast oxidation of the intermediate OOH_{ad} to O_2^{ad} by Co^{III}/Co^{IV} surface redox couple (kinetic process), Figure 13. The apparent OER on Co_3O_4 NSs that proceeded via a four-electron pathway, corresponding to a double two-electron one, ascribed to Co^{III}/Co^{IV} sites acting efficiently to oxidize the generated OOH_{ad}, and facilitating the formation of $O_{2,ad}$.

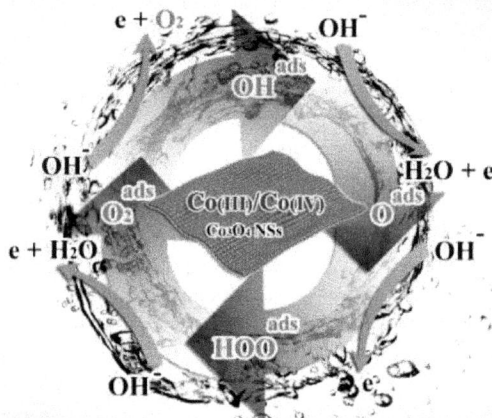

Figure 13. The scheme exemplifying the proposed OER electrocatalytic mechanism on Co_3O_4 NSs. Reproduced from [142], Copyright © Royal Society of Chemistry, 2016.

3.3.2. Chemical Composition

As mentioned above, the electrocatalytic performance of Co_3O_4 nanoparticles mainly depends on the surface area, the oxidation charge of the cobalt atoms, and the oxygen vacancies. Hence, a reasonable tuning of the surface electronic states of undoped Co_3O_4 can provide more electrochemically active sites to favor the OER. The preparation of mesoporous Co_3O_4 nanowires (Co_3O_4 NWs) by Zheng et al. [143], through a facile $NaBH_4$ reduction method led to seven-fold activity enhancement of OER, as compared to pristine Co_3O_4, Figure 14. As shown in Figure 14a, the peak current density observed in the cyclic voltammetry curves increased dramatically, after the chemical reduction process, signifying the existence of more active sites as a result of the oxygen vacancies, confirmed by DFT calculations. The increased Co^{4+}/Co^{3+} redox peaks (1.4–1.5 V) was ascribed to the formation of $[Co^{4+}-O]$ intermediate, involved in the turnover-limiting chemical step of oxygen evolution. The onset potential of the reduced Co_3O_4 NWs was ca. 1.52 V vs. RHE, i.e., ca. 50 mV and 100 mV more negative than pristine Co_3O_4 NWs and Pt/C catalyst, respectively, Figure 14b. Besides reduction post-treatment, Xia et al. [144] fabricated 3D ordered mesoporous cubic Co_3O_4 implementing hard-templating strategies. X-ray photoelectron spectroscopy (XPS) analysis revealed that the molar surface ratio Co^{III+}/Co^{II+} of ordered mesoporous Co_3O_4 was much lower than that of bulk-Co, suggesting more surface oxygen vacancies on the former, which benefited the adsorption and activation of molecular oxygen.

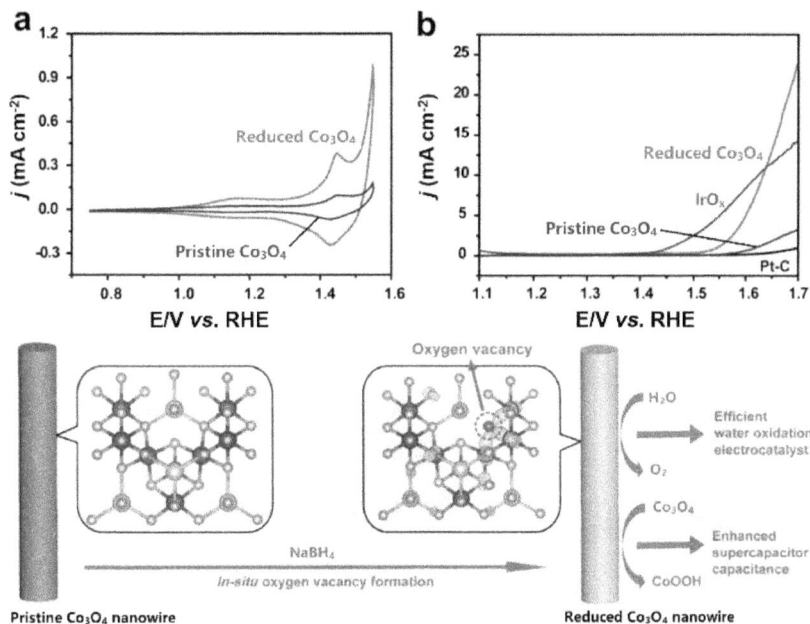

Figure 14. (a) Cyclic voltammograms of reduced Co_3O_4, and pristine Co_3O_4 NWs deposited on glassy carbon electrodes in O_2-saturated 1 M KOH at 5 mV s^{-1}; (b) water oxidation current-potential characteristic of reduced Co_3O_4 NWs (red curve), pristine Co_3O_4 NWs (blue curve), IrO_x (brown curve) and Pt/C (black curve) in O_2-saturated 1 M KOH at a scan rate 5 mV s^{-1} at 1600 rpm (anodic sweep) at 25 ± 1 °C with *iR*-compensation. The catalyst mass loadings in all cases was 0.136 mg cm^{-2}; The bottom panel shows the ins situ chemical reduction via $NaBH_4$ to form oxygen vacancies in Co_3O_4 NWs. Reproduced from [143], Copyright © Wiley, 2014.

3.3.3. Synthesis Strategy

The synthesis of graphene–Co_3O_4 (G–Co_3O_4) composite having a unique sandwich-architecture was reported by Zhao et al. [145]. The large amount of tiny Co_3O_4 nanocrystals, uniformly dispersed on both sides of graphene sheets, allowed for a favorable electron transfer kinetics. The onset potential of G–Co_3O_4 was 0.406 V vs. Ag/AgCl in 1 M KOH, and 0.858 V vs. Ag/AgCl in neutral phosphate buffer solution (PBS). The overpotential at a current density of 10 mA cm^{-2} ($\eta_{@10}$) was 313 mV in 1 M KOH, and 498 mV in PBS, respectively. Kim et al. [146], on the other hand, used dextrose as chemical source to obtain mesoporous carbon, and urea together with $CoCl_2\cdot6H_2O$ in a hydrothermal treatment to obtain carbon–cobalt oxide–nanorods (C–Co_3O_4–NRs). This latter provided an overpotential at a current density of 10 mA cm^{-2} ($\eta_{@10}$) of 415 mV, a value much lower than that of carbon free Co_3O_4 nanorods. The OER activity of the C–Co_3O_4–NRs electrode was significantly increased with a low onset potential of 356 mV and Tafel slope of 53 mV dec^{-1}.

Additionally, materials with high conductivity and mobility, e.g., Co foil, Ni foam and carbon fiber paper, were employed as support grown on non-precious metal electrocatalysts to facilitate the electrolyte diffusion, drive off the as-formed gas bubbles from the electrode surface during the oxygen evolution process, favoring the kinetics, and the chemical stability. In this connection, a simple and reasonable method to synthesize self-supported Co_3O_4 nanocrystal/carbon fiber paper via thermal decomposition of the $[Co(NH_3)_n]^{2+}$–oleic acid complex and subsequent spray deposition was done by Fu et al. [147]. The Co_3O_4 NCs with a loading of 0.35 mg cm^{-2} showed a current density of 16.5 mA cm^{-2} at $\eta_{@10}$ of 350 mV in 1 M KOH. With the same perspective, Wei et al. [148] reported the synthesis of Co_3O_4 nanorods array on Co foil (Co_3O_4 NA/CF) oxygen-evolving catalyst using the so-called in situ self-standing method, Figure 15a. The 1D Co_3O_4 NA/CF OER material only needed η = 308 mV to drive a geometrical current density of 15 mA cm^{-2} in 1 M KOH, exceeding the value reported so far on Co_3O_4-based electrocatalysts. The electrocatalyst delivered an excellent long-term durability of 22 h, and a turnover frequency of 0.646 mol O_2 s^{-1} at η = 410 mV in Figure 15b. The suggested OER high activity of Co_3O_4 NA/CF was thought to be the formation of Co^{III} (in an octahedral environment) on CoOOH as a result of Co_3O_4 oxidation. The surface Co^{III}-containing octahedral forms Co^{III}–OH, which was further oxidized to form the active catalytic center: Co^{IV}–O for OER. A further coupling of Co^{IV}–O with neighboring species formed hydroperoxo (Co^{IV}–OOH) to peroxo (Co^{IV}–OO) species, leading to the release of O_2 and initial Co^{III}, see scheme in Figure 15c.

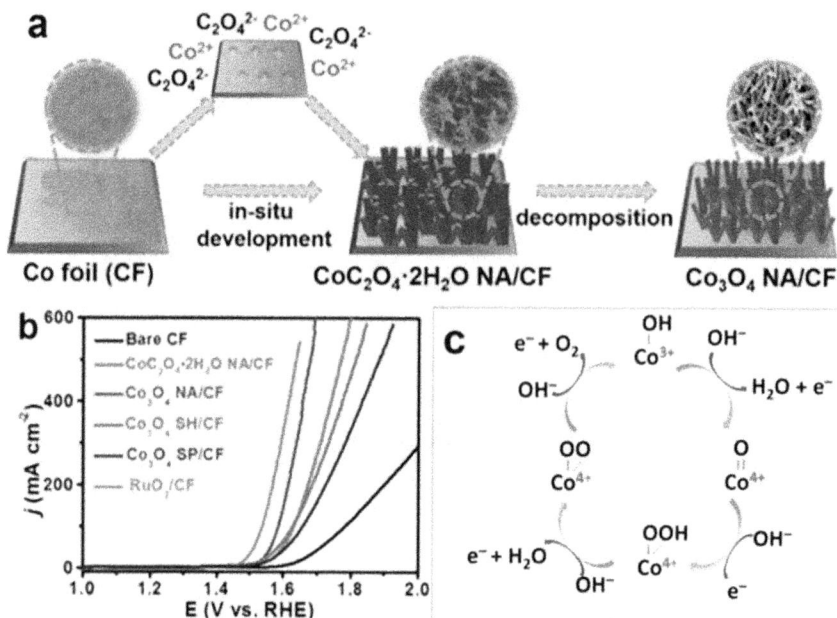

Figure 15. (a) Two-step manufacture of Co_3O_4 NA/CF; (b) LSV curves (anodic sweep) of RuO_2/CF, Co_3O_4 NA/CF, $CoC_2O_4 \cdot 2H_2O$ NA/CF, and bare Co catalysts in O_2-saturated 1 M KOH at 5 mV s^{-1} at a rotation speed of 1600 rpm at 25 °C. The constant catalyst loading was 1.9 mg cm^{-2}; (c) Suggested OER mechanism on Co_3O_4. Reproduced from [148], Copyright © Royal Society of Chemistry, 2018.

3.4. Oxygen Evolution on Cobalt-Based Layered Double-Hydroxides Catalysts

The use of LDH materials for OER has been recently promoted, see Table 2. Cobalt-containing LDHs composed of edge-sharing octahedral MO_6 layers, which are OER active sites, were successfully synthesized and showed an unusual catalytic activity and stability [149–151]. Li et al. [15] presented a strategy for a direct nucleation and growth of CoMn–LDH material on modified multiwall carbon nanotubes (MWCNTs) with three-dimensional hierarchical configuration. This approach afforded an intimate chemical and electrical coupling between LDH nanoplates and carbon materials, to allow for a rapid electron charge transfer from the active sites to the support. By tuning Co/Mn ratio, the Co_5Mn–LDH/MWCNT activated the reaction at a low overpotential of $\eta_{@10}$ ~300 mV in 1 M KOH. Ren et al. [152] designed and synthesized randomly cross-linked CoNi–LDH/CoO via an in situ reduction and interface-directed assembly in air. Owing to the orbital hybridization between metal 3d and O 2p orbitals, and electron transfer between metal atoms through Ni–O–Co, some Co and Ni atoms in the CoNi LDH underwent a high +3 valence. For transition metals, highly oxidized redox couples, e.g., $Co^{4+/3+}$ and $Ni^{4+/3+}$ were considered as active centers for OER [153].

The specific activity of the material for the target reaction is usually highly dependent on the chemical composition and electronic structure. The higher conductivity induced by modifying the electronic structure of Co is applicable for the electrochemical catalysis, such as de-lithiated hexagonal $LiCo_2$ for OER, and spinel $LiCo_2$ for ORR [153]. For the OER, a small amount of Fe doping was effective for enhancing the OER activities of Ni hydroxides or oxides, possibly due to the enhanced structure disorder and conductivity. Inspired by this, Sun et al. [154] systematically investigated the ORR and OER activities of ternary NiCoFe–LDHs and observed that a peroxidation treatment of NiCoFe–LDHs led to obtain o–NiCoFe–LDHs that significantly enhanced the corresponding bifunctional performance of (ORR/OER), as shown in Figure 16a. The XPS results and Zeta potential measurements evidenced that Co^{2+} was partially oxidized to a higher Co^{3+} state, while negligible chemical state change of Ni and Fe elements was observed in o–NiCoFe–LDHs. The partial conversion of Co^{2+} to Co^{3+} state stimulated the charge transfer to the catalyst surface, which could lead to the enhancement of the conductivity. In the Figure 16b, in 6 M KOH medium, the o–NiCoFe–LDHs afforded a high ORR current density of -20 mA cm^{-2} at a required potential of 0.65 V vs. RHE. Moreover, there was a negligible ORR current density degradation of o–NiCoFe–LDHs for 40 h. The higher conductivity induced from the higher valence state of Co, might enhance the electrophilicity of the adsorbed O and thus facilitating the reaction of an OH^- anion with an adsorbed O atom on the catalytic active sites to form adsorbed –OOH species, which was considered as the rate-limiting step for OER [138,155].

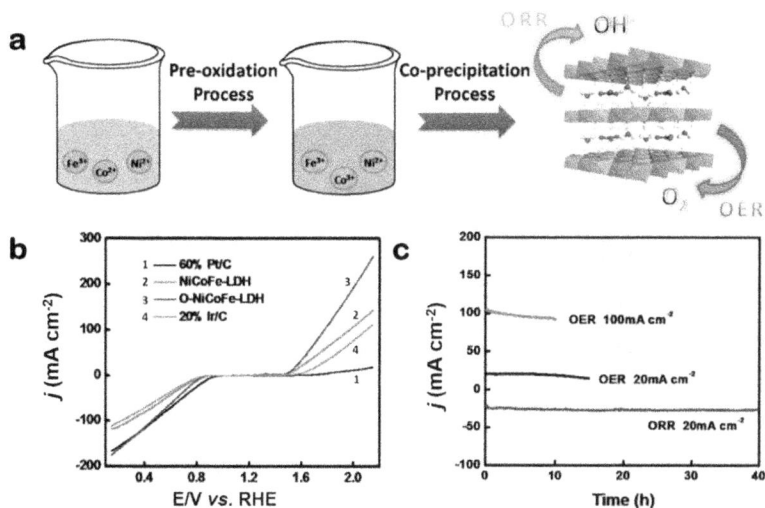

Figure 16. (**a**) Schematic chemical synthesis route and crystal structure of peroxidized ternary LDH bifunctional catalyst; (**b**) The global polarization curves of various catalysts loaded onto Teflon-treated carbon fiber paper (T–CFP) in O_2-saturated 6 M KOH electrolyte with a scan rate of 1 mV s^{-1}, the mass-loading of 1 mg cm^{-2} without iR-compensation; (**c**) ORR (cathodic sweep) and OER (anodic sweep) stability measurements of o–NiCoFe–LDH/Y–CFP electrode in O_2-saturated 6 M KOH electrolyte with a constant potential, the mass-loading of 1 mg cm^{-2}. Reproduced from [154], Copyright © Wiley, 2015.

It is fascinating to design and build a bifunctional oxygen electrode catalyst based on the combination of highly OER-active Ni–Fe hydroxides or oxides, and highly ORR-active Co, Fe-based compounds. Feng's group [33] designed and obtained a bifunctional electrocatalyst based on $NiFeO_x/Co-N_y-C$ by a simple calcination of $Ni_2Fe–CoPcTs–LDH$ precursor based on the intercalation of cobalt phthalocyanine tetrasulfonate (CoPcTs) into $Ni_2Fe–LDHs$ under N_2 atmosphere at 600 °C. The key aspects of this bifunctional electrocatalyst, for the reversible oxygen electrode, was the mutual incorporation of the ORR-active centers (Co-based compound), and OER-active ones (spinel $NiFe_2O_4$). Particularly, the CoPcTs–intercalated structure and $Ni_2Fe–LDH$ host sheet significantly enhanced the immobilization of CoPcTs and improved the dispersion degree of catalytic sites, respectively.

Moreover, the surface area of lamellar architecture materials can be further enlarged by swelling and exfoliating into individually single layers by mechanical, chemical or electrochemical means [156–159]. The exfoliated single-layered nanosheets revealed significantly higher oxygen evolution activity than the corresponding bulk LDH in alkaline conditions. Hu et al. [159] inferred that the higher OER activities of exfoliated LDHs (CoCo, NiCo, and NiFe–LDH) were mainly attributed to the increase in the number of active edge sites and to higher electronic conductivity. Analogously, Jin et al. [160] selected NiCo LDH as a representative material to demonstrate the concept of the amplified influence of exfoliation using a newly developed high-temperature high-pressure hydrothermal continuous flow reactor (HCFR), see Figure 17. The major findings were presented as follows: (1) the utilized HCFR technology effectively maintain the supersaturation to control the morphology and size of the product; (2) the exfoliation not only resulted in thinner layers with reduced size, but it was also caused by a change in the electronic structure; (3) the increase in the number of edge sites to activate OER; and (4) the increase in the electrochemical active surface area (ECSA) upon the exfoliation was not the only important factor that led to the enhanced OER performance. This work provided a general strategy to enhance the electrocatalytic performance of layered materials by chemical exfoliation.

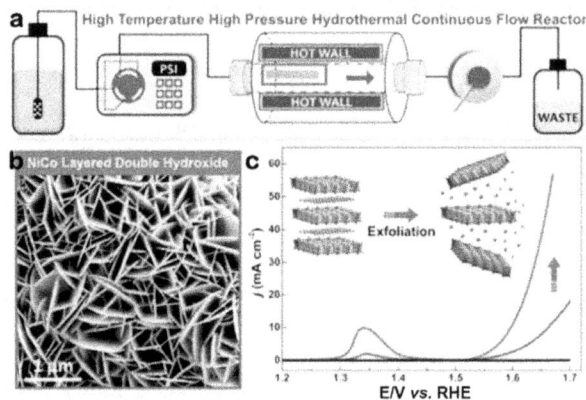

Figure 17. (**a**) The high-temperature high-pressure continuous flow reactor (HCFR) scheme; (**b**) scanning electron microscope (SEM) image of exfoliated NiCo LDH from HCFR; (**c**) *iR*-corrected and background subtracted polarization curves (anodic sweep) of NiCo LDH nanoplates (green curve) made of HCFR with mass catalyst loading of ~0.23 mg cm^{-2}, NiCo LDH nanosheets (red curve) synthesized nanoplates from exfoliated HCFR with a mass catalyst loading of ~0.17 mg cm^{-2}, and carbon paper (black curve) in O_2-saturated 1 M KOH at a scan rate of 0.5 mV s^{-1}. All experiments were conducted at room temperature (25 °C). Reproduced from [160], Copyright © American Chemical Society, 2015.

Table 2. Selected Co-based catalyst towards the oxygen evolution reaction.

Catalysts	Mass Loading (mg cm^{-2})	Electrolyte	RPM (rpm)	E_{onset} mV vs. RHE	η@ 10 mA cm^{-2} (mV)	Tafel Slope (mV dec^{-1})	TOF (s^{-1})	Refs.
Nanobelts CoSe$_2$/N-doped RGO [a]	0.2	0.1 M KOH	1600	-	366	76	-	[135]
Co$_{1-x}$S/N-S-G [b]	0.5	0.1 M KOH	1600	-	371	63	-	[121]
N-Co$_9$S$_8$/G [c]	0.2	0.1 M KOH	1600	1.51	409	83	-	[19]
Co$_2$B/CoSe$_2$	0.4	1 M KOH	-	-	320	56	-	[136]
Co$_3$O$_4$-NSs [d]	1.76	1 M KOH	1600	1.51	330	69	-	[142]
Reduced Co$_3$O$_4$ NWs [e]	0.136	1 M KOH	1000	1.52	400	72	-	[143]
G-Co$_3$O$_4$	0.189	1 M KOH	1600	0.41 vs. Ag/AgCl	313	56	0.45 @ 0.35 V	[145]
C-Co$_3$O$_4$-NRs [f]	0.142	1 M KOH	-	1.59	415	53	-	[146]
Co$_3$O$_4$ NCs [g]	0.35	1 M KOH	-	1.52	350 @ 16.5 mA cm^{-2}	101	-	[147]
OA-Co$_3$O$_4$ NCs	0.35	1 M KOH	-	1.55	-	118	-	[147]
Co$_3$O$_4$ NA/CF [h]	1.9	1 M KOH	-	-	308 @ 15 mA cm^{-2}	71	0.65 @ 0.41 V	[148]
NiCo$_{2.7}$(OH)$_x$	0.2	1 M KOH	1600	1.48	350	65	0.18 @ 0.35 V	[149]
CoFe/C	-	1 M KOH	1600	1.45	300	61	-	[151]
CoNi/C	-	1 M KOH	1600	1.56	360	39	-	[151]
Co$_5$Mn-LDH/MWCNT [i]	0.283	1 M KOH	1600	-	300	74	0.47 @ 0.35 V	[15]
CoNi LDH/CoO	0.265	1 M KOH	1600	1.48	300	123	1.40 @ 0.40 V	[152]
CoCo LDH/CoO	0.265	1 M KOH	1600	1.54	340	123	0.80 @ 0.40 V	[152]
O-NiCoFe-LDH	0.12	0.1 M KOH	1600	-	340	93	0.02 @ 0.30 V	[154]
NiFeO$_x$/Co-N$_y$-C	0.196	1 M KOH	1600	1.47	310	60	0.01 @ 0.30 V	[33]
NiCo-NS [k]	0.07	1 M KOH	-	1.52 @ 1 mA cm^{-2}	334	41	-	[159]
CoCo-NS [k]	0.07	1 M KOH	-	1.54 @ 1 mA cm^{-2}	353	45	-	[159]
Exfoliated NiCo LDH	0.17	1 M KOH	-	-	367	112	-	[160]

E_{onset}, onset potential; η @ 10 mA cm^{-2}, the overpotential (η) located at the current density of 10 mA cm^{-2}; TOF, the turnover frequency. [a] RGO, reduced graphene oxides; [b] N-S-G, N, S-doped graphene; [c] G, graphene; [d] NSs, nanosheets; [e] NWs, nanowires; [f] NRs, nanorods; [g] NCs, nanocrystals; [h] CF, carbon fiber; [i] MWCNT, multiwall carbon nanotubes; [k] NS, nanosheets.

3.5. Cobalt-Based Bifunctional Catalysts in Assembled Unitized Regenerative Fuel Cells

Bifunctional oxygen electrode catalysts play a vital role in the development of unitized regenerative fuel cells (URFCs). URFC, a compact energy storage and conversion device, can simultaneously work in a fuel cell mode to produce electricity, and as a water electrolyzer mode to store off-peak electricity in the form of hydrogen. In this closed-loop system, the essential component is the desired bifunctional oxygen electrode catalysts with high catalytic activity, long-term durability, and strongly resistance to anodic corrosion. To date, several previous studies have reported cobalt-based materials as promising bifunctional ORR/OER electrocatalysts for a URFC system [161]. Scott's group used cobalt-based ORR catalysts as the cathodes and demonstrated high and stable power density performance (>200 mW cm^{-2}) in alkaline anion exchange membrane fuel cells (AAEMFCs) [162]. They further [163] modified the ORR and OER electrode with $Cu_{0.6}Mn_{0.3}Co_{2.1}O_4$ catalyst for regenerative H_2–O_2 fuel cell. In water electrolyzer mode, the onset voltage for water electrolysis (deionized water as electrolyte) was ca. 1.55 V. At 100 mA cm^{-2}, the voltages of fuel cell mode and electrolyzer mode were 0.58 V and 1.82 V, respectively, indicating that the fuel cell to electrolyzer voltage ratio was ca. 31.87%. However, the current densities in both modes were much lower than those cells with KOH as electrolyte, probably because of the large electrolytic resistance.

Xu et al. [161] measured the performance of a unitized regenerative anion exchange membrane fuel cells (UR–AEMFCs), using a bifunctional ORR/OER catalysts, Co_3O_4/oCNT, (oCNT stands for oxidized CNTs) on the oxygen electrode. The obtained performance in fuel cell mode was basically consistent with the result in half-cell mode, while the electrolyzer performance was poor. Likewise, 0.1 M KOH solution in the cell, significantly improved the performance of electrolysis.

Our group designed a self-assembly laminar flow unitized regenerative micro-cell (LFURMC) without gas separator, which was provided with Pt/C as the hydrogen catalyst, bifunctional oxygen NiFeOx/CoNy–C catalyst and 3 M KOH electrolyte [33]. In the fuel cell mode, Figure 18, the NiFeO$_x$/CoN$_y$–C electrode showed the maximum power density of 56 mW cm^{-2}, which was very close to that of Pt/C (63 mW cm^{-2}) and IrO$_x$/C (58 mW cm^{-2}); in the electrolyzer mode, the maximum electrical power consumed on NiFeO$_x$/CoN$_y$–C (237 mW cm^{-2}), was more than three times that on Pt/C (73 mW cm^{-2}). Moreover, the calculated round-trip efficiency (RTE) of NiFeO$_x$/CoN$_y$–C catalyst nearly held a level of ca. 52% during three cycles, evaluating a super reversibility of the NiFeO$_x$/CoN$_y$–C electrode.

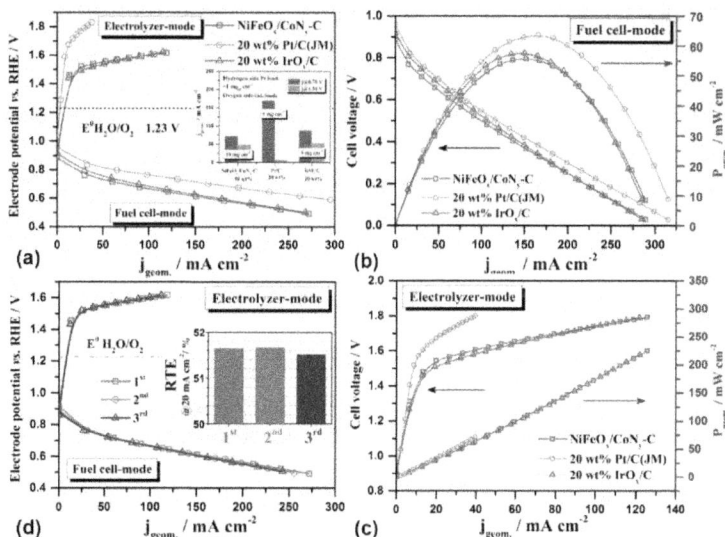

Figure 18. Alkaline laminar flow unitized regenerative micro-cell (LFURMC) experiment. (**a**) Current-potential of the material as anode (electrolyzer mode) and cathode (fuel cell mode). inset: The geometric activities comparison of three LFURMC; (**b**) the cell voltage and power density curves of the fuel cell with platinum as the hydrogen catalyst, and NiFeO$_x$/CoN$_y$–C, 20 wt% Pt/C and 20 wt% IrO$_x$/C cathode catalysts; (**c**) the cell voltage and power density curves of the electrolyzer with platinum as the hydrogen catalyst, and NiFeO$_x$/CoN$_y$–C, 20 wt% Pt/C and 20 wt% IrO$_x$/C cathode catalysts; (**d**) Stability tests of LFURMC based on the NiFeO$_x$/CoN$_y$–C catalyst (three runs), inset: round-trip-efficiency (RTE) for the three cycles. All measurements were conducted at room temperature (25 °C). Remark: no Ohmic-drop correction was made for all determinations. Reproduced from [33], Copyright © Elsevier, 2017.

4. Electrocatalysts for Hydrogen Evolution Reaction (HER)

4.1. Mechanism of HER

The hydrogen evolution reaction (HER) is the half-reaction carried out at the cathode of an electrolyzer, in which protons (acidic environment) or water molecules (alkaline environment) are reduced, accompanied by the subsequent evolution of gaseous hydrogen through the water splitting process. The overall HER proceeds as follow (in all cases, any catalytic site is denoted as "*"):

Acid medium:

$$2H^+ + 2e^- \rightarrow H_{2(g)} \tag{19}$$

Alkaline medium:

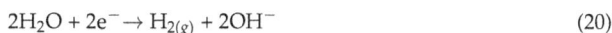

$$2H_2O + 2e^- \rightarrow H_{2(g)} + 2OH^- \tag{20}$$

The standard potentials (E°) are different due to the nature of the active ions in the reaction. As we can expect, the HER as any electrochemical reaction possesses a certain activation energy barrier to promote the reaction, usually denoted as overpotential η. Therefore, the HER usually demands the assistance of electrocatalytic materials to lower η, and consequently to increase the reaction rate and efficiency. The mechanisms to achieve the HER can be Volmer–Heyrovsky or Volmer–Tafel. In acid medium, the HER proceeds according to the following steps:

i. A hydrogen atom adsorption, which is the result of the combination of a proton and an electron on the electrode surface (proton discharge) is the Volmer reaction:

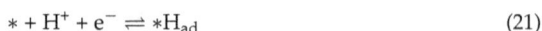

$$* + H^+ + e^- \rightleftharpoons *H_{ad} \qquad (21)$$

ii. The adsorbed hydrogen atom interacting with a proton and an electron leads to an electrochemical desorption. This reaction is the Heyrovsky reaction:

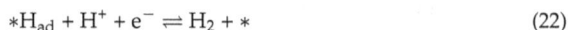

$$*H_{ad} + H^+ + e^- \rightleftharpoons H_2 + * \qquad (22)$$

iii. The coupling of the two adsorbed hydrogen atoms leads to a dissociative desorption of hydrogen, the Tafel reaction:

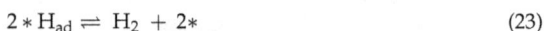

$$2 * H_{ad} \rightleftharpoons H_2 + 2* \qquad (23)$$

In alkaline electrolyte, due to the OH^- abundance, the HER proceeds according to the following steps. For the Volmer reaction, the molecular water couples with an electron, leading in an adsorbed hydrogen atom at the electrode surface:

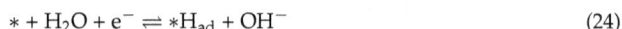

$$* + H_2O + e^- \rightleftharpoons *H_{ad} + OH^- \qquad (24)$$

i. For the Heyrovsky reaction, the adsorbed hydrogen atom combines with molecular water and an electron, allowing the electrochemical desorption of hydrogen:

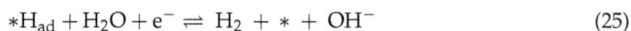

$$*H_{ad} + H_2O + e^- \rightleftharpoons H_2 + * + OH^- \qquad (25)$$

ii. The Tafel reaction is similar to that of the acidic medium.

In acid and alkaline media, for the HER, the hydrogen adsorption starts via the Volmer reaction, Equations (21) or (24). The successive hydrogen desorption can proceed via the Heyrovsky reaction, Equations (22) or (25) or the dissociative desorption via the Tafel reaction, Equation (23). On the other hand, the HER can be proposed by the Tafel slope, as derived from the HER polarization curves. The Tafel slope analysis represents the intrinsic nature of the electrocatalytic material, and the empirical magnitude of the Tafel slope can provide information to distinguish the mechanism. For a Tafel rate determining step (RDS), the slope is 30 mV dec^{-1}, whereas for Heyrovsky and Volmer RDS, slopes are 40, and 120 mV dec^{-1}, respectively.

The HER mechanism on Co-based catalysts follows the Volmer–Heyrovsky mechanism, where normally the Heyrovsky step is the RDS in acid medium [164,165]; whereas the Volmer step is considered the RDS in alkaline medium [165].

4.2. Hydrogen Evolution on Cobalt Chalcogenides Catalysts

4.2.1. Synthesis Strategy

Co-based catalysts are potential materials for the HER. Considerable research efforts were devoted to synthesizing cobalt chalcogenides as catalysts for HER. Those dichalcogenides are in the pyrite/marcasite phase. The metal atoms are octahedrally coordinated to adjacent S/Se atoms. Kong et al. [164] claimed that $CoSe_2$ possessed one of the best HER performance among the first raw transition-metal chalcogenides, because of the partially filled e_g band of $CoSe_2$. With this knowledge, Liu et al. [166] developed $CoSe_2$ nanowire arrays on carbon cloth ($CoSe_2$ NW/CC) through a facile two-step hydrothermal preparative strategy. The hydrogen evolution reaction performance was tested in 0.5 M H_2SO_4. Current densities of 10 and 100 mA cm^{-2} at overpotentials of 130 and 164 mV were maintained for at least 48 h. The excellent HER activity and durability for $CoSe_2$ NW/CC were explained as follows. (1) The nanoarray allows for the exposure of more active sites; (2) The direct growth of $CoSe_2$ on CC offers intimate contact, good mechanical adhesion and excellent electrical connection between them; and (3) The absence of polymer binder for catalyst immobilization.

Vertical-aligned graphene nanoribbons (VA–GNRs) were used as a support for cobalt carbide nanocrystals, e.g., Co_3C/VA–GNRs by Fan et al. [119]. The HER performance in acid medium revealed a Tafel slope of 57 mV dec^{-1} and $\eta_{@10}$ of ca. 0.125 V. Zhang et al. [167] synthesized by a hydrothermal procedure polymorphic $CoSe_2$ supported onto graphite with promising HER performance, Figure 19. These authors observed that the optimized calcination temperature of $CoSe_x$ was 300 °C. This treatment allowed the change of a mixed orthorhombic to cubic phase of $CoSe_2$. The polymorphic $CoSe_2$ catalytic center showed higher HER performance compared to $CoSe_x$, cubic $CoSe_2$ and CoSe synthesized under the same conditions. The high performance observed in polymorphic $CoSe_2$ was attributed to the enhanced chemisorption of H atoms onto the mixed-phase in the Tafel step.

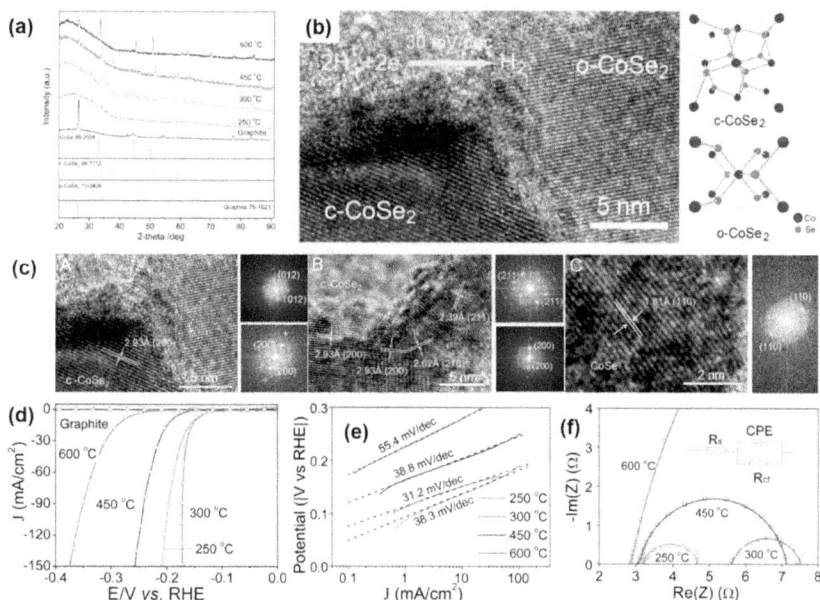

Figure 19. (a) XRD patterns of the as-prepared $CoSe_x$/GD samples calcined at different temperatures; (b) the crystal structures of cubic $CoSe_2$ (c–$CoSe_2$) and orthorhombic $CoSe_2$ (o–$CoSe_2$); (c) HRTEM micrographs and electron-diffraction patterns of c–$CoSe_2$ annealed at 300 °C (A), c–$CoSe_2$ annealed at 450 °C (B) and CoSe annealed at 600 °C (C), respectively; (d) *iR*-corrected hydrogen evolution reaction (HER) polarization curves (cathodic sweep) of $CoSe_x$ (250 °C), c–$CoSe_2$ (300 °C), c–$CoSe_2$ (450 °C) and CoSe (600 °C) in H_2-saturated 0.5 M H_2SO_4 solution with a scan rate of 3 mV s^{-1}; (e) Tafel slopes analysis of polymorphic $CoSe_2$/GD catalysts; (f) Nyquist plots for the HER at an overpotential of −0.5 V vs. SCE from 200 kHz to 50 mHz, the data were fitted by the simplified Randles equivalent circuit (inset). Reproduced from [167], Copyright © American Chemical Society, 2015.

4.2.2. Crystal Structure and Nanostrcuture

Tuning the crystal structure and surface morphology, a catalytic center can enhance its performance for electrochemical energy conversion devices. For example, the combined effect of morphology and phase of $CoSe_2$ was studied by Li et al. [168]. These authors fashioned $CoSe_2$ nanotubes with orthorhombic- (o–$CoSe_2$) and cubic-phases (c–$CoSe_2$) by a facile precursor transformation method, Figure 20a. In the synthetic process, the crystal structure and surface morphology of $CoSe_2$ were adjusted by the calcination temperature. Benefiting from the advantageous tubular structure, including the functional shells and well-defined interior voids, $CoSe_2$ nanotubes showed a clear HER performance in alkaline medium. Interestingly, the authors observed that the orthorhombic phase (o–$CoSe_2$) possessed the highest HER performance in terms of lowest onset overpotential (~54 mV) and smallest Tafel slope (~65.9 mV decade^{-1}) in alkaline medium, see Figure 20b,c. Another approach where the morphology effect towards the HER was observed were given by $CoSe_2$ nanoparticles [21], nanowires [166,169], interwoven [170], nanocomposites [171], hollow microspheres [172], and nanosheets [173].

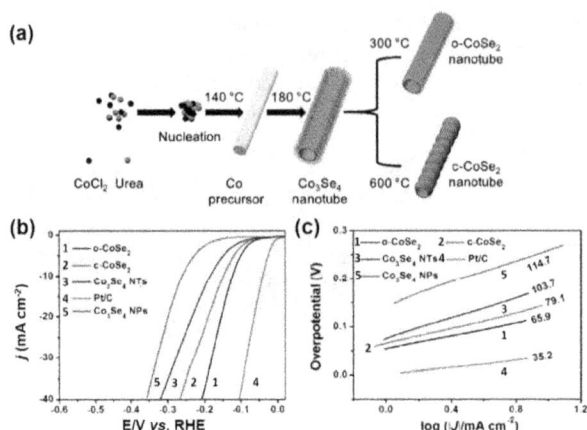

Figure 20. (a) Synthesis route of $CoSe_2$ nanotubes; (b) *iR*-corrected HER polarization curves (cathodic sweep) of o–$CoSe_2$, c–$CoSe_2$, Co_3Se_4 NTs, Pt/C and Co_3Se_4 nanoparticles (NPs) catalysts at a scan rate of 5 mV s^{-1} in 1 M KOH solution; and (c) Tafel slopes analysis of all the catalysts. The mass loading of the catalyst was ~0.283 mg cm^{-2}. Reproduced from [168], Copyright © Royal Society of Chemistry, 2017.

On the other hand, the construction of the ultra-thin two-dimensional (2D) nanostructure have become an promising approach to tailor novel high active electrocatalytic centers, due to the confined charge interaction in the planar dimension with minimum interlayer interaction (which in fact hinders the electrical conductivity), thus improving catalytic properties [174]. An ultrathin 2D structure with abundant low-coordinated surface atoms offers adequate active sites for hydrogen evolution reactions. Liu group [175], prepared the $Mn_{0.05}Co_{0.95}Se_2$ ultrathin nanosheets with 1.2 nm thickness by usual liquid exfoliation of homogeneous Mn doped $CoSe_2$/DETA hybrid nanosheets. The subtle distortion of atomic arrangement was induced after the incorporation of Mn^{2+} into $CoSe_2$, which boosted the exposure of more active edge Se sites, optimizing the HER activity. Since the chalcogenide atoms at the edges of pyrite catalyst have been verified to be HER active sites analogous to MoS_2, the authors proposed that the edge Se sites in the $CoSe_2$ probably were responsible for HER as the active sites. The Mn-doped $CoSe_2$ ultrathin nanosheets displayed outstanding HER performance, with a low overpotential of 174 mV, and a Tafel slope of 36 mV dec^{-1}.

4.3. Hydrogen Evolution on MOFs Catalysts

Regarding derived-MOF electrocatalysts, the porous ZIF-9 was used as a cobalt source and sacrificial template to synthesize CoS_2, see Figure 21a [176]. The ZIF-9 presented very weak HER electrocatalytic performance, the benzimidazole ligands of ZIF-9 were substituted by S ions after adding sulfur sources in the appropriate mention of 200 °C and 1 atm hydrothermal temperature and pressure, the resulting CoS_2 had excellent HER activity. The introduction of graphene to further enhance the conductivity of the material, and improve the dispersion of cobalt disulfide, on RGO (CoS_2/RGO) was necessary. In Figure 21b, the HER performance of the material with an overpotential of 180 mV at 10 mA cm^{-2} was smaller than those of ZIF-9, and ZIF-9/RGO. The Tafel slope of CoS_2/RGO was 75 mV decade^{-1}. This slope was near that obtained on Pt/C electrode. Zhang et al. [177] synthesized the core–shell structure Co/Co$_9$S$_8$ anchored onto S-, N- co-doped porous graphene sheet (Co/Co$_9$S$_8$@SNGS). They employed thiophene-2,5-dicarboxylate (Tdc), and 4,4-bipyridine (Bpy) organic ligands assembled Co-based metal–organic frameworks in situ grown on graphene oxide sheets. The S-containing Tdc, and the N-containing Bpy did not only trigger the growth of Co–MOF nanocrystals but also fixed the S/N atomic ratio of 1:2.4 on graphene oxide sheets. After pyrolysis of Co–MOF at 1000 °C the catalyst showed high bifunctional catalytic activities for the OER, and HER in 0.1 M KOH electrolyte, with an overpotential of 290 mV for OER at a current density of 10 mA cm^{-2}, and 350 mV for HER at a current density of 20 mA cm^{-2}.

Besides the cobalt sulfide, cobalt selenide could also be synthesized using MOFs.CoSe$_2$ nanoparticles embedded in defective carbon nanotubes (CoSe$_2$@DC) was produced by carbon–oxidation–selenization procedure of Co-based MOFs [64], Figure 22. The pre-oxidation treatment was a crucial step in introducing an increasing number of defects into carbon nanotubes, which promoted the reaction between Co@Carbon and selenium. The authors associated the enhanced HER performance to the induced carbon-surface defect density, which favored the diffusion of Se atoms through Co atoms, and made the best use of the exposed active surface area. Lin et al. [178] synthesized MOF-derived cobalt diselenide (MOF–CoSe$_2$) built with CoSe$_2$ nanoparticles anchored into nitrogen-doped graphitic carbon through in situ selenization of Co-based MOFs. The N-doped MOFs derived architecture benefited of the high conductivity provided by the abundant active reaction sites, ensuring a robust contact between CoSe$_2$ nanoparticles and N-doped carbon support.

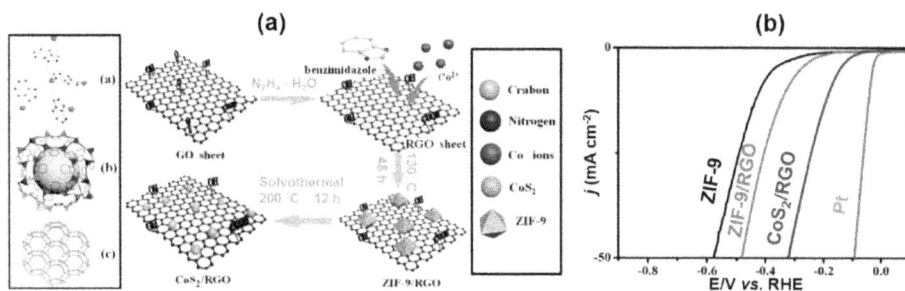

Figure 21. (a) Schematic illustration of the synthesis of CoS$_2$/RGO; and (b) HER polarization curves (cathodic sweep) collected on different catalysts with mass loading of 0.285 mg cm^{-2} in 0.5 M H$_2$SO$_4$ acidic solution at a scan rate of 5 mV s^{-1} at room temperature. Reproduced from [176], Copyright © Elsevier, 2017.

Table 3 summarizes the relevant contributions regarding the HER Co-based catalysts, wherein it is possible to highlight that these catalysts present considerable performance in acid and in alkaline media. One can stress some remarkable parameters regarding the HER electrocatalytic performance, namely, (1) among non-precious catalysts, Co-based catalytic centers are one potential candidate to replace Pt-based electrodes for the water-splitting cathodic reaction, their high activity and stability; (2) since cobalt selenides catalytic centers boost the HER in acidic and in alkaline media, the chemical coordination of Co with sulfides and phosphides showed considerable performance in acid media; (3) the crystalline structure, due to the exposed active sites and electronic properties are crystalline-phase dependent; (4) the surface morphology can expose active sites; (5) the electrical coupling with conductive supports; in as much as the coupling of Co-based catalytic centers with high surface area and high electrical conductive supports modulate the dispersion of the catalytic centers and the interfacial charge-transfer. Doped carbon supports (mainly N-doped graphitic surfaces) could modify the electronic structure of the anchored catalytic centers; and (6) the synthesis route and metal source; e.g., MOFs, could provide hybrid materials which combine the surface properties of the porous carbon network and the catalytic properties of Co domains. Those parts are critical points to be considered in tailoring HER advanced materials.

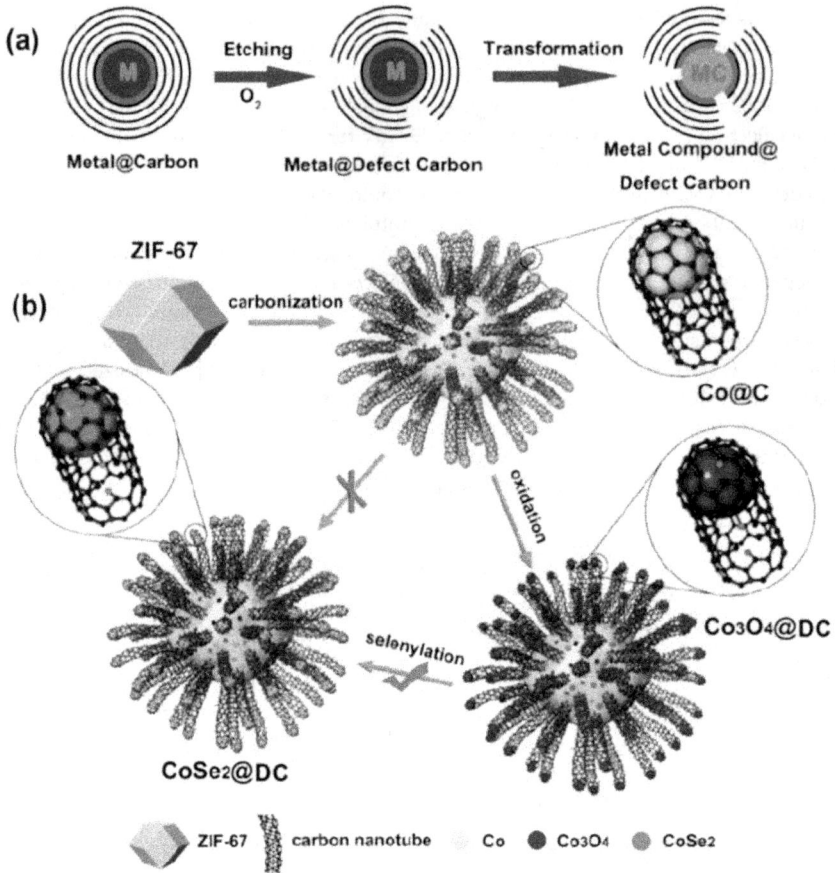

Figure 22. Schematic representation of (**a**) the etching process of metal@carbon by pre-oxidation; (**b**) the synthetic chemical route for CoSe₂@DC. Reproduced from [64], Copyright © Elsevier, 2016.

Table 3. Selected state-of-the-art Co-based catalyst towards the hydrogen evolution reaction.

Catalysts	Mass Loading (mg cm^{-2})	Electrolyte	RPM (rpm)	η@10 mA cm^{-2} (mV)	Tafel Slope (mV dec^{-1})	Refs.
CoSe$_2$@DC [a]	0.357	0.5 M H$_2$SO$_4$	-	132	82	[64]
Co$_3$C–GNRs [b]	0.142	0.5 M H$_2$SO$_4$	1600	125	57	[119]
Co$_2$B/CoSe$_2$	0.4	1 M KOH	-	300	76	[136]
Cubic CoSe$_2$/GD [c]	2.8	0.5 M H$_2$SO$_4$	-	200	42	[179]
Cubic CoSe$_2$ nanoparticles/CF [d]	0.26	0.5 M H$_2$SO$_4$	-	137	42	[21]
Interwoven CoSe$_2$/CNT [e]	0.54	0.5 M H$_2$SO$_4$	-	186	32	[170]
CoSe$_2$–CNT [e]	0.255	0.5 M H$_2$SO$_4$	2000	174	38	[171]
Orthorhombic CoSe$_2$ nanotubes	0.283	1 M KOH	-	124	66	[168]
Cubic CoSe$_2$ nanotubes	0.283	1 M KOH	-	149	79	[168]
Polymorphic CoSe/GD [c]	-	0.5 M H$_2$SO$_4$	-	150	31	[167]
CoSe$_2$ hollow microspheres/rGO	0.277	0.5 M H$_2$SO$_4$	-	250	55	[172]
CoPS/NC [f]	0.17	0.5 M H$_2$SO$_4$	2000	80	68	[180]
CoPS/NC [f]	0.17	1 M KOH	2000	148	78	[180]
Nanowires CoSe$_2$/CF [d]	-	0.5 M H$_2$SO$_4$	-	150	34	[169]
Nanowires CoSe$_2$/CC [g]	-	0.5 M H$_2$SO$_4$	-	150	32	[166]
Nanosheets CoSe$_2$/Ti plate	0.16	0.5 M H$_2$SO$_4$	-	165	39	[173]
CoS/CC [g]	3.77	1 M KOH	-	197	98	[181]
		0.5 M H$_2$SO$_4$	-	212	112	
CoS$_2$/RGO [h]	0.285	0.5 M H$_2$SO$_4$	-	18	75	[176]
CoS$_2$/P	-	0.5 M H$_2$SO$_4$	-	67	50	[182]
		1 M KOH	-	67	60	
Co$_9$S$_8$/NC [e] @MoS$_2$	0.283	0.5 M H$_2$SO$_4$	-	117	69	[183]
		1 M PBS	-	267	126	

η @ 10 mA cm^{-2}, the overpotential (η) located at the current density of 10 mA cm^{-2}; [a] DC, defective carbon nanotubes; [b] GNRs, graphene nanoribbons; [c] GD, graphite disk; [d] CF, carbon fiber; [e] CNT, carbon nanotubes; [f] NC, N-doped carbon, [g] CC, carbon cloth; [h] RGO, reduced graphene oxides.

5. Summary and Outlook

This review mainly focused on novel cobalt-based catalysts (CoCat) for electrochemical processes, e.g., ORR, OER and HER in acid and alkaline electrolyte, categorized into five groups, namely, cobalt chalcogenides (selenides, sulfides), cobalt oxides, Co–LDH, Co–MOFs, and Co–N$_x$/C. Tables 1–3 summarize the corresponding electrocatalytic parameters such as onset potential, half-wave potential, Tafel slope, and so on. Based on those results, we assess that cobalt-based materials have emerged as interesting and potential alternatives because of their activities, superior electrochemical stability, and durability compared with precious metals (e.g., Pt, Ir and Ru). While loading cobalt metal compounds on conducting carbonaceous materials, the cobalt–carbon hybrids showed enhanced electrochemical performance. With respect to the electrocatalytic performance of cobalt-based catalysts, the surface engineering (e.g., doping, etching and edging), the structural properties and morphologies (e.g., specific surface area, porosity, core–shell and hollow), anionic substitution (S, Se and Te), defects (e.g., vacancies, topological defects, lattice defects and edge sites), and support materials (e.g., RGO, CNTs, CNHs, and Vulcan-XC-72) were discussed in detail. Furthermore, these catalysts achieved desired catalytic activities towards ORR, OER and HER, as a result of the following strategies: (1) enhancing the SMSI effect, and the synergistic coupling of catalyst–support, resulting in faster electron transport and charge transfer; (2) modulation of the active sites Co^{3+}/Co^{2+} ratio via the NaBH$_4$-assisted route; (3) elucidation of active sites and building well-engineered architecture under the confinement effect; (4) exfoliating the bulk catalysts into atomic level thickness, boosted the exposure and generation of active sites; (5) the direct growth of cobalt-based catalysts on 3D conductive support, e.g., Co foil, Ni foam, and carbon fiber paper, avoided the use of polymer binders, facilitating the electrolyte diffusion, and driving off the as-formed gas bubbles from the electrode surface.

In search of potentially promising and suitable candidates to replace precious catalyst centers for energy conversion and storage systems, a series of cobalt-based materials has been developed. The last 10 years of research on such materials, there has been a boom in the number of publications, that helped to highlight the assembly of unitized regenerated cells based on bifunctional ORR and OER catalysts by exchanging the fuel cell mode to electrolyzer mode. Taking advantage of the exceptional properties which can be achieved by the rational design of materials, such as a large specific surface area, ultrathin/atomic level thickness, optimized porous structure for the exposure of active sites, high conductivity, and a high uniform dispersion, the Co-based nanomaterials will continue to open new avenues to further enhance electrocatalytic activity and stability. Among these Co-based catalysts, due to the coexistence of the diverse active sites including doping via nitrogen, $Co-N_x/C$ derived from LDH, MOF and other precursor used as templates were considered to substitute Pt, Ru, and Ir-based catalysts in acidic and alkaline medium. Last but not least, in-depth studies and interdisciplinary cooperation are still urgently required. Therefore, the design and development of cobalt-based electrocatalysts should be concentrated on the following aspects in the future: (1) understanding the fundamental reaction mechanisms of ORR, OER and HER by virtue of theoretical prediction and simulation; (2) probing and identifying the ideal active sites, and then integrating the different types of active sites to develop the so-called bifunctional oxygen electrode electrocatalysts or bifunctional HER–OER electrocatalysts; (3) gaining insights into the active sites involved such as metal species, oxygen vacancies, and topological defects, by using various advanced characterization techniques, e.g., X-ray absorption near edge structure spectroscopy (XANES), sub-Ångström-resolution aberration-corrected scanning transmission electron microscopy (HAADF–STEM), and in situ Raman; (4) developing simple and low-cost approaches to synthesize the catalysts with satisfactory activity and stability is expected to achieve the mass-production and high quality required for large-scale applications.

Author Contributions: H.Z. searched the literature, and wrote the main part of the current work; C.A.C.-R. contributed with the data analysis, writing and revision of the current literature; Y.Z. wrote part of this work; S.Z. redacted and corrected the references; Y.F. and N.A.-V. organized, supervised and reviewed the work.

Funding: This research received no external funding.

Acknowledgments: This work is supported by National Key R&D Program of China (No. 2016YFB0301600), the National Natural Science Foundation of China (No. 21571015, 21627813), Program for Changjiang Scholars and Innovative Research Team in University (No. IRT1205), China Scholarship Council (H.Z.), The Priority Academic Program Development of Jiangsu Higher Education Institutions (Y.Z.), and CONACyT-Mexico Scholarship (C.A.C.-R.).

Conflicts of Interest: The authors declare no conflict of interest.

References

1. Park, S.; Shao, Y.Y.; Liu, J.; Wang, Y. Electrocatalysts for Water Electrolyzers and Reversible Fuel Cells: Status and Perspective. *Energy Environ. Sci.* **2012**, *5*, 9331–9344. [CrossRef]
2. Carrette, L.; Friedrich, K.A.; Stimming, U. Fuel Cells –Fundamentals and Applications. *Fuel Cells* **2001**, *1*, 5–39. [CrossRef]
3. Jiao, Y.; Zheng, Y.; Jaroniec, M.; Qiao, S.Z. Design of Electrocatalysts for Oxygen- and Hydrogen-Involving Energy Conversion Reactions. *Chem. Soc. Rev.* **2015**, *44*, 2060–2086. [CrossRef] [PubMed]
4. Suen, N.T.; Hung, S.F.; Quan, Q.; Zhang, N.; Xu, Y.J.; Chen, H.M. Electrocatalysis for the Oxygen Evolution Reaction: Recent Development and Future Perspectives. *Chem. Soc. Rev.* **2017**, *46*, 337–365. [CrossRef] [PubMed]
5. Ge, X.; Sumboja, A.; Wuu, D.; An, T.; Li, B.; Goh, F.W.T.; Hor, T.S.A.; Zong, Y.; Liu, Z. Oxygen Reduction in Alkaline Media: From Mechanisms to Recent Advances of Catalysts. *ACS Catal.* **2015**, *5*, 4643–4667. [CrossRef]
6. Shao, M.H.; Chang, Q.W.; Dodelet, J.P.; Chenitz, R. Recent Advances in Electrocatalysts for Oxygen Reduction Reaction. *Chem. Rev.* **2016**, *116*, 3594–3657. [CrossRef] [PubMed]
7. Dai, L.M.; Xue, Y.H.; Qu, L.T.; Choi, H.J.; Baek, J.B. Metal-Free Catalysts for Oxygen Reduction Reaction. *Chem. Rev.* **2015**, *115*, 4823–4892. [CrossRef] [PubMed]
8. Liu, Q.; Jin, J.T.; Zhang, J.Y. NiCo$_2$S$_4$@Graphene as a Bifunctional Electrocatalyst for Oxygen Reduction and Evolution Reactions. *ACS Appl. Mater. Interfaces* **2013**, *5*, 5002–5008. [CrossRef] [PubMed]
9. Gong, M.; Li, Y.G.; Wang, H.L.; Liang, Y.Y.; Wu, J.Z.; Zhou, J.G.; Wang, J.; Regier, T.; Wei, F.; Dai, H.J. An Advanced Ni−Fe Layered Double Hydroxide Electrocatalyst for Water Oxidation. *J. Am. Chem. Soc.* **2013**, *135*, 8452–8455. [CrossRef] [PubMed]

10. Zhao, A.Q.; Masa, J.; Xia, W.; Maljusch, A.; Willinger, M.G.; Calavel, G.; Xie, K.P.; Schlogl, R.; Schuhmann, W.; Muhlert, M. Spinel Mn−Co Oxide in N-Doped Carbon Nanotubes as a Bifunctional Electrocatalyst Synthesized by Oxidative Cutting. *J. Am. Chem. Soc.* **2014**, *136*, 7551–7554. [CrossRef] [PubMed]

11. Xie, R.F.; Fan, G.L.; Ma, Q.; Yang, L.; Li, F. Facile Synthesis and Enhanced Catalytic Performance of Graphene-Supported Ni Nanocatalyst from a Layered Double Hydroxide-Based Composite Precursor. *J. Mater. Chem. A* **2014**, *2*, 7880–7889. [CrossRef]

12. Sumboja, A.; An, T.; Goh, H.Y.; Lubke, M.; Howard, D.P.; Xu, Y.; Handoko, A.D.; Zong, Y.; Liu, Z. One-Step Facile Synthesis of Cobalt Phosphides for Hydrogen Evolution Reaction Catalysts in Acidic and Alkaline Medium. *ACS Appl. Mater. Interfaces* **2018**, *10*, 15673–15680. [CrossRef] [PubMed]

13. Pramana, S.S.; Cavallaro, A.; Li, C.; Handoko, A.D.; Chan, K.W.; Walker, R.J.; Regoutz, A.; Herrin, J.S.; Yeo, B.S.; Payne, D.J.; et al. Crystal Structure and Surface Characteristics of Sr-Doped GdBaCo$_2$O$_{6-\delta}$ Double Perovskites: Oxygen Evolution Reaction and Conductivity. *J. Mater. Chem. A* **2018**, *6*, 5335–5345. [CrossRef]

14. Lübke, M.; Sumboja, A.; McCafferty, L.; Armer, C.F.; Handoko, A.D.; Du, Y.H.; McColl, K.; Cora, F.; Brett, D.; Liu, Z.L.; et al. Transition-Metal-Doped α-MnO$_2$ Nanorods as Bifunctional Catalysts for Efficient Oxygen Reduction and Evolution Reactions. *ChemistrySelect* **2018**, *3*, 2613–2622. [CrossRef]

15. Jia, G.; Hu, Y.; Qian, Q.; Yao, Y.; Zhang, S.; Li, Z.; Zou, Z. Formation of Hierarchical Structure Composed of (Co/Ni)Mn-LDH Nanosheets on MWCNT Backbones for Efficient Electrocatalytic Water Oxidation. *ACS Appl. Mater. Interfaces* **2016**, *8*, 14527–14534. [CrossRef] [PubMed]

16. Yang, Z.; Yao, Z.; Li, G.; Fang, G.Y.; Nie, H.G.; Liu, Z.; Zhou, X.M.; Chen, X.A.; Huang, S.M. Sulfur-Doped Graphene as an Efficient Metal-Free Cathode Catalyst for Oxygen Reduction. *ACS Nano* **2012**, *6*, 205–211. [CrossRef] [PubMed]

17. Gago, A.S.; Gochi-Ponce, Y.; Feng, J.Y.; Esquivel, J.P.; Sabate, N.; Santander, J.; Alonse-Vante, N. Tolerant Chalcogenide Cathodes of Membraneless Micro Fuel Cells. *ChemSusChem* **2012**, *5*, 1488–1494. [CrossRef] [PubMed]

18. Wang, H.L.; Liang, Y.Y.; Li, Y.G.; Dai, H.J. Co$_{1-x}$S-Graphene Hybrid: A High-Performance Metal Chalcogenide Electrocatalyst for Oxygen Reduction. *Angew. Chem. Int. Ed. Engl.* **2011**, *50*, 10969–10972. [CrossRef] [PubMed]

19. Dou, S.; Tao, L.; Hou, J.; Wang, S.Y.; Dai, L.M. Etched and Doped Co$_9$S$_8$/Graphene Hybrid for Oxygen Electrocatalysis. *Energy Environ. Sci.* **2016**, *9*, 1320–1326. [CrossRef]

20. Chao, Y.S.; Tsai, D.S.; Wu, A.P.; Tseng, L.W.; Huang, Y.S. Cobalt Selenide Electrocatalyst Supported by Nitrogen-Doped Carbon and its Stable Activity toward Oxygen Reduction Reaction. *Int. J. Hydrogen Energy* **2013**, *38*, 5655–5664. [CrossRef]

21. Kong, D.S.; Wang, H.T.; Lu, Z.Y.; Cui, Y. CoSe$_2$ Nanoparticles Grown on Carbon Fiber Paper: An Efficient and Stable Electrocatalyst for Hydrogen Evolution Rreaction. *J. Am. Chem. Soc.* **2014**, *136*, 4897–4900. [CrossRef] [PubMed]

22. Zhang, H.; Wang, H.H.; Sumboja, A.; Zang, W.J.; Xie, J.P.; Gao, D.Q.; Pennycook, S.J.; Liu, Z.L.; Guan, C.; Wang, J. Integrated Hierarchical Carbon Flake Arrays with Hollow P-Doped CoSe$_2$ Nanoclusters as an Advanced Bifunctional Catalyst for Zn-Air Batteries. *Adv. Funct. Mater.* **2018**, *28*, 1804846. [CrossRef]

23. Chen, P.R.; Yang, F.K.; Kostka, A.; Xia, W. Interaction of Cobalt Nanoparticles with Oxygen- and Nitrogen-Functionalized Carbon Nanotubes and Impact on Nitrobenzene Hydrogenation Catalysis. *ACS Catal.* **2014**, *4*, 1478–1486. [CrossRef]

24. Mao, S.; Wen, Z.H.; Huang, T.Z.; Hou, Y.; Chen, J.H. High-Performance Bi-Functional Electrocatalysts of 3D Crumpled Graphene–Cobalt Oxide Nanohybrids for Oxygen Reduction and Evolution Reactions. *Energy Environ. Sci.* **2014**, *7*, 609–616. [CrossRef]

25. Liang, Y.Y.; Li, Y.G.; Wang, H.L.; Zhou, J.G.; Wang, J.; Regier, T.; Dai, H.J. Co$_3$O$_4$ Nanocrystals on Graphene as a Synergistic Catalyst for Oxygen Reduction Reaction. *Nat. Mater.* **2011**, *101*, 780–786. [CrossRef] [PubMed]

26. Chen, Y.Z.; Wang, C.M.; Wu, Z.Y.; Xiong, Y.J.; Xu, Q.; Yu, Y.H.; Jiang, H.L. From Bimetallic Metal-Organic Framework to Porous Carbon: High Surface Area and Multicomponent Active Dopants for Excellent Electrocatalysis. *Adv. Mater.* **2015**, *27*, 5010–5016. [CrossRef] [PubMed]

27. Ma, T.Y.; Dai, S.; Jaroniec, M.; Qiao, S.Z. Metal−Organic Framework Derived Hybrid Co$_3$O$_4$-Carbon Porous Nanowire Arrays as Reversible Oxygen Evolution Electrodes. *J. Am. Chem. Soc.* **2014**, *136*, 13925–13931. [CrossRef] [PubMed]

28. Zhong, H.H.; Lou, Y.; He, S.; Tang, P.G.; Li, D.Q.; Alonso-Vante, N.; Feng, Y.J. Electrocatalytic Cobalt Nanoparticles Interacting with Nitrogen-Doped Carbon Nanotube in Situ Generated from a Metal-Organic Framework for the Oxygen Reduction Reaction. *ACS Appl. Mater. Interfaces* **2017**, *9*, 2541–2549. [CrossRef] [PubMed]
29. Zhang, W.J.; Sumboja, A.; Ma, Y.Y.; Zhang, H.; Wu, Y.; Wu, S.S.; Wu, H.J.; Liu, Z.L.; Guan, C.; Wang, J.; et al. Single Co Atoms Anchored in Porous N-Doped Carbon for Efficient Zinc−Air Battery Cathodes. *ACS Catal.* **2018**, *8*, 8961–8969. [CrossRef]
30. Guan, C.; Sumboja, A.; Zang, W.J.; Qian, Y.H.; Zhang, H.; Liu, X.M.; Liu, Z.L.; Zhao, D.; Pennycook, S.J.; Wang, J. Decorating Co/CoNx Nanoparticles in Nitrogen-Doped Carbon Nanoarrays for Flexible and Rechargeable Zinc-Air Batteries. *Energy Storage Mater.* **2018**, *16*, 243–250. [CrossRef]
31. Fan, G.L.; Li, F.; Evans, D.G.; Duan, X. Catalytic Applications of Layered Double Hydroxides: Recent Advances and Perspectives. *Chem. Soc. Rev.* **2014**, *43*, 7040–7066. [CrossRef] [PubMed]
32. Zhang, S.L.; Zhang, Y.; Jiang, W.J.; Liu, X.; Xu, S.L.; Hou, R.J.; Zhang, F.Z.; Hu, J.S. Co@N-CNTs Derived from Triple-Role CoAl-Layered Double Hydroxide as an Efficient Catalyst for Oxygen Reduction Reaction. *Carbon* **2016**, *107*, 162–170. [CrossRef]
33. Zhong, H.H.; Tian, R.; Gong, X.M.; Li, D.Q.; Tang, P.G.; Alonso-Vante, N.; Feng, Y.J. Advanced Bifunctional Electrocatalyst Generated through Cobalt Phthalocyanine Tetrasulfonate Intercalated Ni2Fe-Layered Double Hydroxides for a Laminar Flow Unitized Regenerative Micro-Cell. *J. Power Sources* **2017**, *361*, 21–30. [CrossRef]
34. Wang, J.H.; Cui, W.; Liu, Q.; Xing, Z.C.; Asiri, A.M.; Sun, X.P. Recent Progress in Cobalt-Based Heterogeneous Catalysts for Electrochemical Water Splitting. *Adv. Mater.* **2016**, *28*, 215–230. [CrossRef] [PubMed]
35. Dai, X.F.; Qiao, J.L.; Zhou, X.J.; Shi, J.J.; Xu, P.; Zhang, L.; Zhang, J.J. Effects of Heat-Treatment and Pyridine Addition on the Catalytic Activity of Carbon-Supported Cobalt-Phthalocyanine for Oxygen Reduction Reaction in Alkaline Electrolyte. *Int. J. Electrochem. Sci.* **2013**, *8*, 3160–3175.
36. Zhao, S.; Rasimick, B.; Mustain, W.; Xu, H. Highly Durable and Active Co3O4 Nanocrystals Supported on Carbon Nanotubes as Bifunctional Electrocatalysts in Alkaline Media. *Appl. Catal. B* **2017**, *203*, 138–145. [CrossRef]
37. Liu, S.Y.; Li, L.J.; Ahnb, H.S.; Manthiram, A. Delineating the Roles of Co3O4 and N-Doped Carbon Nanoweb (CNW) in Bifunctional Co3O4/CNW Catalysts for Oxygen Reduction and Oxygen Evolution Reactions. *J. Mater. Chem. A* **2015**, *3*, 11615–11623. [CrossRef]
38. Wu, Z.S.; Ren, W.C.; Wen, L.; Gao, L.B.; Zhao, J.P.; Chen, Z.P.; Zhou, G.M.; Li, F.; Cheng, H.M. Graphene Anchored with Co3O4 Nanoparticles as Anode of Lithium Ion Batteries with Enhanced Reversible Capacity and Cyclic Performance. *ACS Nano* **2010**, *4*, 3187–3194. [CrossRef] [PubMed]
39. Subramanian, N.P.; Kumaraguru, S.P.; Colon-Mercado, H.; Kim, H.; Popov, B.N.; Black, T.; Chen, D.A. Studies on Co-Based Catalysts Supported on Modified Carbon Substrates for PEMFC Cathodes. *J. Power Sources* **2006**, *157*, 56–63. [CrossRef]
40. Khan, M.; Tahir, M.N.; Adil, S.F.; Khan, H.U.; Siddiqui, M.R.H.; Al-Warthan, A.A.; Tremel, W. Graphene Based Metal and Metal Oxide Nanocomposites: Synthesis, Properties and their Applications. *J. Mater. Chem. A* **2015**, *3*, 18753–18808. [CrossRef]
41. Alonso-Vante, N. Platinum and Non-Platinum Nanomaterials for the Molecular Oxygen Reduction Reaction. *Chemphyschem* **2010**, *11*, 2732–2744. [CrossRef] [PubMed]
42. Tributsch, H.; Alonso-Vante, N. Energy Conversion Catalysis Using Semiconducting Transition Metal Cluster Compounds. *Nature* **1986**, *323*, 431–432.
43. Higgins, D.C.; Hassan, F.M.; Seo, M.H.; Choi, J.Y.; Hoque, M.A.; Lee, D.U.; Chen, Z. Shape-Controlled Octahedral Cobalt Disulfide Nanoparticles Supported on Nitrogen and Sulfur-Doped Graphene/Carbon Nanotube Composites for Oxygen Reduction in Acidic Electrolyte. *J. Mater. Chem. A* **2015**, *3*, 6340–6350. [CrossRef]
44. Delacôte, C.; Lewera, A.; Pisarek, M.; Kulesza, P.J.; Zelenay, P.; Alonso-Vante, N. The Effect of Diluting Ruthenium by Iron in RuxSey Catalyst for Oxygen Reduction. *Electrochim. Acta* **2010**, *55*, 7575–7580. [CrossRef]
45. Lee, K.; Zhang, L.; Zhang, J.J. Ternary Non-Noble Metal Chalcogenide (W–Co–Se) as Electrocatalyst for Oxygen Reduction Reaction. *Electrochem. Commun.* **2007**, *9*, 1704–1708. [CrossRef]

46. Zhao, C.; Li, D.Q.; Feng, Y.J. Size-Controlled Hydrothermal Synthesis and High Electrocatalytic Performance of CoS$_2$ Nanocatalysts as Non-Precious Metal Cathode Materials for Fuel Cells. *J. Mater. Chem. A* **2013**, *1*, 5741–5746. [CrossRef]

47. Alonso-Vante, N.; Malakhov, I.V.; Nikitenko, S.G.; Savinova, E.R.; Kochubey, D.I. The Structure Analysis of the Active Centers of Ru-containing Electrocatalysts for the Oxygen Reduction. An in Situ EXAFS Study. *Electrochim. Acta* **2002**, *47*, 3807–3814. [CrossRef]

48. Behret, H.; Binder, H.; Sandstede, G. Electrocatalytic Oxygen Reduction with Thiospinels and Other Sulphides of Transition Metals. *Electrochim. Acta* **1975**, *20*, 111–117. [CrossRef]

49. Vayner, E.; Sidik, R.A.; Anderson, A.B.; Popov, B.N. Experimental and Theoretical Study of Cobalt Selenide as a Catalyst for O$_2$ Electroreduction. *J. Phys. Chem. C* **2017**, *111*, 10508–10513. [CrossRef]

50. Baresel, D.; Sarholz, W.; Scharner, P.; Schmitz, J. Transition MetalChalcogenides as Oxygen Catalysts for Fuel Cells. *Ber. Bunsenges. Phys. Chem.* **1974**, *78*, 608–618.

51. Sidik, R.A.; Anderson, A.B. Co$_9$S$_8$ as a Catalyst for Electroreduction of O$_2$: Quantum Chemistry Predictions. *J. Phys. Chem. B* **2006**, *110*, 936–941. [CrossRef] [PubMed]

52. Zhu, C.Y.; Aoki, Y.; Habazaki, H. Co$_9$S$_8$ Nanoparticles Incorporated in Hierarchically Porous 3D Few-Layer Graphene-Like Carbon with S, N-Doping as Superior Electrocatalyst for Oxygen Reduction Reaction. *Part. Part. Syst. Charact.* **2017**, *34*, 1700296. [CrossRef]

53. Feng, Y.J.; Alonso-Vante, N. Structure Phase Transition and Oxygen Reduction Activity in Acidic Medium of Carbon-Supported Cobalt Selenide Nanoparticles. *ECS Trans.* **2009**, *25*, 167–173.

54. Feng, Y.J.; Alonso-Vante, N. Carbon-Supported CoSe$_2$ Nanoparticles for Oxygen Reduction Reaction in Acid Medium. *Fuel Cells* **2010**, *10*, 77–83.

55. Wu, G.; Chung, H.T.; Nelson, M.; Artyushkova, K.; More, K.L.; Johnston, C.M.; Zelenay, P. Graphene-Enriched Co$_9$S$_8$-N-C Non-Precious Metal Catalyst for Oxygen Reduction in Alkaline Media. *ECS Trans.* **2011**, *41*, 1709–1717.

56. Apostolova, R.D.; Shembel, E.M.; Talyosef, I.; Grinblat, J.; Markovsky, B.; Aurbach, D. Study of Electrolytic Cobalt Sulfide Co$_9$S$_8$ as an Electrode Material in Lithium Accumulator Prototypes. *Russ. J. Electrochem.* **2009**, *45*, 311–319. [CrossRef]

57. Handoko, A.D.; Deng, S.; Deng, Y.; Cheng, A.W.F.; Chan, K.W.; Tan, H.R.; Pan, Y.; Tok, E.S.; Sow, C.H.; Yeo, B.S. Enhanced Activity of H$_2$O$_2$-Treated Copper(II) Oxide Nanostructures for the Electrochemical Evolution of Oxygen. *Catal. Sci. Technol.* **2016**, *6*, 269–274. [CrossRef]

58. Antolini, E. Structural Parameters of Supported Fuel Cell Catalysts: The Effect of Particle Size, Inter-Particle Distance and Metal Loading on Catalytic Activity and Fuel Cell Performance. *Appl. Catal. B* **2016**, *181*, 298–313. [CrossRef]

59. Wang, J.X.; Inada, H.; Wu, L.J.; Zhu, Y.M.; Choi, Y.M.; Liu, P.; Zhou, W.P.; Adzic, R.R. Oxygen Reduction on Well-Defined Core-Shell Nanocatalysts: Particle Size, Facet, and Pt Shell Thickness Effects. *J. Am. Chem. Soc.* **2009**, *131*, 17298–17302. [CrossRef] [PubMed]

60. Feng, Y.J.; He, T.; Alonso-Vante, N. In Situ Free-surfactant Synthesis and ORR-electrochemistry of Carbon-Supported Co$_3$S$_4$ and CoSe$_2$ Nanoparticles. *Chem. Mater.* **2008**, *20*, 26–28. [CrossRef]

61. Feng, Y.J.; Alonso-Vante, N. Carbon-Supported Cubic CoSe$_2$ Catalysts for Oxygen Reduction Reaction in Alkaline Medium. *Electrochim. Acta* **2012**, *72*, 129–133. [CrossRef]

62. Feng., Y.J.; He, T.; Alonso-Vante, N. Oxygen Reduction Reaction on Carbon-Supported CoSe$_2$ Nanoparticles in an Acidic Medium. *Electrochim. Acta* **2009**, *54*, 5252–5256. [CrossRef]

63. Wood, K.N.; O'Hayre, R.; Pylypenko, S. Recent Progress on Nitrogen/Carbon Structures Designed for Use in Energy and Sustainability Applications. *Energy Environ. Sci.* **2014**, *7*, 1212–1249. [CrossRef]

64. Zhou, W.J.; Lu, L.; Zhou, K.; Yang, L.J.; Ke, Y.T.; Tang, Z.H.; Chen, S.W. CoSe$_2$ Nanoparticles Embedded Defective Carbon Nanotubes Derived from MOFs as Efficient Electrocatalyst for Hydrogen Evolution Reaction. *Nano Energy* **2016**, *28*, 143–150. [CrossRef]

65. Alonso-Vante, N. Photocatalysis an Enhancer of Electrocatalytic Process, Current Opinion in Electrochemistry. *Curr. Opin. Electrochem.* **2018**, *9*, 114–120. [CrossRef]

66. Unni, S.M.; Mora-Hernandez, J.M.; Kurungot, S.; Alonso-Vante, N. CoSe$_2$ Supported on Nitrogen-Doped Carbon Nanohorns as a Methanol-Tolerant Cathode for Air-Breathing Microlaminar Flow Fuel Cells. *ChemElectroChem* **2015**, *2*, 1339–1345. [CrossRef]

67. García-Rosado, I.J.; Uribe-Calderón, J.; Alonso-Vante, N. Nitrogen-Doped Reduced Graphite Oxide as a Support for CoSe Electrocatalyst for Oxygen Reduction Reaction in Alkaline Media. *J. Electrochem. Soc.* **2017**, *164*, F658–F666. [CrossRef]

68. Zhu, A.Q.; Tan, P.F.; Qiao, L.L.; Liu, Y.; Ma, Y.J.; Pan, J. Sulphur and Nitrogen Dual-Doped Mesoporous Carbon Hybrid Coupling with Graphite Coated Cobalt and Cobalt Sulfide Nanoparticles: Rational Synthesis and Advanced Multifunctional Electrochemical Properties. *J. Colloid Interface Sci.* **2018**, *509*, 254–264. [CrossRef] [PubMed]

69. Hu, H.; Han, L.; Yu, M.; Wang, Z.Y.; Lou, X.W. Metal–Organic-Framework-Engaged Formation of Co Nanoparticle-Embedded Carbon@Co$_9$S$_8$ Double-Shelled Nanocages for Efficient Oxygen Reduction. *Energy Environ. Sci.* **2016**, *9*, 107–111. [CrossRef]

70. Chen, B.L.; Li, R.; Ma, G.P.; Gou, X.L.; Zhu, Y.Q.; Xia, Y.D. Cobalt Sulfide/N, S Codoped Porous Carbon Core-Shell Nanocomposites as Superior Bifunctional Electrocatalysts for Oxygen Reduction and Evolution Reactions. *Nanoscale* **2015**, *7*, 20674–20684. [CrossRef] [PubMed]

71. Tao, L.Q.; Zai, J.T.; Wang, K.X.; Zhang, H.J.; Xu, M.; Shen, J.; Su, Y.Z.; Qian, X.F. Co$_3$O$_4$ Nanorods/Graphene Nanosheets Nanocomposites for Lithium Ion Batteries with Improved Reversible Capacity and Cycle Stability. *J. Power Sources* **2012**, *202*, 230–235. [CrossRef]

72. Bai, G.M.; Dai, H.X.; Deng, J.G.; Liu, Y.X.; Wang, F.; Zhao, Z.X.; Qiu., W.G.; Au, C.T. Porous Co$_3$O$_4$ Nanowires and Nanorods: Highly Active Catalysts for the Combustion of Toluene. *Appl. Catal. A* **2013**, *450*, 42–49. [CrossRef]

73. Liu, J.P.; Jiang, J.; Cheng, C.W.; Li, H.X.; Zhang, J.X.; Gong, H.; Fan, H.J. Co$_3$O$_4$ Nanowire@MnO$_2$ Ultrathin Nanosheet Core/Shell Arrays: A New Class of High-Performance Pseudocapacitive Materials. *Adv. Mater.* **2011**, *23*, 2076–2081. [CrossRef] [PubMed]

74. Wang, Y.Y.; Lei, Y.; Li, J.; Gu, L.; Yuan, H.Y.; Xiao, D. Synthesis of 3D-Nanonet Hollow Structured Co$_3$O$_4$ for High Capacity Supercapacitor. *ACS Appl. Mater. Interfaces* **2014**, *6*, 6739–6747. [CrossRef] [PubMed]

75. Singh, S.K.; Dhavale, V.M.; Kurungot, S. Surface-Tuned Co$_3$O$_4$ Nanoparticles Dispersed on Nitrogen-Doped Graphene as an Efficient Cathode Electrocatalyst for Mechanical Rechargeable Zinc-Air Battery Application. *ACS Appl. Mater. Interfaces* **2015**, *7*, 21138–21149. [CrossRef] [PubMed]

76. Jia, X.D.; Gao, S.J.; Liu, T.Y.; Li, D.Q.; Tang, P.G.; Feng, Y.J. Controllable Synthesis and Bi-Functional Electrocatalytic Performance towards Oxygen Electrode Reactions of Co$_3$O$_4$/N-RGO Composites. *Electrochim. Acta* **2017**, *226*, 104–112. [CrossRef]

77. Liu, J.; Jiang, L.H.; Zhang, B.S.; Jin, J.T.; Su, D.S.; Wang, S.L.; Sun, G.Q. Controllable Synthesis of Cobalt Monoxide Nanoparticles and the Size-Dependent Activity for Oxygen Reduction Reaction. *ACS Catal.* **2014**, *4*, 2998–3001. [CrossRef]

78. Wang, J.K.; Gao, R.; Zhou, D.; Chen, Z.j.; Wu, Z.H.; Schumacher, G.; Hu, Z.B.; Liu, X.F. Boosting the Electrocatalytic Activity of Co$_3$O$_4$ Nanosheets for a Li-O$_2$ Battery through Modulating Inner Oxygen Vacancy and Exterior Co^{3+}/Co^{2+} Ratio. *ACS Catal.* **2017**, *7*, 6533–6541. [CrossRef]

79. Xu, J.B.; Gao, P.; Zhao, T.S. Non-Precious Co$_3$O$_4$ Nano-Rod Electrocatalyst for Oxygen Reduction Reaction in Anion-Exchange Membrane Fuel Cells. *Energy Environ. Sci.* **2012**, *5*, 5333–5339. [CrossRef]

80. Odedairo, T.; Yan, X.C.; Ma, J.; Jiao, Y.L.; Yao, X.D.; Du, A.J.; Zhu, Z.H. Nanosheets Co$_3$O$_4$ Interleaved with Graphene for Highly Efficient Oxygen Reduction. *ACS Appl. Mater. Interfaces* **2015**, *7*, 21373–21380. [CrossRef] [PubMed]

81. Chen, Z.Y.; Li, Y.N.; Wang, M.Q.; Xu, M.W.; Bao, S.J. Hollow Co$_3$O$_4$ Nanocages Decorated Graphene Aerogels Derived from Carbon Wrapped Nano-Co for Efficient Oxygen Reduction Reaction. *ChemistrySelect* **2017**, *2*, 6359–6363. [CrossRef]

82. Wu, Z.X.; Wang, J.; Han, L.L.; Lin, R.Q.; Liu, H.F.; Xin, H.L.L.; Wang, D.L. Supramolecular Gel-Assisted Synthesis of Double Shelled Co@CoO@N-C/C Nanoparticles with Synergistic Electrocatalytic Activity for the Oxygen Reduction Reaction. *Nanoscale* **2016**, *8*, 4681–4687. [CrossRef] [PubMed]

83. Wang, D.D.; Chen, X.; Evans, D.G.; Yang, W.S. Well-Dispersed Co$_3$O$_4$/Co$_2$MnO$_4$ Nanocomposites as a Synergistic Bifunctional Catalyst for Oxygen Reduction and Oxygen Evolution Reactions. *Nanoscale* **2013**, *5*, 5312–5315. [CrossRef] [PubMed]

84. Huang, W.; Zhong, H.H.; Li, D.Q.; Tang, P.G.; Feng, Y.J. Reduced Graphene Oxide Supported CoO/MnO$_2$ Electrocatalysts from Layered Double Hydroxides for Oxygen Reduction Reaction. *Electrochim. Acta* **2015**, *173*, 575–580. [CrossRef]

85. Wang, Y.L.; Wang, Z.C.; Wu, X.Q.; Liu, X.W.; Li, M.G. Synergistic Effect between Strongly Coupled CoAl Layered Double Hydroxides and Graphene for the Electrocatalytic Reduction of Oxygen. *Electrochim. Acta* **2016**, *192*, 196–204. [CrossRef]
86. Huo, R.J.; Jiang, W.J.; Xu, S.L.; Zhang, F.Z.; Hu, J.S. Co/CoO/CoFe$_2$O$_4$/G Nanocomposites Derived from Layered Double Hydroxides towards Mass Production of Efficient Pt-Free Electrocatalysts for Oxygen Reduction Reaction. *Nanoscale* **2014**, *6*, 203–206. [CrossRef] [PubMed]
87. Jasinski, R. A New Fuel Cell Cathode Catalyst. *Nature* **1964**, *21*, 1212–1213. [CrossRef]
88. Wu, G.; More, K.L.; Johnston, C.M.; Zelenay, P. High-Performance Electrocatalysts for Oxygen Reduction Derived from Polyaniline, Iron, and Cobalt. *Science* **2011**, *332*, 443–447. [CrossRef] [PubMed]
89. Kothandaraman, R.; Nallathambi, V.; Artyushkova, K.; Barton, S.C. Non-Precious Oxygen Reduction Catalysts Prepared by High-Pressure Pyrolysis for Low-Temperature Fuel Cells. *Appl. Catal. B* **2009**, *92*, 209–216. [CrossRef]
90. Bashyam, R.; Zelenay, P. A Class of Non-Precious Metal Composite Catalysts for Fuel Cells. *Nature* **2006**, *443*, 63–66. [CrossRef] [PubMed]
91. Morozan, A.; Jegou, P.; Jousselme, B.; Palacin, S. Electrochemical Performance of Annealed Cobalt-Benzotriazole/CNTs Catalysts towards the Oxygen Reduction Reaction. *Phys. Chem. Chem. Phys.* **2011**, *13*, 21600–21607. [CrossRef] [PubMed]
92. Chen, X.; Li, F.; Zhang, N.L.; An, L.; Xia, D.G. Mechanism of Oxygen Reduction Reaction Catalyzed by Fe(Co)-N$_x$/C. *Phys. Chem. Chem. Phys.* **2013**, *15*, 19330–19336. [CrossRef] [PubMed]
93. Jia, Q.Y.; Ramaswamy, N.; Tylus, U.; Strickland, K.; Li, J.K.; Serov, A.; Artyushkova, K.; Atanassov, P.; Anibal, J.; Gumeci, C.; et al. Spectroscopic Insights into the Nature of Active Sites in Iron–Nitrogen–Carbon Electrocatalysts for Oxygen Reduction in Acid. *Nano Energy* **2016**, *29*, 65–82. [CrossRef]
94. Wong, W.Y.; Daud, W.R.W.; Mohamad, A.B.; Kadhum, A.A.H.; Loh, K.S.; Majlan, E.H. Recent Progress in Nitrogen-Doped Carbon and its Composites as Electrocatalysts for Fuel Cell Applications. *Int. J. Hydrogen Energy* **2013**, *38*, 9370–9386. [CrossRef]
95. Niwa, H.; Horiba, K.; Harada, Y.; Oshima, M.; Ikeda, T.; Terakura, K.; Ozaki, J.; Miyata, S. X-ray Absorption Analysis of Nitrogen Contribution to Oxygen Reduction Reaction in Carbon Alloy Cathode Catalysts for Polymer Electrolyte Fuel Cells. *J. Power Sources* **2009**, *187*, 93–97. [CrossRef]
96. Hu, C.G.; Xiao, Y.; Zhao, Y.; Chen, N.; Zhang, Z.P.; Cao, M.H.; Qu, L.T. Highly Nitrogen-Doped Carbon Capsules: Scalable Preparation and High-Performance Applications in Fuel Cells and Lithium Ion Batteries. *Nanoscale* **2013**, *5*, 2726–2733. [CrossRef] [PubMed]
97. Kattel, S.; Atanassov, P.; Kiefer, B. Catalytic Activity of Co–N$_x$/C Electrocatalysts for Oxygen Reduction Reaction: A Density Functional Theory Study. *Phys. Chem. Chem. Phys.* **2013**, *15*, 148–153. [CrossRef] [PubMed]
98. Yin, P.Q.; Yao, T.; Wu, Y.; Zheng, L.R.; Lin, Y.; Liu, W.; Ju, H.X.; Zhu, J.F.; Hong, X.; Deng, Z.X.; et al. Single Cobalt Atoms with Precise N-Coordination as Superior Oxygen Reduction Reaction Catalysts. *Angew. Chem. Int. Ed. Engl.* **2016**, *55*, 10800–10805. [CrossRef] [PubMed]
99. Amiinu, I.S.; Liu, X.B.; Pu, Z.H.; Li, W.Q.; Li, Q.D.; Zhang, J.; Tang, H.L.; Zhang, H.N.; Mu, S.C. From 3D ZIF Nanocrystals to Co-N$_x$/C Nanorod Array Electrocatalysts for ORR, OER, and Zn-Air Batteries. *Adv. Funct. Mater.* **2018**, *28*, 1704638–1704646. [CrossRef]
100. Peng, H.L.; Liu, F.F.; Liu, X.J.; Liao, S.J.; You, C.H.; Tian, X.L.; Nan, H.X.; Luo, F.; Song, H.Y.; Fu, Z.Y.; et al. Effect of Transition Metals on the Structure and Performance of the Doped Carbon Catalysts Derived From Polyaniline and Melamine for ORR Application. *ACS Catal.* **2014**, *4*, 3797–3805. [CrossRef]
101. Wiesener, K. N$_4$-Chelates as Electrocatalyst for Cathodic Oxygen Reduction. *Electrochim. Acta* **1986**, *31*, 1073–1078. [CrossRef]
102. Masa, J.; Xia, W.; Muhler, M.; Schuhmann, W. On the Role of Metals in Nitrogen-Doped Carbon Electrocatalysts for Oxygen Reduction. *Angew. Chem. Int. Ed. Engl.* **2015**, *54*, 10102–10120. [CrossRef] [PubMed]
103. Bagotzky, V.S.; Tarasevich, M.R.; Radyushkina, K.A.; Levina, O.A.; Andrusyova, S.I. Electrocatalysis of the Oxygen Reduction Process on Metal Chelates in Acid Electrolyte. *J. Power Sources* **1978**, *2*, 233–240. [CrossRef]
104. Wiesener, K.; Ohms, D.; Neumann, V.; Franke, R. N$_4$ Macrocycles as Electrocatalysts for the Cathodic Reduction of Oxygen. *Mater. Chem. Phys.* **1989**, *22*, 457–475. [CrossRef]

105. Liu, Y.Y.; Yue, X.P.; Li, K.X.; Qiao, J.L.; Wilkinson, D.P.; Zhang, J.J. PEM Fuel Cell Electrocatalysts Based on Transition Metal Macrocyclic Compounds. *Coord. Chem. Rev.* **2016**, *315*, 153–177. [CrossRef]

106. Osmieri, L.; Videla, A.H.A.M.; Specchia, S. Activity of Co–N Multi Walled Carbon Nanotubes Electrocatalysts for Oxygen Reduction Reaction in Acid Conditions. *J. Power Sources* **2015**, *278*, 296–307. [CrossRef]

107. Nallathambi, V.; Lee, J.W.; Kumaraguru, S.P.; Wu, G.; Popov, B.N. Development of High Performance Carbon Composite Catalyst for Oxygen Reduction Reaction in PEM Proton Exchange Membrane Fuel Cells. *J. Power Sources* **2008**, *183*, 34–42. [CrossRef]

108. Lalande, G.; Côté, R.; Tamizhmani, G.; Guay, D.; Dodelet, J.P.; Dignard-Bailey, L.; Weng, L.T.; Bertrand, P. Physical, Chemical and Electrochemical Characterization of Heat-Treated Tetracarboxylic Cobalt Phthalocyanine Adsorbed on Carbon Black as Electrocatalyst for Oxygen Reduction in Polymer Electrolyte Fuel Cells. *Electrochim. Acta* **1995**, *40*, 2635–2646. [CrossRef]

109. Weng, L.T.; Bertrand, P.; Lalande, G.; Guay, D.; Dodelet, J.P. Surface Characterization by Time-of-Flight SIMS of a Catalyst for Oxygen Electroreduction: Pyrolyzed Cobalt Phthalocyanine-on-Carbon Black. *Appl. Surf. Sci.* **1985**, *84*, 9–21. [CrossRef]

110. Jahnke, H.; Schönborn, M.; Zirnmermann, G. Organic Dyestuffs as Catalysts for Fuel Cells. *Top. Curr. Chem.* **1976**, *61*, 133–181. [PubMed]

111. Kong, A.G.; Kong, Y.Y.; Zhu, X.F.; Han, Z.; Shan, Y.K. Ordered Mesoporous Fe (or Co)–N–Graphitic Carbons as Excellent Non-Precious-Metal Electrocatalysts for Oxygen Reduction. *Carbon* **2014**, *78*, 49–59. [CrossRef]

112. Tang, C.; Wang, B.; Wang, H.F.; Zhang, Q. Defect Engineering toward Atomic Co-N$_x$-C in Hierarchical Graphene for Rechargeable Flexible Solid Zn-Air Batteries. *Adv. Mater.* **2017**, *29*, 1703185–1703191. [CrossRef] [PubMed]

113. Zhang, L.L.; Wang, A.Q.; Wang, W.T.; Huang, Y.Q.; Liu, X.Y.; Miao, S.; Liu, J.Y.; Zhang, T. Co–N–C Catalyst for C–C Coupling Reactions: On the Catalytic Performance and Active Sites. *ACS Catal.* **2015**, *5*, 6563–6572. [CrossRef]

114. Hu, F.; Yang, H.C.; Wang, C.H.; Zhang, Y.J.; Lu, H.; Wang, Q.B. Co-N-Doped Mesoporous Carbon Hollow Spheres as Highly Efficient Electrocatalysts for Oxygen Reduction Reaction. *Small* **2017**, *13*, 1602507–1602514. [CrossRef] [PubMed]

115. Cai, S.C.; Meng, Z.H.; Tang, H.L.; Wang, Y.; Tsiakaras, P. 3D Co-N-doped Hollow Carbon Spheres as Excellent Bifunctional Electrocatalysts for Oxygen Reduction Reaction and Oxygen Evolution Reaction. *Appl. Catal. B* **2017**, *217*, 477–484. [CrossRef]

116. Aijaz, A.; Masa, J.; Rosler, C.; Xia, W.; Weide, P.; Botz, A.J.R.; Fischer, R.A.; Schuhmann, W.; Muhler, M. Co@Co$_3$O$_4$ Encapsulated in Carbon Nanotube-Grafted Nitrogen-Doped Carbon Polyhedra as an Advanced Bifunctional Oxygen Electrode. *Angew. Chem. Int. Ed. Engl.* **2016**, *55*, 4087–4091. [CrossRef] [PubMed]

117. Kuang, M.; Wang, Q.; Han, P.; Zheng, G.F. Cu, Co-Embedded N-Enriched Mesoporous Carbon for Efficient Oxygen Reduction and Hydrogen Evolution Reactions. *Adv. Energy Mater.* **2017**, *7*, 1700193–1700200. [CrossRef]

118. Lin, Q.P.; Bu, X.H.; Kong, A.G.; Mao, C.Y.; Bu, F.; Feng, P.Y. Heterometal-Embedded Organic Conjugate Frameworks from Alternating Monomeric Iron and Cobalt Metalloporphyrins and Their Application in Design of Porous Carbon Catalysts. *Adv. Mater.* **2015**, *27*, 3431–3436. [CrossRef] [PubMed]

119. Fan, X.J.; Peng, Z.W.; Ye, R.Q.; Zhou, H.Q.; Guo, X. M$_3$C (M: Fe, Co, Ni) Nanocrystals Encased in Graphene Nanoribbons: An Active and Stable Bifunctional Electrocatalyst for Oxygen Reduction and Hydrogen Evolution Reactions. *ACS Nano* **2015**, *9*, 7407–7418. [CrossRef] [PubMed]

120. Zhu, L.; Teo, M.; Wong, P.C.; Wong, K.C.; Narita, I.; Ernst, F.; Mitchell, K.A.R.; Campbell, S.A. Synthesis, Characterization of a CoSe$_2$ Catalyst for the Oxygen Reduction Reaction. *Appl. Catal. A* **2010**, *386*, 157–165. [CrossRef]

121. Qiao, X.C.; Jin, J.T.; Fan, H.B.; Li, Y.W.; Liao, S.J. In Situ Growth of Cobalt Sulfide Hollow Nanospheres Embedded in Nitrogen and Sulfur Co-doped Graphene Nanoholes as a Highly Active Electrocatalyst for Oxygen Reduction and Evolution. *J. Mater. Chem. A* **2017**, *5*, 12354–12360. [CrossRef]

122. Liang, H.; Li, C.W.; Chen, T.; Cui, L.; Han, J.R.; Peng, Z.; Liu, J.Q. Facile Preparation of Three-Dimensional Co$_{1-x}$S/Sulfur and Nitrogen-Codoped Graphene/Carbon Foam for Highly Efficient Oxygen Reduction Reaction. *J. Power Sources* **2018**, *378*, 699–706. [CrossRef]

123. Zhang, Y.; Li, P.W.; Yin, X.Y.; Yan, Y.; Zhan, K.; Yang, J.H.; Zhao, B. Cobalt Sulfide Supported on Nitrogen and Sulfur Dual-Doped Reduced Graphene Oxide for Highly Active Oxygen Reduction Reaction. *RSC Adv.* **2017**, *7*, 50246–50253. [CrossRef]

124. Ganesan, P.; Prabu, M.; Sanetuntikul, J.; Shanmugam, S. Cobalt Sulfide Nanoparticles Grown on Nitrogen and Sulfur Codoped Graphene Oxide: An Efficient Electrocatalyst for Oxygen Reduction and Evolution Eeactions. *ACS Catal.* **2015**, *5*, 3625–3637. [CrossRef]

125. Mahmood, N.; Zhang, C.Z.; Jiang, J.; Liu, F.; Hou, Y.L. Multifunctional Co_3S_4/Graphene Composites for Lithium Ion Batteries and Oxygen Reduction Reaction. *Chem. Eur. J.* **2013**, *19*, 5183–5190. [CrossRef] [PubMed]

126. Sennu, P.; Christy, M.; Aravindan, V.; Lee, Y.G.; Nahm, K.S.; Lee, Y.S. Two-Dimensional Mesoporous Cobalt Sulfide Nanosheets as a Superior Anode for a Li-Ion Battery and a Bifunctional Electrocatalyst for the LiO_2 System. *Chem. Mater.* **2015**, *27*, 5726–5735. [CrossRef]

127. Arunchander, A.; Peera, S.G.; Giridhar, V.V.; Sahu, A.K. Synthesis of Cobalt Sulfide-Graphene as an Efficient Oxygen Reduction Catalyst in Alkaline Medium and its Application in Anion Exchange Membrane Fuel Cells. *J. Electrochem. Soc.* **2016**, *164*, F71–F80. [CrossRef]

128. Fu, S.F.; Zhu, C.Z.; Song, J.H.; Feng, S.; Du, D.; Engelhard, M.H.; Xiao, D.D.; Li, D.S.; Lin, Y.H. Two-Dimensional N, S-Codoped Carbon/Co_9S_8 Catalysts Derived from $Co(OH)_2$ Nanosheets for Oxygen Reduction Reaction. *ACS Appl. Mater. Interfaces* **2017**, *9*, 36755–36761. [CrossRef] [PubMed]

129. Zhou, Y.X.; Yao, H.B.; Wang, Y.; Liu, H.L.; Gao, M.R.; Shen, P.K.; Yu, S.H. Hierarchical Hollow Co_9S_8 Microspheres: Solvothermal Synthesis, Magnetic, Electrochemical, and Electrocatalytic Properties. *Chemistry* **2010**, *16*, 12000–12007. [CrossRef] [PubMed]

130. Tang, Y.P.; Jing, F.; Xu, Z.X.; Zhang, F.; Mai, Y.Y.; Wu, D.Q. Highly Crumpled Hybrids of Nitrogen/Sulfur Dual-Doped Graphene and Co_9S_8 Nanoplates as Efficient Bifunctional Oxygen Electrocatalysts. *ACS Appl. Mater. Interfaces* **2017**, *9*, 12340–12347. [CrossRef] [PubMed]

131. Mattioli, G.; Giannozzi, P.; Bonapasta, A.A.; Guidoni, L. Reaction Pathways for Oxygen Evolution Promoted by Cobalt Catalyst. *J. Am. Chem. Soc.* **2013**, *135*, 15353–15363. [CrossRef] [PubMed]

132. García-Mota, M.; Bajdich, M.; Viswanathan, V.; Vojvodic, A.; Bell, A.T.; Nørskov, J.K. Importance of Correlation in Determining Electrocatalytic Oxygen Evolution Activity on Cobalt Oxides. *J. Phys. Chem. C* **2012**, *116*, 21077–21082. [CrossRef]

133. Bajdich, M.; Garcia-Mota, M.; Vojvodic, A.; Norskov, J.K.; Bell, A.T. Theoretical Investigation of the Activity of Cobalt Oxides for the Electrochemical Oxidation of Water. *J. Am. Chem. Soc.* **2013**, *135*, 13521–13530. [CrossRef] [PubMed]

134. Liu, Y.W.; Cheng, H.; Lyu, M.J.; Fan, S.J.; Liu, Q.H.; Zhang, W.S.; Zhi, Y.D.; Wang, C.M.; Xiao, C.; Wei, S.Q.; et al. Low Overpotential in Vacancy-Rich Ultrathin $CoSe_2$ Nanosheets for Water Oxidation. *J. Am. Chem. Soc.* **2014**, *136*, 15670–15675. [CrossRef] [PubMed]

135. Gao, M.R.; Cao, X.; Gao, Q.; Xu, Y.F.; Zheng, Y.R.; Jiang, J.; Yu, S. Nitrogen-Doped Graphene Supported $CoSe_2$ Nanobelt Composite Catalyst for Efficient Water Oxidation. *ACS Nano* **2014**, *8*, 3970–3978. [CrossRef] [PubMed]

136. Guo, Y.X.; Yao, Z.Y.; Shang, C.S.; Wang, E.K. Amorphous Co_2B Grown on $CoSe_2$ Nanosheets as a Hybrid Catalyst for Efficient Overall Water Splitting in Alkaline Medium. *ACS Appl. Mater. Interfaces* **2017**, *9*, 39312–39317. [CrossRef] [PubMed]

137. Lyons, M.E.G.; Brandon, M.P. The Oxygen Evolution Reaction on Passive Oxide Covered Transition Metal Electrodes in Alkaline Solution. Part III - Iron. *Int. J. Electrochem. Sci.* **2008**, *3*, 1425–1462.

138. Yeo, B.S.; Bell, A.T. Enhanced Activity of Gold-Supported Cobalt Oxide for the Electrochemical Evolution of Oxygen. *J. Am. Chem. Soc.* **2011**, *133*, 5587–5593. [CrossRef] [PubMed]

139. Simmons, G.W.; Kellerman, E.; Leidheiser, H.J. In Situ Studies of the Passivation and Anodic Oxidation of Cobalt by Emission Moessbauer Spectroscopy. *J. Electrochem. Soc.* **1976**, *123*, 1276–1284. [CrossRef]

140. Liu, J.; Liu, Y.; Liu, N.Y.; Han, Y.Z.; Zhang, X.; Huang, H.; Lifshitz, Y.; Lee, S.T.; Zhong, J.; Kang, Z.H. Metal-Free Efficient Photocatalyst for Stable Visible Water Splitting via A Two-Electron Pathway. *Science* **2015**, *347*, 970–974. [CrossRef] [PubMed]

141. Barkaoui, S.; Haddaoui, M.; Dhaouadi, H.; Raouafi, N.; Touati, F. Hydrothermal Synthesis of Urchin-Like Co_3O_4 Nanostructures and their Electrochemical Sensing Performance of H_2O_2. *J. Solid State Chem.* **2015**, *228*, 226–231. [CrossRef]

142. Du, S.C.; Ren, Z.Y.; Qu, Y.; Wu, J.; Xi, W.; Zhu, J.Q.; Fu, H.G. Co_3O_4 Nanosheets as a High-Performance Catalyst for Oxygen Evolution Proceeding via a Double Two-Electron Process. *Chem. Commun.* **2016**, *52*, 6705–6708. [CrossRef] [PubMed]

143. Wang, Y.C.; Zhou, T.; Jiang, K.; Da, P.M.; Peng, Z.; Tang, J.; Kong, B.A.; Cai, W.B.; Yang, Z.Q.; Zheng, G.F. Reduced Mesoporous Co_3O_4 Nanowires as Efficient Water Oxidation Electrocatalysts and Supercapacitor Electrodes. *Adv. Energy Mater.* **2014**, *4*, 1400696–1400702. [CrossRef]

144. Xia, Y.S.; Dai, H.X.; Jiang, H.Y.; Zhang, L. Three-Dimensional Ordered Mesoporous Cobalt Oxides: Highly Active Catalysts for the Oxidation of Toluene and Methanol. *Catal. Commun.* **2010**, *11*, 1171–1175. [CrossRef]

145. Zhao, Y.F.; Chen, S.Q.; Sun, B.; Su, D.W.; Huang, X.D.; Liu, H.; Yan, Y.M.; Sun, K.N.; Wang, G.X. Graphene-Co_3O_4 Nanocomposite as Electrocatalyst with High Performance for Oxygen Evolution Reaction. *Sci. Rep.* **2015**, *5*, 7629–7635. [CrossRef] [PubMed]

146. Jadhav, A.R.; Bandal, H.A.; Tamboli, A.H.; Kim, H. Environment Friendly Hydrothermal Synthesis of Carbon–Co_3O_4 Nanorods Composite as an Efficient Catalyst for Oxygen Evolution Reaction. *J. Energy Chem.* **2017**, *26*, 695–702. [CrossRef]

147. Du, S.C.; Ren, Z.Y.; Zhang, J.; Wu, J.; Xi, W.; Zhu, J.Q.; Fu, H.G. Co_3O_4 Nanocrystal Ink Printed on Carbon Fiber Paper as a Large-Area Electrode for Electrochemical Water Splitting. *Chem. Commun.* **2015**, *51*, 8066–8069. [CrossRef] [PubMed]

148. Wei, Y.C.; Ren, X.; Ma, H.M.; Sun, X.; Zhang, Y.; Kuang, X.; Yan, T.; Ju, H.X.; Wu, D.; Wei, Q. $CoC_2O_4\cdot2H_2O$ Derived Co_3O_4 Nanorods Array: A High-Efficiency 1D Electrocatalyst for Alkaline Oxygen Evolution Reaction. *Chem. Commun.* **2018**, *54*, 1533–1536. [CrossRef] [PubMed]

149. Nai, J.W.; Yin, H.J.; You, T.T.; Zheng, L.R.; Zhang, J.; Wang, P.X.; Jin, Z.; Tian, Y.; Liu, J.Z.; Tang, Z.Y.; et al. Efficient Electrocatalytic Water Oxidation by Using Amorphous Ni-Co Double Hydroxides Nanocages. *Adv. Energy Mater.* **2015**, *5*, 1401880–1401887. [CrossRef]

150. Surendranath, Y.; Lutterman, D.A.; Liu, Y.; Nocera, D.G. Nucleation, Growth, and Repair of Acobalt-Based Oxygen Evolving Catalyst. *J. Am. Chem. Soc.* **2012**, *134*, 6326–6336. [CrossRef] [PubMed]

151. Ni, B.; Wang, X. Edge Overgrowth of Spiral Bimetallic Hydroxides Ultrathin-Nanosheets for Water Oxidation. *Chem. Sci.* **2015**, *6*, 3572–3576. [CrossRef] [PubMed]

152. Wu, J.; Ren, Z.Y.; Du, S.C.; Kong, L.J.; Liu, B.W.; Xi, W.; Zhu, J.Q.; Fu, H.G. A Highly Active Oxygen Evolution Electrocatalyst: Ultrathin CoNi Double Hydroxide/CoO Nanosheets Synthesized via Interface-Directed Assembly. *Nano Res.* **2016**, *9*, 713–725. [CrossRef]

153. Maiyalagan, T.; Jarvis, K.A.; Therese, S.; Ferreira, P.J.; Manthiram, A. Spinel-Type Lithium Cobalt Oxide as a Bifunctional Electrocatalyst for the Oxygen Evolution and Oxygen Reduction Reactions. *Nat. Commun.* **2014**, *5*, 3949–3956. [CrossRef] [PubMed]

154. Qian, L.; Lu, Z.Y.; Xu, T.H.; Wu, X.C.; Tian, Y.; Li, Y.P.; Huo, Z.Y.; Sun, X.M.; Duan, X. Trinary Layered Double Hydroxides as High-Performance Bifunctional Materials for Oxygen Electrocatalysis. *Adv. Energy Mater.* **2015**, *5*, 1500245–1500250. [CrossRef]

155. Grimaud, A.; May, K.J.; Carlton, C.E.; Lee, Y.L.; Risch, M.; Hong, W.T.; Zhou, J.G.; Shao-Horn, Y. Double Perovskites as a Family of Highly Active Catalysts for Oxygen Evolution in Alkaline Solution. *Nat. Commun.* **2013**, *4*, 2439–2745. [CrossRef] [PubMed]

156. Lukowski, M.A.; Daniel, A.S.; Meng, F.; Forticaux, A.; Li, L.S.; Jin, S. Enhanced Hydrogen Evolution Catalysis from Chemically Exfoliated Metallic MoS_2 Nanosheets. *J. Am. Chem. Soc.* **2013**, *135*, 10274–10277. [CrossRef] [PubMed]

157. Voiry, D.; Yamaguchi, H.; Li, J.W.; Silva, R.; Alves, D.C.B.; Fujita, T.; Chen, M.W.; Asefa, T.; Shenoy, V.B.; Eda, G.; et al. Enhanced Catalytic Activity in Strained Chemically Exfoliated WS_2 Nanosheets for Hydrogen Evolution. *Nat. Mater.* **2013**, *12*, 850–855. [CrossRef] [PubMed]

158. Yu, L.; Yan, Y.X.; Liu, Q.; Wang, J.; Yang, B.; Wang, B.; Jing, X.Y.; Liu, L.H. Exfoliation at Room Temperature for Improving Electrochemical Performance for Supercapacitors of Layered MnO_2. *J. Electrochem. Soc.* **2013**, *161*, E1–E5. [CrossRef]

159. Song, F.; Hu, X.L. Exfoliation of Layered Double Hydroxides for Enhanced Oxygen Evolution Catalysis. *Nat. Commun.* **2014**, *5*, 4477–4485. [CrossRef] [PubMed]

160. Liang, H.F.; Meng, F.; Caban-Acevedo, M.; Li, L.S.; Forticaux, A.; Xiu, L.C.; Wang, Z.C.; Jin, S. Hydrothermal Continuous Flow Synthesis and Exfoliation of NiCo Layered Double Hydroxide Nanosheets for Enhanced Oxygen Evolution Catalysis. *Nano Lett.* **2015**, *15*, 1421–1427. [CrossRef] [PubMed]

161. Zhao, S.; Yan, L.T.; Luo, H.; Mustain, W.; Xu, H. Recent Progress and Perspectives of Bifunctional Oxygen Reduction/Evolution Catalyst Development for Regenerative Anion Exchange Membrane Fuel Cells. *Nano Energy* **2018**, *47*, 172–198. [CrossRef]

162. Mamlouk, M.; Kumar, S.M.S.; Gouerec, P.; Scott, K. Electrochemical and Fuel Cell Evaluation of Co Based Catalyst for Oxygen Reduction in Anion Exchange Polymer Membrane Fuel Cells. *J. Power Sources* **2011**, *196*, 7594–7600. [CrossRef]

163. Wu, X.; Scott, K. A Non-Precious Metal Bifunctional Oxygen Electrode for Alkaline Anion Exchange Membrane Cells. *J. Power Sources* **2012**, *206*, 14–19. [CrossRef]

164. Kong, D.S.; Cha, J.J.; Wang, H.T.; Lee, H.R.; Cui, Y. First-Row Transition Metal Dichalcogenide Catalysts for Hydrogen Evolution Reaction. *Energy Environ. Sci.* **2013**, *6*, 3553–3558. [CrossRef]

165. Anantharaj, S.; Ede, S.R.; Sakthikumar, K.; Karthick, K.; Mishra, S.; Kundu, S. Recent Trends and Perspectives in Electrochemical Water Splitting with an Emphasis on Sulfide, Selenide, and Phosphide Catalysts of Fe, Co, and Ni: A Review. *ACS Catal.* **2016**, *6*, 8069–8097. [CrossRef]

166. Liu, Q.; Shi, J.L.; Hu, J.M.; Asiri, A.M.; Luo, Y.L.; Sun, X.P. $CoSe_2$ Nanowires Array as a 3D Electrode for Highly Efficient Electrochemical Hydrogen Evolution. *ACS Appl. Mater. Interfaces* **2015**, *7*, 3877–3881. [CrossRef] [PubMed]

167. Zhang, H.X.; Yang, B.; Wu, X.L.; Li, Z.J.; Lei, L.C.; Zhang, X.W. Polymorphic $CoSe_2$ with Mixed Orthorhombic and Cubic Phases for Highly Efficient Hydrogen Evolution Reaction. *ACS Appl. Mater. Interfaces* **2015**, *7*, 1772–1779. [CrossRef] [PubMed]

168. Li, H.M.; Qian, X.; Zhu, C.L.; Jiang, X.X.; Shao, L.; Hou, L.X. Template Synthesis of $CoSe_2/Co_3Se_4$ Nanotubes: Tuning of their Crystal Structures for Photovoltaics and Hydrogen Evolution in Alkaline Medium. *J. Mater. Chem. A* **2017**, *5*, 4513–4526. [CrossRef]

169. Wang, K.; Xi, D.; Zhou, C.J.; Shi, Z.Q.; Xia, H.Y.; Liu, G.W.; Qiao, G.J. $CoSe_2$ Necklace-Like Nanowires Supported by Carbon Fiber Paper: A 3D Integrated Electrode for the Hydrogen Evolution Reaction. *J. Mater. Chem. A* **2015**, *3*, 9415–9420. [CrossRef]

170. Yue, H.H.; Yu, B.; Qi, F.; Zhou, J.H.; Wang, X.Q.; Zheng, B.J.; Zhang, W.L.; Li, Y.R.; Chen, Y.F. Interwoven $CoSe_2/CNTs$ Hybrid as a Highly Efficient and Stable Electrocatalyst for Hydrogen Evolution Reaction. *Electrochim. Acta* **2017**, *253*, 200–207. [CrossRef]

171. Kim, J.K.; Park, G.D.; Kim, J.H.; Park, S.K.; Kang, Y.C. Rational Design and Synthesis of Extremely Efficient Macroporous $CoSe_2$-CNT Composite Microspheres for Hydrogen Evolution Reaction. *Small* **2017**, *13*, 1700068. [CrossRef] [PubMed]

172. Dai, C.; Tian, X.K.; Nie, Y.L.; Tian, C.; Yang, C.; Zhou, Z.X.; Li, Y.; Gao, X.Y. Successful Synthesis of 3D $CoSe_2$ Hollow Microspheres with High Surface Roughness and its Excellent Performance in Catalytic Hydrogen Evolution Reaction. *Chem. Eng. J.* **2017**, *321*, 105–112. [CrossRef]

173. Xiao, H.Q.; Wang, S.T.; Wang, C.; Li, Y.Y.; Zhang, H.R.; Wang, Z.J.; Zhou, Y.; An, C.H.; Zhang, J. Lamellar Structured $CoSe_2$ Nanosheets Directly Arrayed on Ti Plate as an Efficient Electrochemical Catalyst for Hydrogen Evolution. *Electrochim. Acta* **2016**, *217*, 156–162. [CrossRef]

174. Di, J.; Yan, C.; Handoko, A.D.; Seh, Z.W.; Li, H.M.; Liu, Z. Ultrathin Two-Dimensional Materials for Photo- and Electrocatalytic Hydrogen Evolution. *Mater. Today* **2018**, *21*, 749–770. [CrossRef]

175. Liu, Y.W.; Hua, X.M.; Xiao, C.; Zhou, T.F.; Huang, P.C.; Guo, Z.P.; Pan, B.C.; Xie, Y. Heterogeneous Spin States in Ultrathin Nanosheet Inducing Subtle Lattice Distortion for Efficiently Triggering Hydrogen Evolution. *J. Am. Chem. Soc.* **2016**, *138*, 5087–5092. [CrossRef] [PubMed]

176. Yang, Y.Y.; Li, F.; Li, W.Z.; Gao, W.B.; Wen, H.; Li, J.; Hu, Y.P.; Luo, Y.T.; Li, R. Porous CoS_2 Nanostructures Based on ZIF-9 Supported on Reduced Graphene Oxide: Favourable Electrocatalysis for Hydrogen Evolution Reaction. *Int. J. Hydrogen Energy* **2017**, *42*, 6665–6673. [CrossRef]

177. Zhang, X.; Liu, S.W.; Zang, Y.P.; Liu, R.R.; Liu, G.Q.; Wang, G.Z.; Zhang, Y.X.; Zhang, H.M.; Zhao, H.J. $Co/Co_9S_8@S$, N-Doped Porous Graphene Sheets Derived from S, N Dual Organic Ligands Assembled Co-MOFs as Superior Electrocatalysts for Full Water Splitting in Alkaline Media. *Nano Energy* **2016**, *30*, 93–102. [CrossRef]

178. Lin, J.; He, J.R.; Qi, F.; Zheng, B.J.; Wang, X.Q.; Yu, B.; Zhou, K.R.; Zhang, W.L.; Li, Y.R.; Chen, Y.F. In-Situ Selenization of Co-based Metal-Organic Frameworks as a Highly Efficient Electrocatalyst for Hydrogen Evolution Reaction. *Electrochim. Acta* **2017**, *247*, 258–264. [CrossRef]

179. Zhang, H.X.; Lei, L.C.; Zhang, X.W. One-Step Synthesis of Cubic Pyrite-Type CoSe$_2$ at Low Temperature for Efficient Hydrogen Evolution Reaction. *RSC Adv.* **2014**, *4*, 54344–54348. [CrossRef]

180. Li, Y.Z.; Niu, S.Q.; Rakov, D.; Wang, Y.; Caban-Acevedo, M.; Zheng, S.J.; Song, B.; Xu, P. Metal Organic Framework-Derived CoPS/N-doped Carbon for Efficient Electrocatalytic Hydrogen Evolution. *Nanoscale* **2018**, *10*, 7291–7297. [CrossRef] [PubMed]

181. Li, N.; Liu, X.; Li, G.D.; Wu, Y.Y.; Gao, R.Q.; Zou, X.X. Vertically Grown CoS Nanosheets on Carbon Cloth as Efficient Hydrogen Evolution Electrocatalysts. *Int. J. Hydrogen Energy* **2017**, *42*, 9914–9921. [CrossRef]

182. Ouyang, C.B.; Wang, X.; Wang, S.Y. Phosphorus-Doped CoS$_2$ Nanosheet Arrays as Ultra-Efficient Electrocatalysts for the Hydrogen Evolution Reaction. *Chem. Commun.* **2015**, *51*, 14160–14163. [CrossRef] [PubMed]

183. Li, H.M.; Qian, X.; Xu, C.; Huang, S.W.; Zhu, C.L.; Jiang, X.C.; Shao, L.; Hou, L.X. Hierarchical Porous Co$_9$S$_8$/Nitrogen-Doped Carbon@MoS$_2$ Polyhedrons as pH Universal Electrocatalysts for Highly Efficient Hydrogen Evolution Reaction. *ACS Appl. Mater. Interfaces* **2017**, *9*, 28394–28405. [CrossRef] [PubMed]

![catalysts logo] *catalysts*

MDPI

Article

A Novel Metal–Organic Framework Route to Embed Co Nanoparticles into Multi-Walled Carbon Nanotubes for Effective Oxygen Reduction in Alkaline Media

Hong Zhu *, Ke Li, Minglin Chen, Hehuan Cao and Fanghui Wang

State Key Laboratory of Chemical Resource Engineering, Institute of Modern Catalysis, Department of Organic Chemistry, School of Science, Beijing University of Chemical Technology, Beijing 100029, China; like200933@163.com (K.L.); chml24@163.com (M.C.); 2016410013@mail.buct.edu.cn (H.C.); fhwang@mail.buct.edu.cn (F.W.)
* Correspondence: zhuho128@126.com; Tel.: +86-10-6444-4919

Received: 1 November 2017; Accepted: 23 November 2017; Published: 27 November 2017

Abstract: Metal–organic framework (MOF) materials can be used as precursors to prepare non-precious metal catalysts (NPMCs) for oxygen reduction reaction (ORR). Herein, we prepared a novel MOF material (denoted as Co-bpdc) and then combined it with multi-walled carbon nanotubes (MWCNTs) to form Co-bpdc/MWCNTs composites. After calcination, the cobalt ions from Co-bpdc were converted into Co nanoparticles, which were distributed in the graphite carbon layers and MWCNTs to form Co-bpdc/MWCNTs. The prepared catalysts were characterized by TEM (Transmission electron microscopy), XRD (X-ray diffraction), XPS (X-ray photoelectron spectroscopy), BET (Brunauer–Emmett–Teller), and Raman spectroscopy. The electrocatalytic activity was measured by using rotating disk electrode (RDE) voltammetry. The catalysts showed higher ORR catalytic activity than the commercial Pt/C catalyst in alkaline solution. Co-bpdc/MWCNTs-100 showed the highest ORR catalytic activity, with an initial reduction potential and half-wave potential reaching 0.99 V and 0.92 V, respectively. The prepared catalysts also showed superior stability and followed the 4-electron pathway ORR process in alkaline solution.

Keywords: metal–organic framework; non-precious metal catalyst; oxygen reduction reaction; Co-bpdc; Co-bpdc/MWCNTs composites

1. Introduction

In recent years, as a result of the global energy crisis and environmental pollution, the search for cheap, efficient, and environmentally friendly alternative energy conversion and storage systems has aroused huge and sustained interest [1,2]. Hydrogen is a clean energy source, and fuel cells are the best way to utilize it. Fuel cells are considered one of the most efficient clean power-generating technologies of the 21st century. However, the high cost of fuel cells limits their commercialization; the precious metal catalysts make up a large part of the cost. To realize commercialization of the fuel cells as soon as possible, we urgently need to develop high-efficiency and low-cost non-precious metal catalysts to replace the high-cost, platinum-based noble metal catalysts [3–7].

Oxygen reduction reaction (ORR) plays a pivotal role in fuel cells and metal–air batteries [8–10]. Among the various non-noble metal ORR catalysts, transition metal–nitrogen–carbon (M/N/C) complexes or composites have attracted much attention due to their low price, high activity, long durability, and high resistance to methanol exchange [11,12]. Over the past several decades, non-noble metal catalysts (mainly Fe or Co) have been the most studied and have made great progress [13–17]. Jasinski first reported that nitrogen-containing metal macrocyclic compounds were able to catalyze oxygen reduction reaction, and thus created a new era of non-precious metals and non-metallic catalysts [10]. Subsequent studies have found that the activity and stability of the catalyst can be improved by pyrolyzing the metal salt compound and the nitrogen macrocyclic compound, which contain transition metal, nitrogen source, and carbon source [17–19]. Furthermore, the heteroatom-doped carbon material is a kind of promising material that can be used to replace the noble metal catalyst, and the doping of the carbon material with heteroatoms can effectively improve electrocatalytic activity and stability [20–22]. Among the heteroatom-doped carbon materials, nitrogen-doped carbon nanomaterials have been extensively studied because of their low cost, high electrocatalytic activity, high durability, and good environmental protection [23–25]. Many results have shown that the introduction of nitrogen atoms can induce the ionization of adjacent carbon atoms, promote the adsorption of oxygen, and lead to the improvement of catalytic activity. Nitrogen-doped carbon materials can also work with metal compounds to significantly improve catalytic efficiency [21,26].

Metal–organic framework (MOF) materials are composed of inorganic metal centers (metal ions or metal clusters) and bridged organic ligands through self-assembly interconnection, forming a porous crystalline material with a cyclical network structure [27–30]. MOF materials have a great prospect in non-precious metal catalysts because MOF materials have many advantages, such as structural diversity, adjustable aperture, and large specific surface area. These advantages have attracted the attention of researchers, and make MOF materials widely used in the catalytic field [29,31–33]. Since MOF-derived catalysts have many fully exposed nanostructures, resulting in their excellent electron transport properties, these catalysts have high electrocatalytic activity in fuel cells, in both acidic and alkaline electrolytes [29,34]. We can modify a MOF material to make it rich in $Fe-N_4$ or $Co-N_4$ coordination structure, so that the MOF material contains transition metal, nitrogen source, and carbon source. Specifically, we can introduce nitrogen atoms to form nitrogen-containing organic ligands, and then coordinate the modified ligands with the transition metal ions, to form a new kind of MOF material that contains nitrogen [30,33,34].

2. Results

Herein, we reported a novel MOF material (denoted as Co-bpdc) and the preparation of Co-bpdc/MWCNTs composites. The Co-bpdc and Co-bpdc/MWCNTs composites were heated at a high temperature in a N_2 atmosphere to form the catalysts. After calcination, the structure of the Co-bpdc/MWCNTs composites was destroyed, and the organic ligands from the Co-bpdc were carbonized to the graphite carbon layer. The central cobalt ions from the Co-bpdc were converted into Co nanoparticles, which were then distributed in the graphite carbon layers and MWCNTs. The MWCNTs worked as carriers and helped to disperse the cobalt atoms, and the addition of MWCNTs also improved the conductivity and electrocatalytic activity of the catalysts. For all the above reasons, the prepared catalysts have a higher ORR catalytic activity than the commercial Pt/C catalyst in alkaline media.

Formatting of Mathematical Components. The electron transfer number (n) during the oxygen reduction reaction was calculated from the Koutecky–Levich equation:

$$\frac{1}{J} = \frac{1}{J_L} + \frac{1}{J_K} = \frac{1}{B\omega^{1/2}} + \frac{1}{J_K} \tag{1}$$

$$B = 0.62nFC_0(D_0)^{2/3}v^{-1/6} \tag{2}$$

$$J_K = nFkC_0 \tag{3}$$

Here, J is the current density, J_L and J_K are the diffusion- and kinetic-limiting current densities, respectively, ω is the angular velocity, F is the Faraday constant (96,485 C mol^{-1}), C_0 is the O_2 bulk concentration ($C_0 = 1.26 \times 10^{-6}$ mol cm^{-3}), D_0 is the diffusion coefficient of the O_2 in 0.1 M KOH or

$HClO_4$ ($D_0 = 1.9 \times 10^{-5}$ cm s^{-1}), v is the kinematic viscosity of the electrolyte ($v = 0.1$ m^2 s^{-1} in 0.1 M KOH or $HClO_4$) [35].

Rotating ring-disk electron (RRDE) measurement can verify the ORR catalytic pathways of catalysts. We performed the measurement using a glassy carbon disk with a Pt ring at 0.73 V at a rotating speed of 1600 rpm. The electron transfer number (n) and the peroxide yield (HO^{2-}) during the ORR process can be calculated according the following equation:

$$n = \frac{4I_d}{I_d + \frac{I_r}{N}} \tag{4}$$

$$\%(H_2O_2) = \frac{\frac{200I_r}{N}}{I_d + \frac{I_r}{N}} \tag{5}$$

Here, I_d and I_r are the disk current and ring current, respectively, and N is the current collection efficiency of the Pt ring (0.37) [35].

3. Discussion

Our preparation process of the Co-bpdc/MWCNTs composites is based on the Co-bpdc and MWCNTs. Firstly, phenanthroline was oxidized to 2,2'-bipyridine-3,3'-dicarboxylic acid (H$_2$bpdc) by potassium permanganate under alkaline conditions. Secondly, the Co-bpdc was synthesized with the cobalt center and 2,2'-bipyridine-3,3'-dicarboxylic acid under acidic conditions with the hydrothermal synthesis. Co-bpdc works as MOF precursors, which provided the transition metal and nitrogen sources. The coordination between cobalt ions and H$_2$bpdc is shown in Figure 1, and Co-bpdc can also be expressed as [Co(bpdc)(H$_2$O)$_2$]$_n$. Secondly, the MWCNTs were introduced during the coordination reaction to obtain Co-bpdc/MWCNTs composites. Finally, the Co-bpdc and Co-bpdc/MWCNTs composites were pyrolyzed by heating at a high temperature in N$_2$ atmosphere, and then the catalysts were prepared.

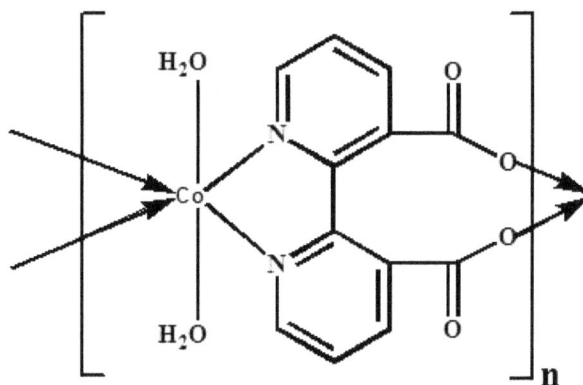

Figure 1. The coordination diagram between cobalt ions and H$_2$bpdc.

The formation of Co-bpdc and Co-bpdc/MWCNTs composites after calcination was confirmed by X-ray diffraction (XRD), as shown in Figure 2. The characteristic diffraction peak at 26°/2θ belongs to the (002) plane of the graphitic carbon in the MWCNTs, consistent with the results in the literature [35]. The (100) plane of the graphitic carbon in the prepared catalysts disappeared, and the (002) plane of graphitic carbon intensity in all the prepared catalysts was lower than that of graphitic carbon intensity in MWCNTs, indicating that the introduction of Co-bpdc decreased the graphitization of MWCNTs. In addition, the characteristic diffraction peaks at 44°, 51°, and 75°/2θ belong to the (111),

(200), and (220) planes of Co nanoparticles (JCPDS no. 15-0806), indicating that the central cobalt ions from the Co-bpdc and Co-bpdc/MWCNTs have been converted into Co nanoparticles after calcination. This result proves that the prepared catalysts contain the single crystal face-centered cubic structure [36]. The diffraction peak for Co (220) is used to evaluate the particle size of Co by Scherrer's equation:

$$D = \frac{0.9\lambda}{B cos\theta} \qquad (6)$$

Here, D is average particle size, the wavelength λ is close to 0.154056 nm, θ is the angle of Co (220) peak and B is the full width at half-maximum in radians (FWHM) [37]. The calculated average particle size of Co nanoparticles is 16.54 nm.

To some extent, with the increase of MWCNTs content, more dispersed cobalt particles result in more active sites being exposed, because of the greater dispersion of cobalt particles. When the content of MWCNTs is excessive, the excessive MWCNTs will cover part of active sites, resulting in the reduction of active sites in catalysts.

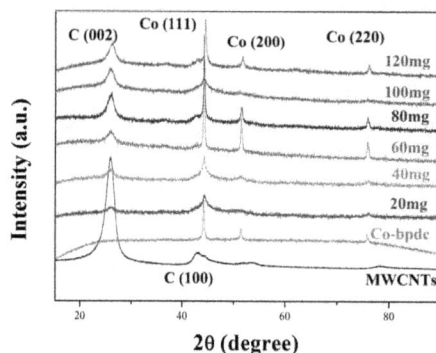

Figure 2. XRD patterns of Co-bpdc/MWCNTs compositions with different contents of MWCNTs.

Figure 3 shows the Raman spectra of MWCNTs and Co-bpdc/MWCNTs composites after calcination. The Raman spectra display the characteristic D band (at about 1345 cm^{-1}) and G band (at about 1590 cm^{-1}) of the carbon in the catalysts [38]. Because the incorporation of heteroatoms (mainly nitrogen) leads to a certain degree of defects in the carbon material, the ratio of the intensity of the D band to that of the G band (I_D/I_G) is used to characterize the defects of the carbon material. Here, the ratio for MWCNTs, reaching 1.09, is the lowest and the ratio for Co-bpdc/MWCNTs-100, reaching 1.64, is the highest among the prepared catalysts. These results indicate that the introduction of Co-bpdc increases the defects of MWCNTs.

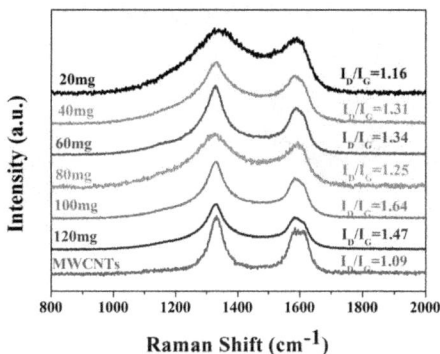

Figure 3. Ramanspectra of Co-bpdc/MWCNTs compositions with different contents of MWCNTs.

Figure 4 exhibits the typical morphology of Co-bpdc and Co-bpdc/MWCNTs composites. The Co-bpdc crystal has a rectangular structure in Figure 4a. The MWCNTs were successfully combined with the Co-bpdc as observed in Figure 4b. After the calcination, Figure 4c,d demonstrates the microstructure of Co-bpdc and Co-bpdc/MWCNT composites. From Figure 4c,d, we found that the structure of Co-bpdc was destroyed and a degree of agglomeration took place. The structure of Co-bpdc/MWCNTs composites was destroyed, and the organic ligands from the Co-bpdc carbonized to the graphite carbon layer. The central cobalt ions from the Co-bpdc were converted into Co nanoparticles, and then the Co nanoparticles were distributed in the graphite carbon layers and MWCNTs. The average particle size of the Co nanoparticles was calculated to be 16.93 nm (Figure 4e), which is close to the XRD result. This result indicates that the addition of MWCNTs can make the cobalt nanoparticles distribute more uniformly, and this structure can contribute to the transmission of oxygen and electrolytes to the catalyst surface. A HRTEM (High-resolution transmission electron microscopy) image of the Co-bpdc/MWCNTs-100 is shown in Figure 4f. As shown in Figure 4f, the metal particle is coated by carbon layers, and the lattice distance of 0.244 nm corresponds to the lattice fringe (111) of Co and is consistent with the XRD results [39].

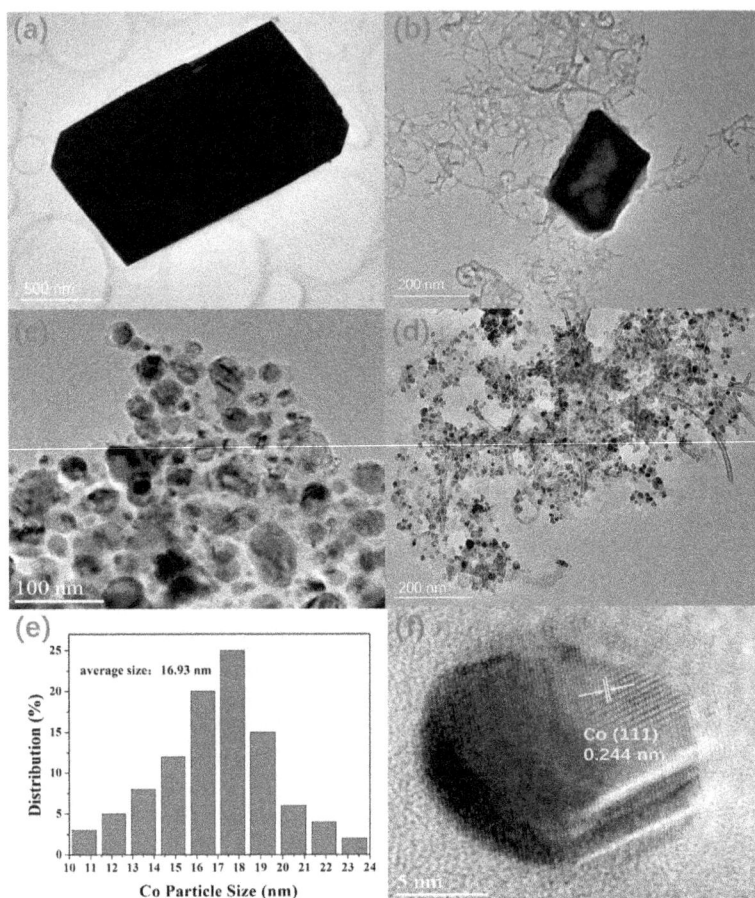

Figure 4. (a) TEM image of Co-bpdc; (b) TEM image of Co-bpdc/MWCNTs-100; (c) TEM image of Co-bpdc after calcination; (d) TEM image of Co-bpdc/MWCNTs-100 after calcination; (e) particle size distribution of Co-bpdc/MWCNTs-100; and (f) HRTEM image of Co-bpdc/MWCNTs-100.

The elemental compositions and valence states of these catalysts were studied by X-ray photoelectron spectroscopy (XPS). Figure 5a reveals the presence of C, N, O, and Co in Co-bpdc/MWCNTs-100. For example, the high-resolution C 1s spectrum is deconvoluted in Figure 5b. The peaks at 284.6 and 285.6 eV are attributed to sp^2-hybridized graphite-like carbon (C–C sp^2) and sp^3-hybridized diamond-like carbon (C–C sp^3), respectively. The peaks at 286.5 and 287.7 eV correspond to C–O/C–N and C=O/C=N, respectively, on the surface of the catalyst, indicating successful N-doping [35,40]. The peak at 290.4 eV belongs to the p-p* shake up satellites of sp^2 graphite-like carbon, indicating the occurrence of aromatization during the carbonization of the catalyst. The N element in the catalysts comes from the organic ligands in the Co-bpdc. The N element in Co-bpdc/MWCNTs-0 is the highest among the prepared catalysts, reaching 3.54 wt %, followed by that in Co-bpdc/MWCNTs-100 (Table 1). The fitted N 1s spectrum of Co-bpdc/MWCNTs-100 in Figure 5c can be deconvoluted to four N species: pyridinic N at 398.2 eV, pyrrolic N at 399.8 eV, graphitic N at 401.2 eV, and oxidized N at 402.5 eV [35]. Figure 5d shows the deconvoluted Co 2p spectrum. The two peaks at 780.5 eV and 795.6 eV correspond to Co $2p_{3/2}$ and Co $2p_{1/2}$, in accordance with the report on Co_3O_4 [35,39]. These results indicate that the element of cobalt appears in the catalysts mainly as cobalt metal and in a small amount as cobalt oxide.

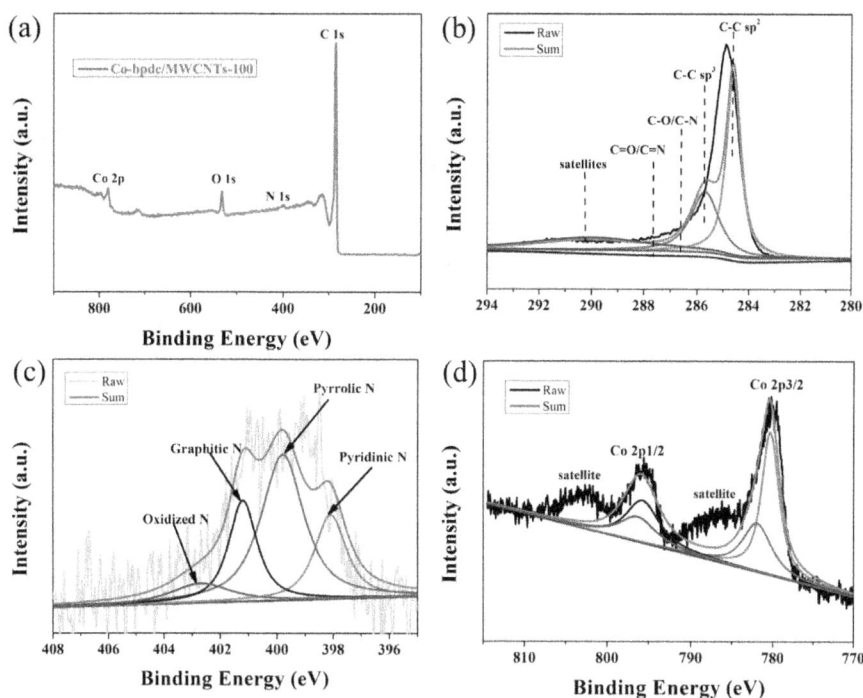

Figure 5. XPS spectra of Co-bpdc/MWCNTs-100: (**a**) Survey scan spectrum; (**b**) deconvoluted C 1s spectrum; (**c**) deconvoluted N 1s spectrum; and (**d**) deconvoluted Co 2p.

Table 1. The proportion of different atoms in Co-bpdc/MWCNTs composites with different contents of MWCNTs.

Elements	0	20	40	60	80	100	120
C	66.65	88.01	89.75	91.05	92.51	91.94	92.67
O	23.79	7.24	6.06	5.65	4.39	4.9	4.79
Co	6.01	3.12	2.35	1.44	1.24	1.14	1.3
N	3.54	1.63	1.84	1.86	1.87	2.02	1.24

From Table 1, we found that the relative content of O, Co and N in the catalysts containing MWCNTs decreases, compared with Co-bpdc. The main reasons are as follows. During the formation of Co-bpdc/MWCNTs composites, MWCNTs are incorporated into Co-bpdc, resulting in Co-bpdc/MWCNTs composites becoming a porous material. After calcination, Co-bpdc/MWCNTs composites were carbonized and the structure of Co-bpdc/MWCNTs composites was destroyed, resulting in a substantial loss of elements in the catalysts. The relative contents of O, Co and N in Co-bpdc were higher than the other catalysts, due to the large amount of agglomeration occurring during calcination. The presence of MWCNTs make Co particles distribute on the MWCNTs more uniformly. Excluding Co-bpdc, with increasing content of MWCNTs, the content of nitrogen in the catalysts first increases, reaching a maximum of 2.02% at 100 mg of MWCNTs, and then drops. However, the trend of Co elements is opposite to the trend of nitrogen elements.

Figure 6 depicts the nitrogen adsorption–desorption isotherm and the corresponding pore-size distribution curve of Co-bpdc/MWCNTs-100. The Brunauer–Emmett–Teller (BET) surface area and the cumulative pore volume of Co-bpdc/MWCNTs-100 are 221.02 m^2 g^{-1} and 0.47 cm^3 g^{-1}, respectively. The isotherm is a type IV BET isotherm with a H_3-type hysteresis loop ($P/P_0 > 0.4$), and the pore-size distribution curve shows a wide pore size distribution, ranging from micropores to mesopores, indicating the mesoporous characteristic of the Co-bpdc/MWCNTs-100. This characteristic contributes to the diffusion of oxygen onto the catalyst surface [35]. As a comparison, the BET surface area of Co-bpdc is 92.92 m^2 g^{-1}, which is lower than Co-bpdc/MWCNTs-100. The higher specific surface area means that the catalysts contain more active sites in unit volume, leading to the higher catalytic activity of the corresponding catalyst.

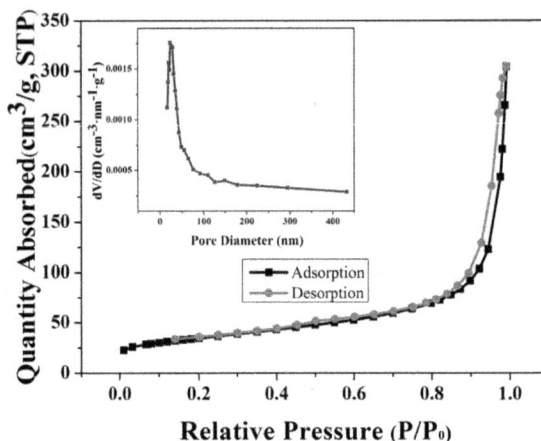

Figure 6. Nitrogen adsorption–desorption isotherm of Co-bpdc/MWCNTs-100.

The electrocatalytic activities of the as-prepared catalysts toward oxygen reduction reaction (ORR) were measured by cyclic voltammetry (CV) and linear sweep voltammetry (LSV) with a rotating disk electrode (RDE) in 0.1 M KOH solution. In Figure 7a, between 0.1 V and 1.6 V, all the prepared catalysts show a featureless voltammogram current curve in N_2-saturated solution, but a voltammogram with well-defined cathodic peaks in an O_2-saturated solution indicates that the catalysts have good oxygen reduction electrocatalytic activity [19]. The polarization curves of the prepared catalysts and the commercial Pt/C catalyst are shown in Figure 7b. Of all the prepared catalysts, Co-bpdc/MWCNTs-100 exhibits the best electrocatalytic activity toward ORR, and its initial reduction potential and half-wave potential reach 0.99 V and 0.92 V, respectively. The electrocatalytic activity of Co-bpdc/MWCNTs-100 toward ORR is much higher than that of the commercial Pt/C catalyst. Moreover, Co-bpdc/MWCNTs-100 also shows a larger current density than the other catalysts. On the other hand, Co-bpdc/MWCNTs-0 exhibits the worst electrocatalytic activity, and its initial reduction potential and half-wave potential reaches 0.90 V and 0.87 V, respectively, indicating that the introduction of MWCNTs can enhance the electrocatalytic activity. This result confirms that the higher specific areas of catalysts mean higher catalytic activity.

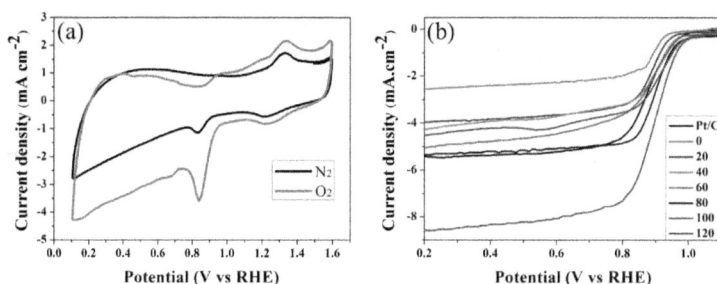

Figure 7. (**a**) Cyclic voltammetries (CVs) of Co-bpdc/MWCNTs-100; and (**b**) Oxygen reduction reaction (ORR) polarization curves of 20% Pt/C and Co-bpdc/MWCNTs catalysts with different contents of MWCNTs in 0.1 M KOH (Scan rate: 10 mV s^{-1}; rotation rate: 1600 rpm).

To better understand the role of cobalt atoms in these catalysts, we acidified the Co-bpdc/MWCNTs-100 catalyst to retain the cobalt atoms in the catalyst. Figure 8a shows that the catalytic activity of the acidified catalyst is not as good as that of Co-bpdc/MWCNTs-100, an indication that the cobalt atoms in the catalyst affect the oxygen reduction reaction. In addition, based on the ORR polarization curves of Co-bpdc/MWCNTs-100 and commercial Pt/C catalyst, the corresponding Tafel plots are shown in Figure 8b. The Tafel plot of Co-bpdc/MWCNTs-100 was determined to be 99.6 mV dec^{-1}, which is close to commercial Pt/C catalyst (96.6 mV dec^{-1}), indicating a similar reaction mechanism to the commercial Pt/C catalyst [41].

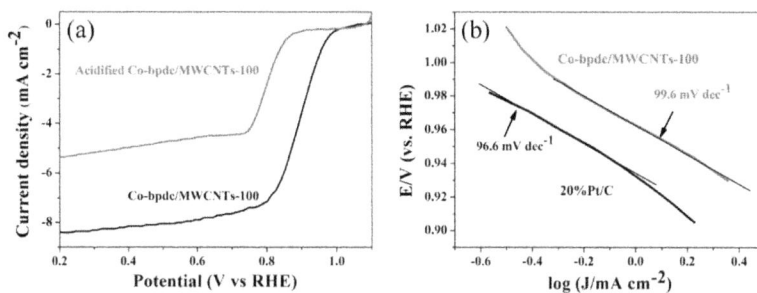

Figure 8. (**a**) ORR polarization curves of Co-bpdc/MWCNTs-100 and acidified Co-bpdc/MWCNTs-100 in 0.1 M KOH (Scan rate: 10 mV s^{-1}; rotation rate: 1600 rpm); (**b**) Tafel plots of 20% Pt/C and Co-bpdc/MWCNTs-100 in 0.1 M KOH.

It can be deduced that the nitrogen content is not the only factor that can affect the electrocatalytic activity of the prepared catalysts. The addition of MWCNTs allows the cobalt nanoparticles to distribute more uniformly, thus allowing the transmission of oxygen and electrolytes to the catalyst surface, thereby increasing the electrocatalytic activity towards the oxygen reduction reaction.

To further understand the reaction mechanism toward ORR in alkaline solution, the polarization curves of Co-bpdc/MWCNTs-100 at different rotating speeds are plotted in Figure 9a, and the corresponding Koutecky–Levich (K–L) plots are shown in Figure 9b. The polarization curves of Co-bpdc/MWCNTs-100 at different rotating speeds have the same initial reduction potential, but the current density increases with the increase of rotating speed. The K–L plots show excellent linearity and near coincidence in alkaline solution, indicating that the reactions on Co-bpdc/MWCNTs-100 follow first-order reaction kinetics and the electron transfer numbers (*n*) at different potentials are similar. The electron transfer numbers of the standard commercial Pt/C catalyst should be 4 [37]. The electron transfer numbers of Co-bpdc/MWCNTs-100 at different potentials were calculated to be in the range of 3.92–4.00 in an alkaline solution, with an average value of 3.97, which is very close to 4. These results indicate that the catalysts we prepared favor the 4-electron ORR process.

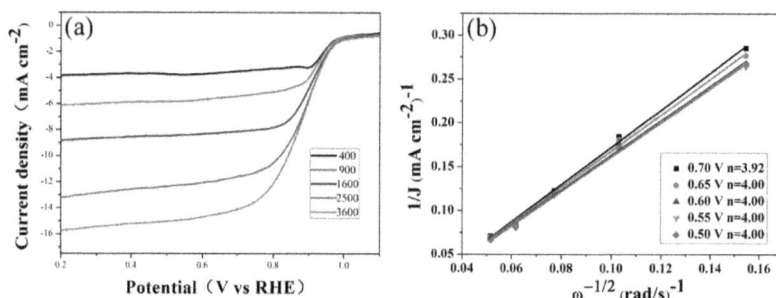

Figure 9. (a) ORR polarization curves of Co-bpdc/MWCNTs-100 at various rotating rates in 0.1 M KOH (scan rate: 10 mV s^{-1}); and (b) Koutecky–Levich plots of Co-bpdc/MWCNTs-100 at various potentials.

The RRDE measurements are shown in Figure 10. The measured H$_2$O$_2$ yields of Co-bpdc/MWCNTs-100 are below ~5% over the potential range of 0.15 V–1.0 V, and the electron transfer numbers are close to 4.0, consistent with the results calculated from the Koutecky–Levich equation based on the RDE measurements. The above results indicate that the prepared catalysts mainly favor the 4-electron ORR process.

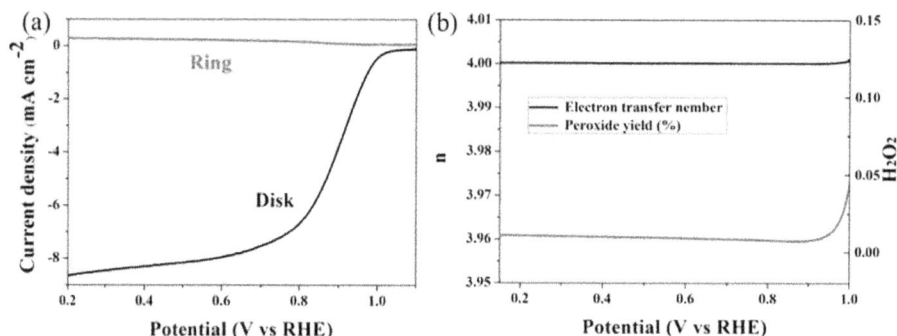

Figure 10. (a) Rotating ring-disk electron (RRDE) curve of Co-bpdc/MWCNTs-100 at 1600 rpm; and (b) electron transfer number and peroxide yield obtained at various potentials from RRDE curves.

The specific surface activity (SSA) values of Co-bpdc/MWCNTs-100 in the alkaline solution are obtained by the following equation:

$$SSA = \frac{j_k}{m_{EC} - A_{BET}} \qquad (7)$$

Here, j_k is the ORR kinetic current obtained on K–L plots at 0.90 V vs. RHE. m_{EC} is the Co-bpdc/MWCNTs-100 mass deposited on the RRDE tip and the A_{BET} is the BET area of Co-bpdc/MWCNTs-100 [42]. It was calculated that j_k is 0.511 mA cm^{-2} and m_{EC} of Co-bpdc/MWCNTs-100 is 0.408 mg cm^{-2}. The SSA of Co-bpdc/MWCNTs-100 is 0.57 μg cm^{-2}. The SSA of Co-bpdc/MWCNTs-100 is almost three orders of magnitude lower than the Pt/C Refs. [42,43].

The probable intrinsic performance of the active sites is as follows: the ORR process in the alkaline solution can be rationalized considering that (a) O_2 is absorbed on the Co nanoparticles; (b) O_2 is absorbed on the surface of Co-bpdc/MWCNTs-100. At the low ORR overpotentials, (b) the process is inhibited and (a) the process takes place on the Co nanoparticles of Co-bpdc/MWCNTs-100, which favor the 4-electron ORR process. When the ORR overpotentials exceed 9.0 V, Co-bpdc/MWCNTs-100 favor the 4-electron ORR process and 2-electron ORR process simultaneously, which could produce the H_2O_2. Finally, H_2O_2 is reduced to water on the surface of Co-bpdc/MWCNTs-100.

The electrocatalytic stability of Co-bpdc/MWCNTs-100 was assessed by the repetitive cycles in the range 0.5–1.5 V (at a scan rate of 100 mV s^{-1}). The polarization curve of Co-bpdc/MWCNTs-100 after 5000 cycles and the initial polarization curve of Co-bpdc/MWCNTs-100 are shown in Figure 11. The initial reduction potential does not decrease significantly and the half-wave potential of Co-bpdc/MWCNTs-100 shows only a negative shift of 5 mV. In addition, in the potential range of 0.4–0.7 V, the polarization curve after 5000 cycles shows an even higher current density than the initial polarization curve. These results indicate that Co-bpdc/MWCNTs-100 has a good ORR stability. The outstanding catalytic activity and stability of Co-bpdc/MWCNT composites make it a promising ORR electrocatalyst.

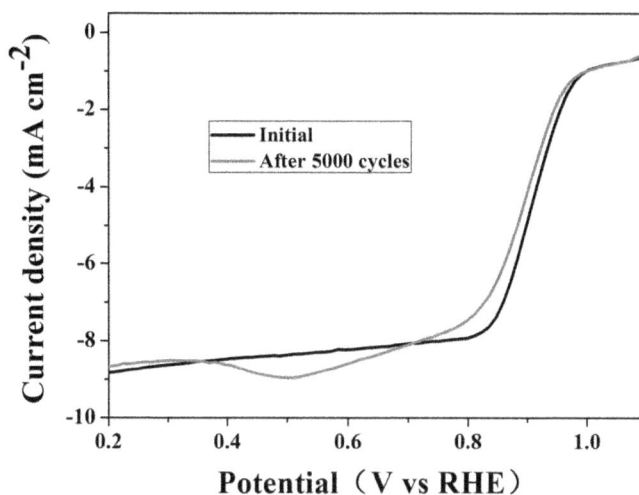

Figure 11. ORR polarization curves at 0 and 5000 cycles for Co-bpdc/MWCNTs-100.

4. Materials and Methods

4.1. Catalyst Preparation

Preparation of H$_2$bpdc. Phenanthroline (8 g), potassium permanganate (19 g) and sodium hydroxide (3.2 g) were added into 300 mL of deionized water in a flask. The mixture was heated at 98 °C for 3 h to obtain a purple black turbid liquid and a dark brown precipitate. A dark yellow transparent liquid was obtained by filtration. The filtrate was concentrated by evaporation to 150 mL and hydrochloric acid was added to adjust the pH to 2.0–2.5, resulting in a large amount of white precipitate after 24 h. Finally, the precipitate was collected by filtration and washed with deionized water to give the organic ligand (H$_2$bpdc).

Preparation of Co-bpdc. Co(NO$_3$)$_2$.6H$_2$O (0.291 g) and H$_2$bpdc (0.244 g) were ultrasonically dissolved in 40 mL of a solution of ethanol (V $_{water}$:V $_{ethanol}$ = 2:1), NaOH was added to adjust the pH to 6, and the mixture was heated and stirred for 5 h. The resulting orange crystals were collected by filtration, washed with deionized water, and dried in a dryer.

Preparation of Co-bpdc/MWCNTs composites. $Co(NO_3)_2 \cdot 6H_2O$ (0.291 g) and H_2bpdc (0.244 g) were ultrasonically dissolved in 40 mL of a solution of ethanol in water (V $_{water}$:V $_{ethanol}$ = 2:1), and then a NaOH solution was added to adjust the pH to 6. Different amounts of acid-treated MWCNTs (20, 40, 60, 80, 100, and 120 mg) were added the mixture solution, heated, and stirred for 5 h. The resulting crystals were collected by filtration, washed with deionized water, and dried in a dryer.

Preparation of catalysts. The above prepared Co-bpdc and Co-bpdc/MWCNTs composites were transferred into a tube furnace and exposed to a flow of nitrogen for 30 min. The furnace was heated to 600 °C at a rate of 4 °C min^{-1} for 4 h and then cooled down to room temperature. Finally, the catalysts were ground and sealed.

4.2. Materials Characterization

X-ray diffraction (XRD) was performed on a Bruker D8 Advanced diffractometer (BrukerAXSGmbH, Karlsruhe, Germany) with Cu Kα radiation (λ = 1.5406 Å) to obtain the crystalline structures of catalysts. Raman spectra were obtained on a Raman Spectrometer (Thermo Fisher Scientific, Waltham, MA, USA) equipped with a Leica DMLM microscope and a 514 nm Ar$^+$ ion laser as an excitation source. The microstructure and morphology of the catalysts were observed under a JEOL JEM-3010HR and a JEM-2100 microscope (Japan electronics Co., Ltd., Tokyo, Japan). X-ray photoelectron spectroscopy (XPS) was performed on a Thermo Fisher Scientific ESCALAB 250 imaging photoelectron spectrometer (Thermo Fisher Scientific, Waltham, MA, USA). The specific surface area and the pore size distribution are measured by the BET model and Barrette–Joynere–Halenda (BJH) model, respectively, on an ASAP 2460 instrument (Quantachrome, South San Francisco, CA, USA).

4.3. Electrochemical Test

The electrochemical measurements were obtained at room temperature on an IM6ex electrochemical workstation manufactured by ZAHNER ENNIUM (Kronach, Germany) with a standard three-electrode system. The glassy carbon rotating disk electrode (5 mm in diameter) was used as the working electrode. A platinum electrode and an $Hg/HgSO_4$ electrode with saturated KCl solution were used as the counter and reference electrodes, respectively. All the potentials measured were converted to reversible hydrogen electrode (RHE) scale. Cyclic voltammetry (CV) and linear sweeping voltammetry (LSV) were performed with a rotating disk electrode (RDE) and a rotating ring disk electrode (RRDE) on samples in 0.1 M KOH solution saturated with N_2 or O_2 (versus RHE). The preparation process of catalyst ink is as follows: 4 mg of the catalysts and 10 μL of 0.5 wt % Nafion solution were dispersed in 1 mL of isopropanol solution (V $_{isopropanol}$:V $_{H2O}$ = 2:3) ultrasonically. We took 20 μL of the catalyst ink (containing 80 μg of catalysts) and dropped it on the surface of the glassy carbon electrode, and then the catalyst ink was naturally dried at room temperature before the measurement.

5. Conclusions

We have synthesized a novel MOF material (named as Co-bpdc) and applied it to a series of non-precious metal catalysts for fuel cells. The prepared catalysts show higher ORR catalytic activity than the commercial Pt/C catalyst in alkaline solution, and Co-bpdc/MWCNTs-100 shows the highest ORR catalytic activity of the series, with its initial reduction potential and half-wave potential reaching 0.99 V and 0.92 V, respectively, in 0.1 M KOH solution. Moreover, the catalysts favor the 4-electron ORR process in alkaline solution, and exhibit higher stability in alkaline solution than the commercial Pt/C catalyst. The addition of MWCNTs can enhance the electrocatalytic activity. In addition, this work provides a novel idea for the synthesis of MOF materials as precursors to prepare ORR electrocatalysts. Our results indicate that MOF materials have great potential in non-precious catalysts for full cell applications.

Acknowledgments: The authors gratefully acknowledge the financial support from the National Key Research and Development Program of China (No. 2016YFB0101203), Beijing Municipal Science and Technology program (No. Z171100000917019), the National Natural Science Foundation of China (No. 21376022 and 21776014), the International S&T Cooperation Program of China (No. 2013DFA51860) and the Fundamental Research Funds for the Central Universities (No. JC1504).

Author Contributions: Ke Li performed a literature review, categorized the electrocatalysts in the introduction, did the research work and the main part of data analysis, and wrote the main part of the current work. Minglin Chen and Hehuan Cao corrected the English, helped with the part of data analysis and wrote part of this work. Fanghui Wang and Hong Zhu wrote part of this work, corrected the English grammar and syntax and supervised the whole work.

Conflicts of Interest: The authors declare no conflict of interest.

References

1. Qiao, X.C.; Liao, S.J.; Zheng, R.P.; Deng, Y.J.; Song, H.Y.; Du, L. Cobalt and nitrogen co-doped graphene with inserted carbon nanospheres as an efficient bifunctional electrocatalyst for oxygen reduction and evolution. *ACS Sustain. Chem. Eng.* **2016**, *4*, 4131–4136. [CrossRef]
2. Chu, S.; Majumdar, A. Opportunities and challenges for a sustainable energy future. *Nature* **2012**, *488*, 294–303. [CrossRef] [PubMed]
3. Lai, Q.X.; Su, Q.; Gao, Q.W.; Liang, Y.Y.; Wang, Y.X.; Yang, Z.; Zhang, X.G.; He, J.P.; Tong, H. In situ self-scrificed template synthesis of Fe-N/G catalysts for enhanced oxygen reduction. *ACS Appl. Mater. Interfaces* **2015**, *7*, 18170–18178. [CrossRef] [PubMed]
4. Xia, B.Y.; Yan, Y.; Wang, X.; Lou, X.W. Recent progress on graphene-based hybrid electrocatalysts. *Mater. Horiz.* **2014**, *1*, 379–399. [CrossRef]
5. Chen, A.; Holt-Hindle, P. Platinum-based nanostructured materials: Synthesis, properties, and applications. *Chem. Rev.* **2010**, *110*, 3767–3804. [CrossRef] [PubMed]
6. Zhang, Y.; Huang, L.B.; Jiang, W.J.; Zhang, X.; Chen, Y.Y.; Wei, Z.D.; Wan, L.J.; Hu, J.S. Sodium chloride-assisted green synthesis of a 3D Fe–N–C hybrid as a highly active electrocatalyst for the oxygen reduction reaction. *J. Mater. Chem. A* **2016**, *4*, 7781–7787. [CrossRef]
7. Wang, Y.; Nie, Y.; Ding, W.; Chen, S.G.; Xiong, K.; Qi, X.Q.; Zhang, Y.; Wang, J.; Wei, Z.D. Unification of catalytic oxygen reduction and hydrogen evolution reactions: Highly dispersive Co nanoparticles encapsulated inside Co and nitrogen co-doped carbon. *Chem. Commun.* **2015**, *51*, 8942–8945. [CrossRef] [PubMed]
8. Shang, L.; Yu, H.J.; Huang, X.; Bian, T.; Shi, R.; Zhao, Y.F.; Waterhouse, G.I.N.; Wu, L.Z.; Tung, C.H.; Zhang, T.R. Well-dispersed ZIF-derived Co,N-co-doped carbon nanoframes through mesoporous-silica-protected calcination as efficient oxygen reduction electrocatalysts. *Adv. Mater.* **2016**, *28*, 1668–1674. [CrossRef] [PubMed]
9. Shao, M.H.; Chang, Q.W.; Dodelet, J.P.; Chenitz, R. Recent advances in electrocatalysts for oxygen reduction reaction. *Chem. Rev.* **2016**, *116*, 3594–3657. [CrossRef] [PubMed]
10. Hu, K.; Tao, L.; Liu, D.D.; Huo, J.; Wang, S.Y. Sulfur-doped Fe/N/C nanosheets as highly efficient electrocatalysts for oxygen reduction reaction. *ACS Appl. Mater. Interfaces* **2016**, *8*, 19379–19385. [CrossRef] [PubMed]
11. Yao, Y.; Xiao, H.; Wang, P.; Su, P.P.; Shao, Z.G.; Yang, Q.H. CNTs@Fe–N–C core–shell nanostructures as active electrocatalyst for oxygen reduction. *J. Mater. Chem. A* **2014**, *2*, 11768–11775. [CrossRef]
12. Zhu, Y.S.; Zhang, B.S.; Liu, X.; Wang, D.W.; Su, D.S. Unravelling the structure of electrocatalytically active Fe-N complexes in carbon for the oxygen reduction reaction. *Angew. Chem. Int. Ed.* **2014**, *53*, 10673–10677. [CrossRef] [PubMed]
13. Choi, C.H.; Baldizzone, C.; Polymeros, G.; Pizzutilo, E.; Kasian, O.; Schuppert, A.K.; Sahraie, N.R.; Sougrati, M.T.; Mayrhofer, K.J.J.; Jaouen, F. Minimizing operando demetallation of Fe-N-C electrocatalysts in acidic medium. *ACS Catal.* **2016**, *6*, 3136–3146. [CrossRef]
14. Brouzgou, A.; Song, S.Q.; Liang, Z.X.; Tsiakaras, P. Non-precious electrocatalysts for oxygen reduction reaction in alkaline media: Latest achievements on novel carbon materials. *Catalysts* **2016**, *6*, 159. [CrossRef]
15. Kadam, P.D.; Chuan, H.H. Erratum to: Rectocutaneous fistula with transmigration of the suture: A rare delayed complication of vault fixation with the sacrospinous ligament. *Int. Urogynecol. J.* **2016**, *27*, 505. [CrossRef] [PubMed]

16. Proietti, E.; Jaouen, F.; Lefèvre, M.; Larouche, N.; Tian, J.; Herranz, J.; Dodelet, J.P. Iron-based cathode catalyst with enhanced power density in polymer electrolyte membrane fuel cells. *Nat. Commun.* **2011**, *2*, 416. [CrossRef] [PubMed]

17. Nabae, Y.; Nagata, S.; Hayakawa, T.; Niwa, H.; Harada, Y.; Oshima, M.; Isoda, A.; Matsunaga, A.; Tanaka, K.; Aoki, T. Pt-free carbon-based fuel cell catalyst prepared from spherical polyimide for enhanced oxygen diffusion. *Sci. Rep.* **2016**, *6*, 23276. [CrossRef] [PubMed]

18. Zhang, S.L.; Zhang, Y.; Jiang, W.J.; Liu, X.; Xu, S.L.; Huo, R.J.; Zhang, F.Z.; Hu, J.S. Co@N-CNTs derived from triple-role CoAl-layered double hydroxide as an efficient catalyst for oxygen reduction reaction. *Carbon* **2016**, *107*, 162–170. [CrossRef]

19. Jahnke, H.; Schönborn, M.; Zimmermann, G. Organic dyestuffs as catalysts for fuel cells. *Top. Curr. Chem.* **1976**, *61*, 133–181. [PubMed]

20. Ferrero, G.A.; Preuss, K.; Marinovic, A.; Jorge, A.B.; Mansor, N.; Brett, D.J.L.; Fuertes, A.B.; Sevilla, M.; Titirici, M.M. Fe-N-doped carbon capsules with outstanding electrochemical performance and stability for the oxygen reduction reaction in both acid and alkaline conditions. *ACS Nano* **2016**, *10*, 5922–5932. [CrossRef] [PubMed]

21. Zhou, D.; Yang, L.P.; Yu, L.H.; Kong, J.H.; Yao, X.Y.; Liu, W.S.; Xua, Z.C.; Lu, X.H. Fe/N/C hollow nanospheres by Fe(iii)-dopamine complexation-assisted one-pot doping as nonprecious-metal electrocatalysts for oxygen reduction. *Nanoscale* **2015**, *7*, 1501–1509. [CrossRef] [PubMed]

22. Yu, H.Y.; Fisher, A.; Cheng, D.J.; Cao, D.P. Cu,N-codoped hierarchical porous carbons as electrocatalysts for oxygen reduction reaction. *ACS Appl. Mater. Interfaces* **2016**, *8*, 21431–21439. [CrossRef] [PubMed]

23. Fu, X.G.; Choi, J.Y.; Zamani, P.Y.; Jiang, G.P.; Hoque, M.A.; Hassan, F.M.; Chen, Z.W. Co-N decorated hierarchically porous graphene aerogel for efficient oxygen reduction reaction in acid. *ACS Appl. Mater. Interfaces* **2016**, *8*, 6488–6495. [CrossRef] [PubMed]

24. Zhang, G.; Lu, W.T.; Cao, F.F.; Xiao, Z.D.; Zheng, X.S. N-doped graphene coupled with Co nanoparticles as an efficient electrocatalyst for oxygen reduction in alkaline media. *J. Power Sources* **2016**, *302*, 114–125. [CrossRef]

25. Zhu, H.; Sun, Z.N.; Chen, M.L.; Cao, H.H.; Li, K.; Cai, Y.Z.; Wang, F.H. Highly porous composite based on tungsten carbide and N-doped carbon aerogels for electrocatalyzing oxygen reduction reaction in acidic and alkaline media. *Electrochim. Acta* **2017**, *236*, 154–160. [CrossRef]

26. Zhou, X.J.; Bai, Z.Y.; Wu, M.J.; Qiao, J.L.; Chen, Z.W. 3-Dimensional porous N-doped graphene foam as a non-precious catalyst for the oxygen reduction reaction. *J. Mater. Chem. A* **2015**, *3*, 3343–3350. [CrossRef]

27. Hao, J.H.; Yang, W.S.; Zhang, Z.; Tang, J.L. Metal-organic frameworks derived CoxFe1-xP nanocubes for electrochemical hydrogen evolution. *Nanoscale* **2015**, *7*, 11055–11062. [CrossRef] [PubMed]

28. Huskić, I.; Pekov, I.V.; Krivovichev, S.V.; Friščić, T. Minerals with metal-organic framework structures. *Sci. Adv.* **2016**, *2*, e1600621. [CrossRef] [PubMed]

29. You, S.J.; Gong, X.B.; Wang, W.; Qi, D.P.; Wang, X.H.; Chen, X.D.; Ren, N.Q. Enhanced cathodic oxygen reduction and power production of microbial fuel cell based on noble-metal-free electrocatalyst derived from metal-organic frameworks. *Adv. Energy Mater.* **2016**, *6*, 1501497. [CrossRef]

30. Zhu, Q.L.; Xia, W.; Akita, T.; Zou, R.Q.; Xu, Q. Metal-organic framework-derived honeycomb-like open porous nanostructures as precious-metal-free catalysts for highly efficient oxygen electroreduction. *Adv. Mater.* **2016**, *28*, 6391–6398. [CrossRef] [PubMed]

31. Zhang, Y.F.; Bo, X.J.; Nsabimana, A.; Han, C.; Li, M.; Guo, L.P. Electrocatalytically active cobalt-based metal–organic framework with incorporated macroporous carbon composite for electrochemical applications. *J. Mater. Chem. A* **2015**, *3*, 732–738. [CrossRef]

32. Zhao, H.X.; Zou, Q.; Sun, S.K.; Yu, C.S.; Zhang, X.J.; Lia, R.J.; Fu, Y.Y. The ranostic metal–organic framework core–shell composites for magnetic resonance imaging and drug delivery. *Chem. Sci.* **2016**, *7*, 5294–5301. [CrossRef]

33. Song, Y.H.; Li, X.; Sun, L.L.; Wang, L. Metal/metal oxide nanostructures derived from metal–organic frameworks. *RSC Adv.* **2015**, *5*, 7267–7279. [CrossRef]

34. Mahmood, A.; Guo, W.H.; Tabassum, H.; Zou, R.Q. Metal-organic framework-based nanomaterials for electrocatalysis. *Adv. Energy Mater.* **2016**, *6*, 1600423. [CrossRef]

35. Li, X.Z.; Fang, Y.Y.; Lin, X.Q.; Tian, M.; An, X.C.; Fu, Y.; Li, R.; Jin, J.; Ma, J. MOF derived Co$_3$O$_4$nanoparticles embedded in N-doped mesoporous carbon layer/MWCNT hybrids: Extraordinary bi-functional electrocatalysts for OER and ORR. *J. Mater. Chem. A* **2015**, *3*, 17392–17402. [CrossRef]

36. Maghsodi, A.; Milani, H.M.; Dehghani, M.M.; Kheirmand, M.; Samiee, L.; Shoghi, F.; Kameli, M. Exploration of bimetallic Pt-Pd/C nanoparticles as an electrocatalyst for oxygen reduction reaction. *Appl. Surf. Sci.* **2011**, *257*, 6353–6357. [CrossRef]

37. Yi, L.H.; Liu, L.; Liu, X.; Wang, X.Y.; Yi, W.; He, P.Y.; Wang, X.Y. Carbon-supported Pt-Co nanoparticles as anode catalyst for direct borohydride-hydrogen peroxide fuel cell: Electrocatalysis and fuel cell performance. *Int. J. Hydrog. Energy* **2012**, *37*, 12650–12658. [CrossRef]

38. Geng, D.S.; Chen, Y.; Chen, Y.G.; Li, Y.L.; Li, R.Y.; Sun, X.L.; Ye, S.Y.; Knights, S. High oxygen-reduction activity and durability of nitrogen-doped graphene. *Energy Environ. Sci.* **2011**, *4*, 760–764. [CrossRef]

39. Xia, W.; Zou, R.Q.; An, L.; Xia, D.G.; Guo, S.J. A metal–organic framework route to in situ encapsulation of Co@Co$_3$O$_4$@C core@bishell nanoparticles into a highly ordered porous carbon matrix for oxygen reduction. *Energy Environ. Sci.* **2015**, *8*, 568–576. [CrossRef]

40. Jiang, H.L.; Yao, Y.F.; Zhu, Y.H.; Liu, Y.Y.; Su, Y.H.; Yang, X.L.; Li, C.Z. Iron carbide nanoparticles encapsulated in mesoporous Fe-N-doped graphene-like carbon hybrids as efficient bifunctional oxygen electrocatalysts. *ACS Appl. Mater. Interfaces* **2015**, *7*, 21511–21520. [CrossRef] [PubMed]

41. Zhao, P.P.; Xu, W.; Hua, X.; Luo, W.; Chen, S.L.; Cheng, G.Z. Facile synthesis of a N-doped Fe3C@CNT/porous carbon hybrid for an advanced oxygen reduction and water oxidation electrocatalyst. *J. Phys. Chem. C* **2016**, *120*, 11006–11013. [CrossRef]

42. Vezzù, K.; Delpeuch, A.B.; Negro, E.; Polizzi, S.; Nawn, G.; Bertasi, F.; Pagot, G.; Artyushkova, K.; Atanassov, P.; Noto, V.D. Fe-carbon nitride "core-shell" electrocatalysts for the oxygen reduction reaction. *Electrochim. Acta* **2016**, *222*, 1778–1791. [CrossRef]

43. Negro, E.; Polizzi, S.; Vezzu, K.; Toniolo, L.; Cavinato, G.; Noto, V.D. Interplay between morphology and elecrtrochemical performance of "core-shell" electrocatalysts for oxygen reduction reaction based on a PtNix carbon nitride "shell" and a pyrolyzed polyketone nanoball "core". *Int. J. Hydrog. Energy* **2014**, *39*, 2828–2841. [CrossRef]

MDPI

St. Alban-Anlage 66

4052 Basel

Switzerland

Tel. +41 61 683 77 34

Fax +41 61 302 89 18

www.mdpi.com

Catalysts Editorial Office

E-mail: catalysts@mdpi.com

www.mdpi.com/journal/catalysts

www.ingramcontent.com/pod-product-compliance
Lightning Source LLC
Chambersburg PA
CBHW051754200326
41597CB00025B/4551